AF518306

Problems in Aquaculture and its Solutions

THE AUTHOR

Dr. Gadadhar Dash born on 10.02.1969 is presently working as Professor and Head in the Department of Aquatic Animal Health, Faculty of Fishery Sciences, West Bengal University of Animal and Fishery Sciences, Kolkata, West Bengal, India. He has completed his M.F.Sc., degree as a Gold Medallist from Odisha University of Agriculture and Technology and did his Ph.D. from Kalyani University, West Bengal. He has total 26 years of experience out of which, 23Years of service in teaching, research and extension at UG, PG and Ph.D. level in the present University. Also he has served as an Officer to Dept. Fisheries, Govt. of Odisha and Krisha Vigyan Kendra of Odisha University of Agriculture and Technology, Odisha before joining to this University. He has published two books on Fisheries Parasitology, 3 laboratory Manuals and 7 Training Manuals. He has published 09 Chapters in different books, has guided 27 nos. of Master degree students,2 Ph.D students and handled 3 Research Projects and conducted many Workshops, Training Programmes on Fish health Management and Disease diagnosis. He has trained approximately 50,000 fish and Shrimp farmers of West Bengal and other states through various funding agencies such as NSPAAD, DST, DRDA, DBT, NFDB, DRDC,CADC, ATMA, SDB,NABARD and ICAR (NAE). He is well-known in popularizing scientific fish culture in West Bengal and adjacent States of India. He is a reviewer of many Reputed Journals and Expert Member of Several Committee. He has published more than 107 Research papers of National and International Repute and more than 60 Popular articles. He has good abroad exposure with regard to fisheries activity. He is a Fellow of Zoological Society of India and received many prestigious Awards such as Best Teacher Award, Dr S.P. Gupta Gold Medal Award, Best Performer Award, Best Parenting Award and Reviewer Excellence Award *etc.*

Problems in Aquaculture and its Solutions

Gadadhar Dash

Professor and Head,
Department of Aquatic Animal Health,
Faculty of Fishery Sciences,
West Bengal University of Animal and Fishery Sciences,
Kolkata, West Bengal

2022

Daya Publishing House®
A Division of
Astral International Pvt. Ltd.
New Delhi – 110 002

ISBN: 978-93-90435-05-0 (HB)

Published by : **Daya Publishing House®**
A Division of
Astral International Pvt. Ltd.
– ISO 9001:2015 Certified Company –
4736/23, Ansari Road, Darya Ganj,
New Delhi-110 002
Ph. 011-43549197, 23278134
E-mail: info@astralint.com
Website: www.astralint.com

Digitally Printed at : **Replika Press Pvt. Ltd.**

WEST BENGAL UNIVERSITY OF ANIMAL AND FISHERY SCIENCES

68- Kshudiram Bose Sarani, Belgachia, Kolkata-700 037
Phone: 2556 3450, Resi.: (033) 2550 2229, Fax: 91-33-2557-1986
Web: www.wbuafscl.ac.in, E-mail: drpbiswas56@gmail.com / drpbiswas1956@gmail.com

Prof. Purnendu Biswas, Ph.D.
Ex-Vice-Chancellor, WBUAFS

Foreword

India is privileged with enormous wealth of aquatic resources. Although aquaculture in India started as backyard pond culture of fish for food but gradually with the changing scenario of world trade and globalization, it transformed into a lucrative, revenue generating and export based industry. The demand for fish, due to its unique nutritive value, is bound to create shortages all over the world. The practice of aquaculture with a view to enhance production will probably be a right approach for meeting future demands.

Recently our Fish Industry faced a serious turbulence due to outbreaks of various diseases which caused significant loss of revenue and resources. The practical and need based solutions to various types of Aquaculture Problems will show an array of light to farming communities for not only saving their crops, but also raising their farm income for sustainable economic growth.

India also had a setback due to different varieties of problems/issues in aquaculture. To keep a balance between problems and its solutions, this book will be very much useful for not only the scientists, students, teachers but also for farmers and traders. It will serve as a management tool for the aquatic industry.

I am extremely happy to introduce the book entitled "*Problems in Aquaculture and its Solutions*". I congratulate the author for his endeavour and valuable contribution. I am confident that this book will go a long way in providing solutions to aquaculture crisis in future.

Purnendu Biswas

Preface

India is blessed with the enormous wealth of aquatic resources. However, problems in aquaculture have been striking both in aquaculture industry and the cultivators in general. The numbers of articles published in various reputed journals has increased regarding this topic and it is not possible for one person to read all articles and find the solution. This book entitled "Problems in Aquaculture and its solutions" is a complete package and a comprehensive approach to solve various aquatic problems in ecofriendly manner.

The concept of writing this book is approximately more than seven years old and I have finally completed the rough draft, but each time I saw a new problem in aquaculture and felt that this book requires revision. After too many additions and revisions, I have completed the draft of this book.

I would like to express my appreciation to Various Aqua Industries like M/S. Zymonutrients Pvt. Ltd, M/S. Vetquinal, M/S. Utopia Ventures, M/S. Marsco Neutraceuticals Pvt. Ltd., M/S. Brickcells Technology Pvt. Ltd. and M/S. Altech Pvt. Ltd. *etc* for their support in rendering various data and information on aqua problems and special help to the farming community at the time of their need. I appreciate the time to time advice of Prof. B.B. Hossetti, Kuvempu University, Karnataka, Dr. K.P. Biswas, Ex-Joint Director of Dept. of Fisheries, Govt. of Odisha who have encouraged me for finalizing the manuscript in specific time. Also I appreciate the effort taken by Daya Publishing house A division of M/S Astral International Pvt Ltd in publishing this book.

I am extremely grateful to the Authorities of my University who have tremendous trust and belief on me, for allowing in moving different corners of West Bengal and adjacent states in solving their aquaculture related problems.

I express my sincere gratitude to Prof. Chandrasekar Chakraborty Ex-Vice Chancellor, WBUAFS, Prof. Purnendu Biswas, Ex-Vice Chancellor, WBUAFS for their generous support and encouragement in completing this manuscript.

The help rendered by Miss Beryl Julyianta, Ph.D. Scholar, Miss Anwesha Roy, Ph.D Scholar, Mr. Tausif Ahmad PG student Mrs. (Dr.), Koel Bhattacharya Sanyal, SRF, NSPAAD, Mr. Debapriyo Mukherjee SRF, NSPAAD are gratefully acknowledged. I am duely acknowledging the authors whose opinions and texts are being referred in this manuscript.

I am also thankful to my family members, friends and relatives who have encouraged and supported me to complete the book.

Last but not the least, I am extremely grateful to all the fish and shrimp farmers of the country for their generous help, which leads me to complete this publication.

Gadadhar Dash

Contents

Abbreviation

AchE: Acetyl cholinesterase
APHND: Acute Hepatopancreatic Necrosis Disease
Bigha: 1 ha=2.5Acre=3 bigha
BIS: Biological Insecticides
BKC: Benzal Konium Chloride
BOD: Biological Oxygen Demand
COD: Chemical Oxygen Demand
CO_2: Carbon dioxide
Cfu: Colony Forming Unit
Cucm: Cubic Centimeter
°C: Degree Centigrade
DDT: Dichloro diphenyl trichloro ethane
EDTA: Ethylene Diamine Tetra Acetate
Eh: Redox Potential
EMS: Early Mortality Syndrome:
EUS: Epizootic Ulcerative Syndrome
FAO: Food and Agricultural Organization
FAWC: Farm Animal Welfare Council
FCR: Food Conversion Ratio
FZT: Fish Borne Zoonotic Trematode Infection
GMO: Genetically Modified Organisms
ha: Hectare
HABS: Harmful Algal Blooms

IUU: Illegal, unreported and unregulated fishing
IUCN: International Union for Conservation of Nature
Kg: Killogram
M: Meter
$M^{2:}$ Meter Square
MBV: Monodon Baculo Virus
µM: Micrometer
Mg: Milli gram
Ml: Milli Litre
MPEDA: Marine Products Export Development Agency
N_2: Nitrogen
NH_3: Ammonia
NSP: Neurotoxic Shell Fish Poisoning
O_2: Oxygen
OC: Organo Chlorines
OPS: Organo phosphates
PPM: Parts Per Million
PPT: Parts Per Thousand
PYS: Pyrethroids
RVC: Rhabdo Virus Carpio
SPF: Specific Pathogen Free
SPR: Specific Pathogen Resistant
SPT: Specific Pathogen Tolerant
SSP: Single Super Phosphate
SVC: Spring Viremia of Carp
TAN: Total Ammonia-Nitrogen
TSP: Triple Super Phosphate

Chapter 1
Problems of Stocking Density and its Solution

Introduction

Aquaculture is developing, expanding and intensifying in almost all regions of the world. Globally production from capture fisheries has leveled off and most of the main fishing areas have reached their maximum potential. Capture fisheries will therefore not be able to meet the growing global demand for aquatic food. Aquaculture has the potential to make a significant contribute to this increasing demand. To ensure a good production in the aquaculture sector it is very much dependent on the stocking density of the culture system. Low stocking density is not that much profitable for the farmers. But a culture system with high stocking density will always face too many problems. Though there are some kinds of solution to get rid of this problem, but it is good to stock the fish's upto the optimal level of stocking density.

Stocking Density

Stocking density is the term normally used to refer to the weight of fish per unit volume or per unit volume in unit time of water flow through the holding environment (Ellis, 2001).

☆ It is the weight of fish in a unit volume of water.

Optimum Stocking Density of different Ponds

Type of Pond	*Stocking Density*
Nursery pond	3 – 5 million/hac
Rearing pond	0.2 – 0.3 million/hac
Stocking pond	7.5 – 10 thousand/hac
Breeding pond	1000 kg/hac

Stocking Density and Fish Growth

The relationship between stocking density and fish growth has been shown to be positive, negative, density independent or dependent on different experimental density ranges. Stocking density might vary as a function of behavioral adjustments, food availability and water quality. Fish confined at high densities may present chronic stress, potentially resulting in lower growth rates, increased susceptibility to disease and increased mortality rates. For species that live naturally in large shoals, low rather than high densities may be harmful, whereas for territorial species the opposite may be the case. Reductions in stocking density may reduce the physical spread of disease, but stocking density will also have an effect on many other aspects such as the water quality, particulate matter, and fish to fish interaction, (Ellis *et al.,* 2002).

Effect of Stocking Density on different Species

The effect of stocking density varies between species. For example, Sea bass showed higher stress levels at high densities, as indicated by cortisol, innate immune response, and expression of stress related genes. High stocking densities in juvenile gilthead sea bream also produce a chronic stress situation, reflected by high cortisol levels, immunosupression, altered metabolism (Montero *et al.,* 1999). In contrast, Arctic charr feed and grow well when stocked at high densities but show depressed food intake and growth rates at low densities (Jorgensen *et al.,* 1993). In halibut tolerance for high stocking density appears to be stage dependent. Small juvenile halibut had better growth and fewer injuries from aggressive encounters at the highest densities whereas abnormal activity and surface swimming, reduced food consumption, and reduced growth rates with increased stocking density in adult halibut. The effects of high density in rainbow trout and Atlantic cod include decreased growth, reductions in food intake, food conversion efficiency and nutritional status, fin erosion, gill damage, a reduced immune capacity, and alterations in swimming behaviour. High densities are associated both with increased competition, aggression, and physical injury (due to increased contact between fish and between fish and the housing/net), and with degradation in water quality. However, it seems that excessive aggressive behaviour and a poor feeding response may also occur when trout are held at very low densities.

Effect of Different Stocking Densities on the MBV Infected Seeds of *Penaeus monodon*

A three and half month study was undertaken by Centre of Advanced Study in Marine Biology, Annamalai University to evaluate the effect of different stocking densities on the MBV infected *Penaeus monodon* seeds.

The stocking densities of 6 PL/m2, 10 PL/m^2 and 15 PL/m2 were considered as D1, D2 and D3 respectively. All the experimental ponds were supplied with commercial pelleted feed of different grades and the amount was based on the age and growth of shrimps.

The water quality parameters and microbial population was properly maintained by the application of probiotic for all the experimental ponds. A significant higher production, survival and profit were noticed from the pond D1.

From the results of the present study it could be confirm that low stocking (6 PL/m2) density with MBV infected seeds can be successfully cultured by proper water quality and feed management.

Growth of Cultured White Leg Shrimp *Litopenaeus vannamei* in Different Stocking Density

A study was conducted on the culture of *L. vannamei* in the brackish water shrimp farm in west Godavari district of Andhra Pradesh, in four culture ponds at different stocking densities.

After 110th days of pond culture, the mean average weights of the shrimp at harvest were 21.2, 18.9, 19.6, and 17.5; survivals were 82, 92, 81, and 80 per cent; FCR was 1.4, 1.34, 1.38, and 1.35. The average production was 8750, 9813, 8138, and 8591 kg/ha for P1, P2, P5, and P6, respectively.

Two ways ANOVA was attempted between stocking density and Average body weight (ABW). Its shows the relationship between stocking density and average body weight is highly significant.

So ABW increases with less stocking density and vice versa. The investigation concluded that *L.vannamei* culture is successful in brackish water environment and the growth is directly related to stocking density.

Problems

A high stocking density can encourage fishermen, but only if the fish are healthy. Too many fish in a culture system can cause too many problems:

- **Spread of disease:** Help the spread of disease by creating more chances for a pathogen to find and infect a new host.
- **Water quality:** Reduce water quality by lowering oxygen levels and increasing the levels of toxins and suspended solids within the water.
- Damage habitat by reducing the number of plants.
- **Stress:** Increase fish stress through competition for food and space and also by overcrowding. Stressed fishes response rapidly to the diseases

than the healthy fishes. Retarded growth, delayed breeding time, delayed maturation, mortality *etc.* can be caused by stress.

- **Aggressive behaviour:** Produces aggressive behaviour in the fishes due to the competition for food and space. Injuries to the eyes, tails and pectoral fins, causing secondary infections and mortality are also seen.
- **Scale erosion, fin erosion** *etc.* can be happen due to overcrowding these wounds can act as the secondary space for disease occurrence.
- **Deterioration of food:** When the stocking density is high, obviously it will need more feeding. Along with stocking density high rate of feeding causes several problems in the water quality. Increased ammonia level, obnoxious gas formation, rancidity, acidosis, decreased oxygen level, destruction of habitat *etc.* are the common problems associated to high stocking density.

Photo courtesy TFH Publications

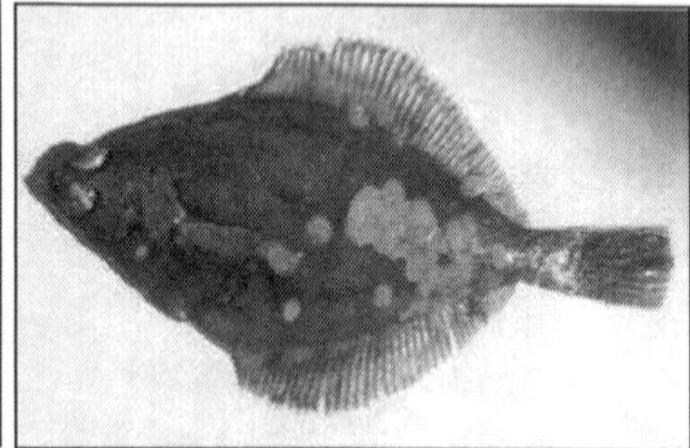

- **Feacal matter:** High stocking density means high number of fishes. So there is a high amount of feacal matter released by the fishes. These feacal matters also deteriorate the water quality of the culture system. Increased ammonia level, obnoxious gas formation, rancidity, acidosis, decreased oxygen level are the result of excessive feacal matter.
- **Shortage of space:** to take a sufficient amount of food and to reflect the natural behaviour, stress response, fear *etc.* it needs a sufficient space. But if the stocking density is too high then a fish is in trouble.
- *Trichodina infestations are typically* caused by high stocking densities and generous feeding rates.
- High feeding rates can lead to high ammonia concentrations, creating an ideal environment for the reproduction of *Trichodina and a* greater chance of a full-blown infection. If the fish are also crowded (as in cage-raised fish) the infestation can spread very rapidly among the fish.
- **Anoxia:** Due to high stocking density the oxygen level of the culture system decreases. Due to this reason the fishes may suffer from anoxia which may cause mass mortality of the stock.
- **Anorexia:** Anorexia is the loss of appetite due to low feeding. Where there is a very high stocking density there is a high competition for food and if the feeding is not sufficient, the fishes suffer from anorexia. It leads to retarded growth, loss of appetite and other problems.

☆ **EUS:** Epizootic ulcerative syndrome is the ulcer formation in the fish body. High stocking density is one of the reasons behind it.

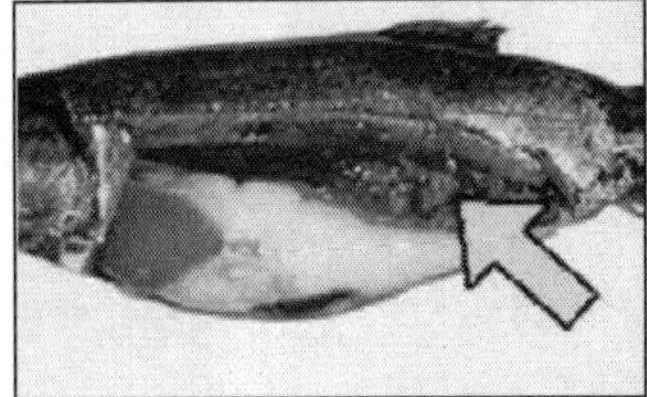

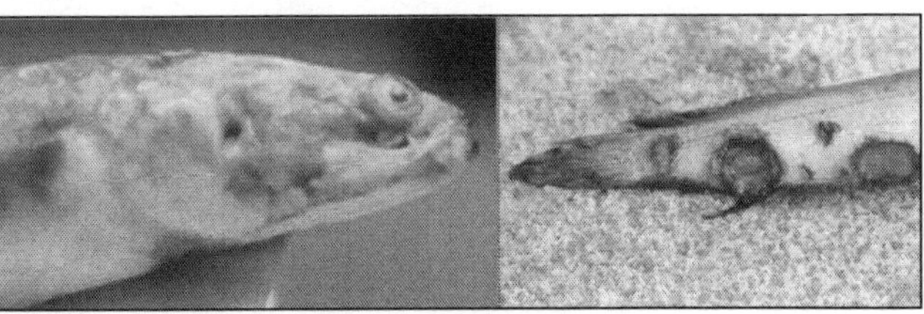

☆ Besides these, high stocking density also leads to chronic stress, reduced feeding, changes in metabolism, down regulation of digestive function and suppression of growth. Also, the high levels of activity associated with agonistic behaviour and chronic stress may result in greater energy expenditure, elevated metabolic rates, decreased food utilization efficiency and impaired immune function.

Categories of stillwater fisheries	Average number of fish	Risk
Natural (fish populations maintained by natural recruitment)	~200lb of fish/acre	Low
Improved (fish populations maintained by natural recruitment and some stocking)	~325lb of fish/acre	↓
Intensive (fish populations maintained by stocking to maximise stock density)	~750lb of fish/acre	High

A guide to different fish stocking densities. The risk increases with the more fish you have.

How high Stock Densities help the Spread of Disease

A mixed species culture with a very high stock density started to lose large numbers of roach, rudd and bream. The deaths started during the summer. The fish were infected with high numbers of the parasite, *I. multifiliis* (white spot). The high stock density meant that it was easier for the parasites to find a fish to live on. It also meant that the fish were more stressed, making it harder for them to fight off the infection.

In a fishery with a low stock density, it is harder for a parasite to find a fish to infect. The parasites that don't find a fish will die, so the number left in the fishery is kept low.

High stock densities can therefore increase the number of pathogens surviving within a fishery. Regulating stock densities is therefore very important in controlling parasite numbers.

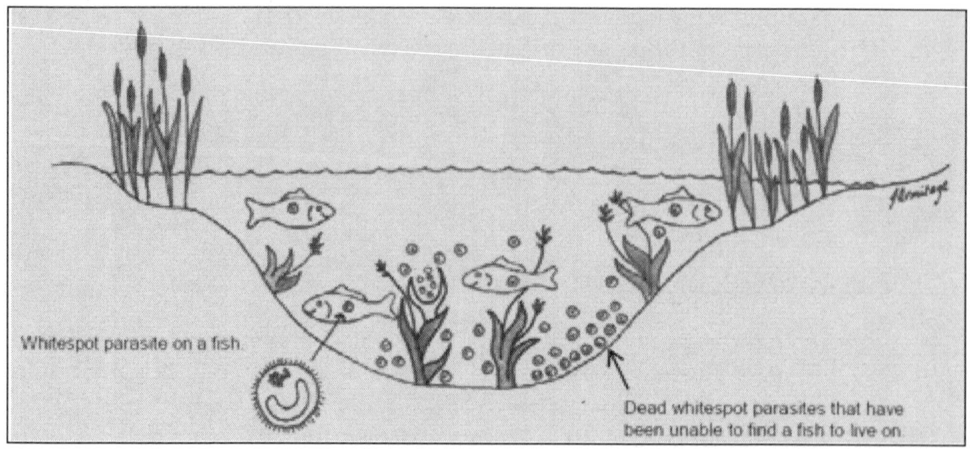

In a fishery with a high stock density, it is much easier for a parasite to find a fish to infect. The number of parasites that die because they can't find a fish is therefore lower.

Solution

A sustainable aquaculture industry has to be ethical but also environmentally and economically sustainable. Unnecessary reductions in the number of fish may render their business unprofitable and as explained above, low stocking densities can also be bad for fish.

What the aquaculture industry requires is a set of guidelines that allows the development of a sustainable industry is good for their fish and is also acceptable to processors, retailers and consumers.

Strategies to reduce disease susceptibility, minimize stress responses, and avoid aggression are, therefore, vital.

However, every fishery is different and can hold varying numbers of fish. There are no set limits as to how many fish can live in water before disease problems occur.

But obviously, "higher the stocking density, greater the risk". However, there are steps to reduce the risk of disease occurring by having good production.

Water Quality

Water quality is most important to resist the diseases and other problems associated with stocking density and other factors. If water quality is good enough then the pathogens can not grow in the system. So there is no intensity of infection. To have a good water quality the parameters should be optimum.

Water Parameters	*Optimum Level*
Dissolved Oxygen	6-8 ppm
Water Temperature	23-27 (°C)
pH	6.8-7.8
Alkalinity	50 – 400 ppm
Total ammonia	less than 1.0 ppm
Hardness	50 – 400 ppm
Iron	Less than 0.05 ppm
Transparency	35 cm
Carbon dioxide (less than)	10 ppm
Calcium	10 -160 ppm
Obnoxious gases	nil

Supplementary Feeding

If stocking density is high then it need supplementary feeding. This will reduce stress levels caused by competition for natural food sources. Care must be taken for choosing the type of food to be feed, as high-fat diets may be bad for the long-term health of fish. A balanced food source is always recommended.

To survive in a stressed condition loaded with microbes and full of diseases sufficient nutrition is required.

To maintain the growth rate at an optimal level, in the highly stocked competitive condition, supplementary feeding is much needed. It also reduces the aggression between the fishes.

Aeration

If the stocking density is more, more oxygen will be taken from the water. Heavily stocked fisheries may need to aerate the water to provide more oxygen for the fish.

There are many different ways to aeration. This allows more oxygen to pass into the water and will keep fish healthy. It is therefore very important to regularly monitor the water quality.

Due to high stocking density, the DO level of water decreases and fishes may suffer from anoxia. To get rid of that aeration is most needed.

Maintaining Optimum Stocking Density

Maintaining optimum stocking density in a culture system the problem of excessive and low stocking density can get rid of.

Water Exchange

Changing the water at least 50 per cent every day or when required will give good result to high stocking density ponds.

Removing Fish

If stocking density is high in the culture system and the fish are reproducing, then it need to consider removing some of them.

The number of fishes removed and their size won't necessarily change the current stock density. But it will stop it from getting any higher as the fish grow and start reproducing themselves.

This is good fisheries management and will enable to maintain the desired stocking density without any further problems.

Water Quality Management in Aquaculture

Aquaculture can be defined as the high-density production of fish, shellfish and plant forms in a controlled environment. Stocking rates for high- density aquaculture are typically thousand fold greater than wild environments. Modem fish culturists employ both open and close systems to raise fish. Open systems, such as, the raceways (used in hatcheries of both finfish and shellfish and also in eel, trout culture) are characterized by rapid turnover of water. Closed systems are commonplace in pond culture of carps, catfishes, tilapia, sea bass, prawn and shrimp among others. Closed aquaculture systems do not have rapid turnover of water, but do not have a high surface to volume ratio facilitating exchange of gases, nutrients, energy *etc.* with the surroundings. Such closed system, intensified, high-density aquaculture forms the basis of concern.

The different forms of high density, intensive aquaculture is quite similar because they all obey the same ~t of physical and chemical principles. These principles compose the subject of water chemistry and its net result *i.e.* the water quality. Poor water chemistry leads to deteriorate water quality, which causes stress to the organisms being raised. Efficient feed conversion, growth and marketability of the final product cannot occur unless the pond system is balanced or in harmony with nature. Therefore the overriding concern of the fish culturist is to maintain, 'balance' or 'equilibrium conditions' with respect to water chemistry and its natural consequence, good water quality.

Water quality for aquaculturists refers to the quality of water that enables successful propagation of the desired organisms. The required water quality is determined by the specific organisms to be cultured and has many components that are interwoven. Sometimes a component can be dealt with separately, but because of the complex interaction between components, the composition of the total array must be addressed. Growth and survival, which together determine the ultimate yield, are influenced by a number of ecological parameters and managerial practices. High stocking density of fish or crustaceans in ponds usually exacerbates problems with water quality and sediment deterioration. Wastes generated by aquaculture activity (faeces and unconsumed feed) first settle in the bottom, as a consequence of organic waste and metabolites of degraded organic matter is accumulated in sediment and water. Part of the waste is flushed out of the ponds immediately or late after the organic matter has been degraded. Low dissolved oxygen level is the major limiting water quality parameter in aquaculture systems. A critically low dissolved oxygen level occurs in ponds particularly when algal blooms die- off and subsequent decomposition of algal blooms and can cause stress or mortality of prawns in ponds. Chronically low dissolved oxygen levels can reduce growth, feeding and moulting frequency.

Culture of penaeids has become intensified since 1986. In an intensive system of *Penaeus monodon* culture salinity, pH and dissolved oxygen were the main parameters, which demonstrated fluctuations of 19.5-27.5 *per cent 0,* 7.4-8.2 and 4.66-8.25 mg/l respectively. In addition, ammonia - N (un- ionized plus ionized ammonia as nitrogen) increased exponentially with culture period, and jumped to 6.5 mg/l after 75 days of cultivation.

Another major consequence of aquaculture production is a high degree of variability in the concentration of dissolved nitrates, nitrites and ammonia. High feeding rates observed in prawn farms lead to eutrophic conditions characterised by substantial phytoplankton blooms. These blooms ultimately senesce and cause rapid increasing in ammonia levels in the ponds. The environmental conditions that create high ammonia concentrations may also cause increases in nitrite concentration. Both ammonia and nitrite can be directly toxic to culture organisms or can induce to sublethal stress in culture populations that results in lowered resistance to diseases.

Ammonia accumulates in culture systems following microbial decomposition of organic material and with some fertilization practices. Microbial decomposition leads to low oxygen concentrations. Low dissolved oxygen concentration increases the toxicity of ammonia to culture organisms. In an aqueous ammonia solution unionized ammonia exists in equilibrium with ionized ammonia and hydroxide ions. The unionized form is usually toxic as it has high lipid solubility and it is able to diffuse quite readily across the cell membrane. Ammonia is utilized as energy source by nitrifying bacteria (*Nitrasomonas* and *Nitrobactor*) and oxidised it to nitrite and nitrate. The production of brackishwater shrimps, which showed a strong increase in the 70's and 80's, showed a declining trend during 90' s in most parts of the world. In general the decrease in shrimp production of aquaculture is attributed to over intensification, leading to the deterioration of the surrounding environment and of pond water and pond sediment quality.

Stress reduced shrimps' resistance to pathogenic diseases, resulting in mass mortality.

Development of aquaculture activities at a particular site cannot be carried out only by considering planned facilities and the quality of water on the site at its origin but also on the aspects of water quality management.

There is a strong relationship between the quality of the water in the pond and that in the water-surrounding environment. Degradation of surrounding water quality will be faster unless proper water quality management techniques are not implemented in the ever-increasing aquaculture system.

Aquaculture Pond Dynamics

Aquaculture ponds are a living dynamic systems they exhibits continuous and constant fluctuations. The pond undergoes a vast collection of both chemical reactions and physical changes. Exchange of atmospheric gases including Oxygen (O_2), nitrogen (N_2) and Carbon dioxide (CO_2) with the pond water are vital to the process of fish metabolism and plant photosynthesis. Inorganic substances (minerals) dissolve from the pond walls and bottom while precipitation of dissolved minerals occurs. Physicals exchanges between the pond its surroundings include absorption of sunlight (radiant energy) to fuel photosynthesis and supply oxygen with in the pond, heat exchange and volume changes caused by evaporation and precipitation (rain). Changes in the volume of a pond are very important as they affect the concentration of dissolved substances and correspondingly requirements for treatment. Hence, the pond dynamics not only depend on its own characters and conditions but also on the surrounding atmospheric weather conditions.

Good production from aquaculture ponds can be achieved when the pond and surroundings make chemical and physical exchanges at a steady state. When all of the processes balance, a state of equilibrium is achieved. Pond equilibrium is the optimum set of conditions for aquaculture, a state completely in harmony with nature.

Water Chemistry

A guiding principle of aquaculture is that water quality and hence efficient production are a direct consequence of good water chemistry. Water may be considered as a 'binder' or 'matrix' in which the dissolved gases, inorganic substances (minerals), as well as organic matter prevails. In addition to dissolved substance, the water matrix gives support to microorganisms, plant and animal life forms and provides a medium for chemical exchange among these populations. However, water is itself relatively chemically inert, physically water has a high heat capacity (holds heat efficiently), is relatively 'polar' affording it the ability to act as an excellent solvent and is also quite dense. Its boiling point is quite high compared to similar molecules and its freezing point quite low. Therefore, water exists as a liquid over a rather broad range of temperature making it a most suitable medium for the support of life forms.

The maintenance of good water quality is essential for both survival and optimum growth of culture organisms. The levels of metabolites in pond water that can have an adverse effect on growth are generally an order of magnitude lower than those tolerated by fishes/prawns/shrimps for survival.

Good water quality is characterized by adequate oxygen and limited levels of metabolites. The culture organisms, algae and microorganisms such as bacteria produce metabolites in a pond. The major source of nutrients in aquaculture is the feed. Because large quantities of feed are loaded in ponds, excess feed, fecal matter and other metabolites become available in large quantities for the growth of algae and microorganisms.

At one point, the increase in population of algae and microorganisms IS exponential. This usually occurs during the second half of the culture period because of available nutrients. About 30 per cent of the total feed consumption is loaded into the pond during the third quarter of the culture period and about 50 per cent is loaded during the last quarter. The algae and microbial population increases until a factor required for growth becomes limiting, after which a sudden decrease in the population can occur. This is referred to as a "collapse" or a "die-off'. The sudden increase and decrease in algal and microbial population can cause drastic changes in water quality parameters, which may affect growth.

By realizing the overriding significance of water chemistry, it IS important to have a firm grasp of some basic concepts. Like:

Temperature

Aquaculture organisms are cold-blooded animals. They can modify their body temperature to the environment in normal condition, unlike the warm- blooded animals, which can react to maintain the optimum body temperature. For ego the optimum range of temperature for the Black Tiger shrimp is between 28°C-30°C. Increase in temperature beyond 30°C increases the activity level and the metabolism. This also increases the growth rate. If the temperature still increases then the shrimp reaches a threshold of physical and nutritional tolerance, which is 33°C in poor quality water or 35°C in good quality water and remains stationary at the pond bottom.

If the environment does not improve the culture organisms may get infected by germs, swim in a disoriented way to the surface or due to exhaustion. If the temperature falls below 28°C, the metabolism reduces and so does the active behaviour and growth rate. Below 20°C, the shrimp will take less feed. Shrimps cannot tolerate a temperature less than B°e.

In the semi intensive culture system, shrimps are more sensmve to temperature than in the extensive one because of the higher biomass and less water volume. During the rainy season, there is a greater possibility of occurrence of thermal stratification in pond water column, as well as the salinity (density) and dissolved oxygen stratification.

Water depth and water volume affect the thermal capacity of the pond and the extent of light penetration. It is related to fluctuation of planktonic algae and benthic algae.

It also influences the volume of the pond and therefore the ponds capacity to support the dissolved oxygen, influencing productivity, biomass and production yield.

Salinity

Salinity plays an important role in the growth of culture orgamsms through osmoregulations of body minerals from that of the surrounding water. For ego the optimum range of salinity for black iger shrimp is between 10 and 25 ppt, although the shrimp will accept salinity between 5 and 38 ppt. since its eurihaline character. The early life stages of both shrimp and prawn requires standard seawater salinities but while growing they can with stand to brackishwater or even to freshwater. However, for better survival and growth optimum range of salinity should be maintained in the aquaculture ponds.

Dissolved Oxygen

Atmospheric oxygen crosses the air-water boundary and dissolves in the water matrix. The only way that oxygen can be introduced from air to water is by diffusion. Atmosphere contains vast amount of oxygen, some of which diffuse into pond waters when they are unsaturated with oxygen. Likewise, oxygen is lost to the atmosphere when pond water have supersaturated with oxygen. The driving force causing net transfer of oxygen between air and water is the difference in the tension between oxygen in the atmosphere and oxygen in the water. Once equilibrium is reached *i.e.* oxygen tensions in air and water are the same, the net oxygen transfer ceases. Oxygen must enter or leave a body of water at the air-water interface and for the very thin film of water in contact with air, the greater the deficit or surplus, the faster oxygen will enter or leave the film, For undisturbed water, the net transfer of oxygen will depend upon deficit or surplus, the area of the air-water interface, the temperature and the time of contact. In general, the rate of diffusion of oxygen depends primarily on the oxygen deficit in water, the amount of water surface exposed to the air and the degree of turbulence.

Typically, dissolved oxygen is measured either in mg. per litre (mgl) or parts per million (ppm) with 0 ppm representing total oxygen depletion and 15 ppm representing the maximum or saturation concentration.

The solubility of oxygen in water decreases as the water temperature increases. It is interesting to note that oxygen appears to operate in a cyclic fashion. Having crossed air-water boundary, dissolved oxygen is utilized by aquatic organism to accommodate metabolism and is excreted as carbon dioxide (CO_2). The liberated CO_2 is used by, photosynthetic plant forms to regenerate oxygen within the pond. The aquatic organism again consumes much of this oxygen and some is returned to the environment. There appears to be a symbiotic relationship between the aquatic organisms and photosynthetic plant forms.

The oxygen cycle and hence oxygen balance can be affected by, what is known as the biochemical oxygen demand (ROD.) of the pond. Decaying plant and animal matter consume substantial amounts of oxygen in the decaying process. It is important to realize that the oxygen cycle and hence dissolved oxygen levels can be affected by changes in the surroundings; a cloudy day with little sunlight will reduce the photosynthetic oxygen contribution to dissolved oxygen.

Similarly, unusually high temperatures will lower the solubility of oxygen in water and hence low dissolved oxygen. When a pond is in 'balance' dissolved oxygen will not vary erratically.

Oxygen is one environmental parameter that exerts a tremendous effect on growth and production through its direct effect on feed consumption and metabolism and its indirect effect on environmental conditions. Oxygen affects the solubility and availability of many nutrients. Low levels of dissolved oxygen can cause changes in oxidation state of substances from the oxidized to the reduced form. Lack of dissolved oxygen can be directly harmful to culture organisms or cause a substantial increase in the level of toxic metabolites. It is therefore important to continuously maintain dissolved oxygen at optimum levels of above 3.5 ppm.

There is above 21 per cent oxygen content in the air. Air acts a big reservoir for oxygen concentration in water is limited by its solubility. The solubility of oxygen:

- Decreases as the temperature increases.
- Decreases exponentially with increase in salinity.
- Decreases with lower atmospheric pressure and higher humidity.
- Increases with depth.

Strategies to maintain optimum levels of DO would be to take advantage of major factors that increase DO and put into check the factors that decrease DO. Photosynthesis plays a major role in oxygen production; respiration of all living organisms in the pond is the major factor involved in oxygen consumption. Oxygen concentration in pond water exhibits a diurnal pattern, with the maximum occurring during the peak of photosynthesis in the afternoon and the minimum occurring at dawn due to nighttime respiration. The magnitude of DO fluctuation is small and occurs around the level of saturated DO when plankton density is low and increases as plankton density increases. Supplemental aeration is generally provided during nighttime when DO increases to levels below 4.0 ppm.

Photosynthesis of phytoplankton is the major contributor of DO during the day and diffusion accounts for increases when DO is below saturation at night. Diffusion at night can be tremendously facilitated with the use of aerators, which exposes more water surface to equilibrate with atmospheric oxygen. Through reverse diffusion, an aerator operated during the day will tend to remove supersaturated DO. The net effect is a milder diurnal fluctuations of DO similar to the conditions of low phytoplankton density. Such conditions are favorable for semi-intensive culture of prawn and shrimp. Photosynthetic oxygen production is also significantly limited when a plankton die-off occurs. The phenomenon is commonly observed when a cyclone occurs. Under these conditions, flushing out decaying plankton, providing for additional aerators and aerating for additional hours may be necessary to maintain DO at optimum levels. When plankton density is high, it has a shading effect which limits the penetration of sunlight in water thereby reducing photosynthetic oxygen production in the bottom of the water column.

High plankton density often results from high nutrient loads and other these conditions, large quantities of feed and fecal wastes are found on the pond bottom. This causes an increase in bacterial population and metabolic activity in the bottom sediments, which are several orders of magnitude higher than that in the water column. Consequently, DO consumption is much greater in the bottom sediment. Limited light penetration and increased DO consumption in the bottom may cause significantly lower DO compared to the top layer of the water column. If this causes DO to deplete to lower than critical levels, disastrous effects on the prawns may happen. Limited light penetration (low secchi disc reading) can also cause differences in the temperature of the top and bottom layer. Temperature stratification usually occurs during calm and warm afternoons. Pond managers should avoid temperature differences of greater than 1°C. It appears that the occurrence of cramps (curved-stiff) in prawns and shrimps, which may cause mortalities, is associated with sudden temperature changes. Circulating the pond water helps remove or minimize stratification by agitators.

It was found that the paddle vheel aerator is capable of elevating the dissolved oxygen level from 0.05 to 4.9 mg/l with in 4 hours in 0.5 ha. Pond. It was also suggested that the low dissolved oxygen values in the aquaculture ponds be improved rapidly by combination of aeration and water exchange.

pH (measure of acidity or alkalinity)

pH or the concentrations of hydrogen ions (W) present in pond water is a measure of acidity or alkalinity. The pH scale extends from 0 to 14 with 0 being the most acidic and 14 the most alkaline. pH 7 is a condition of neutrality and routine aquaculture occurs in the range 7.0 to 9.0 (optimum is 7.5 to 8.5). Exceedingly alkaline water (greater than pH 9) is dangerous as ammonia toxicity increases rapidly. At higher temperatures fish are more sensitive to pH changes.

It is an important chemical parameter to consider because it affects the metabolism and other physiological processes of culture organisms. A certain range of pH (pH 6.8 8.7) should be maintained for acceptable growth and production. But in semi- intensive culture, re-optimum range is better maintained between pH 7.4 - 8.5. pH 7 is the neutral point and water is acidic below pH 7 and basic above pH 7. pH changes in pond water are mainly influenced by carbon dioxide and ions in equilibrium with it. PH can also be altered by a) Organic acids, these are produced by anaerobic bacteria ("acid formers") from protein, carbohydrates and fat from feed wastes, b) Mineral acids such sulfuric acid (acid-sulfate soils), which may be washed down from dikes during rains and c) Lime application.

Like DO, a diurnal fluctuation pattern that is associated with the intensity of photosynthesis, occurs for pH. This is because carbon dioxide is required for photosynthesis and accumulates through nighttime respiration. It peaks before dawn and is at its minimum when photosynthesis is intense. All organisms respire and produce Carbon dioxide (CO_2) continuously, so that the rate of CO_2 production depends on the density of organisms.

The rate of CO_2 consumption depends on phytoplankton density. Carbon dioxide is acidic and it decreases the pH of water. Also, at lower pH, CO_2 becomes the dominant form of carbon and the quantity of bicarbonate and carbonate would decrease. The consumption of CO_2 during photosynthesis causes pH to peak in the afternoon and the accumulation of CO_2 during dark causes pH to be at its minimum before dawn.

The pH should be monitored before dawn for the low level and in the afternoon for the high level. The magnitude of diurnal fluctuation is dependent upon the density of organisms producing and consuming C~ and on the buffering capacity of pond water (greater buffer capacity at higher alkalinity). *i.e.*, Diurnal fluctuation of pH is not great in pond water of higher alkalinity. An alkalinity above 20 ppm CaC03 is preferred in prawn/shrimp ponds. Intervention, such as flushing of ponds to reduce the pH, is advisable when the magnitude of diurnal fluctuation in pH is great.

Nevertheless, one should notice that the drastic fluctuation of pH would cause stress to culture organisms. Normally, it should maintain the daily fluctuation within a range of 0.4 difference. Control of pH is essential for minimizing ammonia and H_2S toxicity.

Ammonia

Ammonia is the second gas of importance in fish culture; its significance to good fish production is overwhelming. High ammonia levels can arise from overfeeding, protein rich, excess feed decays to liberate toxic ammonia gas, which in conjunction with the fishes, excreted ammonia may accumulate to dangerously high levels under certain conditions. Fortunately, ammonia concentrations are partially 'curbed' or 'buffered' by conversion to nontoxic nitrate (NO_3-) ion by nitrifying bacteria. Additionally, ammonia is converted from toxic ammonia (NH_3) to nontoxic ammonium ion CNH_4^+) at pH below 8.0.

Hardness

Numerous inorganic (mineral) substances are dissolved in water. Among these, the metals calcium and magnesium, along with their counter ion carbonate (C03-^{2}) comprise the basis for the measurement of 'hardness'. Optimum hardness for aquaculture is in the range of 40 to 400 ppm of hardness. Hard waters have the capability of buffering the effects of heavy metals such as copper or zinc which are in general toxic to fish. The hardness is a vital factor in maintaining good pond equilibrium.

Turbidity

Water turbidity refers to the quantity of suspended material, which interferes with light penetration in the water column. In prawn ponds, water turbidity can result from planktonic organisms or from suspended clay particles. Turbidity limits light penetration, thereby limiting photosynthesis in the bottom layer. Higher turbidity can cause temperature and DO stratification in prawn ponds.

Planktonic organisms are desirable when not excessive, but suspended clay particles are undesirable. It can cause clogging of gills or direct injury to tissues of prawns. Erosion or the water itself can be the source of small (1-100 nm) colloidal particles responsible for the unwanted turbidity. The particles repel each other due to negative-charges: this can be neutralized by electrolytes resulting in coagulation. It is reported that alum and ferric sulfate are more effective than hydrated lime and gypsum in removing clay turbidity. Both alum and gypsum have acid reactions and can depress pH and total alkalinity, so the simultaneous application of lime is recommended to maintain the suitable range of pH. Treatment rates depend on the type of soil.

Redox Potential (Oxidation-Reduction Eh)

Redox Potential is an index indicating the status of oxidation or reduction. It is correlated with chemical substances, such as Q, CO_2 and mineral composed of aerobic layer, whereas H_2S, CO_2, NH_3, H_2SO_4 and others comprise of anaerobic layer. Microorganisms are correlated with the status of oxidation or reduction. With the degree of Eh, it is indicative of one of the parameters that show the supporting ability of water and soil to the prawn biomass.

In semi intensive culture photosynthetic bacteria (PSB) plays an important role through absorption and conversion of organic matter into the minerals and nutrients as a secondary production, compared to the primary production of algal population. PSB exist particularly due to low oxygen level and high intensity of light and can significantly improve the culture environment.

Water Quality Management

Water quality parameters should be monitored to serve as guide for managing a pond so that conditions that can adversely affect the growth of prawns can be avoided. In cases where problems are encountered, these parameters can help in the diagnosis, so a remedy can be formulated. Individual parameters usually do not tell much, but several parameters put together can serve as indicators of dynamic processes occurring in the pond.

The population of phytoplankton and microorganisms are major determinants of the level of oxygen and metabolites in the pond. The diet fluctuation of DO (including its vertical profile), pH and CO_2 serve as indicators of their population. Since C~ is the major factor affecting the dial fluctuation of pH, monitoring pH fluctuation may be adequate. Also, CO_2 is more difficult to measure.

Daily measurements are conducted at early hours *i.e.* 5-6 am and after noon measurements in the *i.e.* 2- 3 pm. This represents the period before the start of photosynthesis and the peak of photosynthesis, respectively. Thus the maximal and the minima of these parameters occur during this period. The other parameters do not have a distinct dial pattern and therefore can be monitored only once a day, preferable at a common time.

Feed and growth data need to be presented with water quality parameters, side by side. This is because algal blooms are consequences of nutrients from feeds and excess feed can cause the rapid deterioration of water quality.

Careful monitoring and data collection will remain useless unless it influences decisions regarding water management. This becomes more important as cost to implement various management schemes (aeration, water exchange, inputs) increase.

Most of the water quality problems can be solved with adequate water exchange. Thus, if large quantities of water suitable for aquaculture were available, monitoring would not be as critical and high production levels can be targeted. If water is limited, the risk of encountering water quality and disease problems increases as one goes for more intensive culture.

Water Exchange

When stocking density increases, It IS of primary importance to have a dependable water supply and to maintain good water quality. So far, besides aeration, water exchange is still the most effective and widely employed method to maintain good water quality besides water quality enhancers like sanitizers, zeolite *etc.* Generally, water exchange is used to adjust salinity as desired, to remove excess metabolites, to keep algae healthy and producing ample oxygen and to regulate pond water temperature. The exchange rate varies with the production period, stocking density and total biomass, levels of natural productivity, turbidity and source and volume of water. The principle of water exchange is to change the water in a way such that the water quality changes gradually instead of abruptly. In semi- intensive systems, frequent and sometimes even continuous water exchange at a small flowing rate is employed. Abrupt addition of large quantities of water in small ponds may result in sudden environment change, which subsequently can stress in culture organism. Therefore, massive water replacement is not recommended unless there is sudden die-off of plankton, critical low oxygen or after the application of chemicals.

Continuous water exchange should be accompanied with running of paddlewheels to have the pond water fully mixed. Otherwise, it will cause great differences of water quality within a pond and heterogeneous distribution of culture organisms on the pond bottom. Lowering the water level first and adding new water is not recommended, especially during daytime in summer. Increasing water temperature, while lowering water level, can reduce the capability of water to hold oxygen and hasten the degeneration of the pond bottom, leading to oxygen depletion. It is better to add new water first according to the predetermined exchange rate, have the paddlewheel running to homogenize water throughout the pond and then discharge water. Both bottom and surface water discharge can be considered. Water quality in the bottom layer is generally worse than near the surface. Surface water should be discharged when there are scums, shrimp or prawn faeces and floating dead plankton are present. During and after showers, the lower density freshwater at the surface layer should be discharged to avoid salinity change.

The lowering of salinity more than 5 ppt, at each time, by the water exchange is not recommended. Drastic changes of salinity may alter the phytoplankton fauna and their population densities and lead to instability of the ecosystem.

Water exchange is the first method to improve pond management, except when

- ☆ Good quality of water is not available.
- ☆ A drastic change to the pond environment should be avoided.
- ☆ Culture organisms have been greatly weakened as a result of nutrition depletion and diseases.
- ☆ Ponds are being treated with chemicals and medicine.

While doing water exchange several parameters should be noticed such as pH value, salinity, temperature, turbidity and other parameters related to defining good pond environment.

In the first month of culture, the water cultivation is most important and therefore only low level of exchange water addition to fill up pond to take care of evaporation and seepage needs to be done instead of heavy water exchange. As the culture proceeds the degree of contamination is different and the purpose of water exchange is to decrease the contamination that can influence shrimp/prawn growth. Basically, the water exchange rate is determined by maintaining the optimum range of parameters discussed above and particularly on the amount of biomass.

Acceptable water quality can be achieved using the following pond management techniques:

- ☆ Controlled water exchange. Water exchange is done daily, as routine or extra exchange may be recommended to prevent onset of a problem or be recommended as integral to a crisis management.
- ☆ Co- ordinating water exchange and fertilization to maintain populations of algae.
- ☆ Avoid overfeeding through proper management of feeding trays.
- ☆ Aeration.

☆ Continual or periodic removal of accumulated organic material from the pond bottom.

☆ Maintenance of a high density bacterial flock *i.e.* probiotics combined with water circulation and aeration.

Water exchange alone does not in prove dissolved oxygen unless there is a signifIcant difference in DO levels between intake and pond water. In farms that practice the stepwise water exchange method, it is beneficial to schedule filling of pond and incoming water is the maximum. An effective method of achieving desirable water quality during culture is a periodic reduction of 25- 50 ern in the water level of the pond followed by heavy simultaneous exchange of water for 12- 24 hours a the reduced depth. Since the water volume in the pond is reduced a greater percentage exchange is achieved in a shorter period.

This is recommended to achieve the following:

Temporary improvements in DO caused by cropping algal populations below the "compensation depth" of the pond, which is the depth at which oxygen production and respiration by phytoplankton are equal. Light -limited phytoplankton below the compensation depth are net consumers of oxygen. The compensation depth roughly corresponds to twice the secchi disk depth. Prevent an increase of undesirable algal species. Water exchange alone can reduce the density of undesirable species. If fertilization and inoculation is done simultaneously with water exchange, culture of appropriate species can be achieved. The reduced pond volume amplifies the effects of fertilization.

Routine water exchange requirements are determined by shrimp/prawn! biomass, feeding rates, phytoplankton and microbial densities and the presence or absence of aerators and sediment disposal systems. Daily water exchange is correlated with shrimp/prawn biomass and adjusted in accordance with DO and secchi disk readings. Plotting morning and afternoon DO and secchi disk readings facilitates early detection of trends that would require adjustments in the basic water exchange program. Nighttime DO can be predicted by simply plotting DO readings at dusk and during the night, and extrapolating the results to dawn.

Although it is possible to maintain large biomass of shrimp with low rates of water exchange, pumping costs are generally less than 10 per cent of the total cost of production and unless water resources limited, it is not advisable to reduce water exchanges rates in semi- intensive systems simply to minimize costs. Higher water exchange reduces the risk of water quality related disease problems. It is also one of the few management tools available for preserving survival in diseases animals that does not rely on use of chemicals.

Aeration

Padd1ewhee1s are commonly used in serm- mtensive shrimp/prawn culture and is one of the major capital cost item in the farm. The padd1ewhee1 aerators are used to increase contact surface of water with air thereby increasing the area through which oxygen is absorbed by the water and to create a circular movement of the pond water. This has the following advantages:

- It increases the dissolved oxygen level of the water and prevents oxygen depletion during the night.
- It accelerates the diffusion effect of not only the oxygen, but also enables the capture or release of carbon dioxide. Carbon dioxide is important for culture of algae and therefore for maintenance of appropriate watercolor.
- It facilitates the volatilization of undesirable gases such as N_2, NH_3, C~ and H_2S.
- It reduces the daily fluctuation range of pH value.
- It accelerates the decomposition and mineralization of organic matter ill water and soil and helps in the release of nutritive value of fertilizers.
- It diminishes the possible stratification of pH, DO, salinity and temperature ill the pond water.
- It helps in mixing the pond water and maintenance of ideal conditions all over the pond.
- It increases turbidity when necessary.

As the pond water moves in a circular fashion the pond bottom is cleaned and the waste matter gets accumulated in the center and the comers. By this method most of the pond bottom is kept clean.It is very important during application of chemicals or medical treatment with respect to distribution of chemicals/medicines throughout the pond and ensure to be careful in proper water exchange to reduce any side effects. Use of paddlewheels also has the advantage of accelerating the evaporation rate thereby increasing the salinity. This is however, an advantage in the condition of low salinity.

There is no hard and fast rule regarding the number of paddlewheels to be used in a pond because it is difficult to define the relation between biomass in the pond and the interaction of various water parameters.

The arrangement of paddlewheels in the pond is done to maxmuze circulation efficiency and minimize dead comer areas. The arrangement of paddlewheels should be to ensure anticlockwise movement of water in the pond in the Northern Hemisphere and clockwise movement of water in the pond in the Southern Hemisphere. This ensures that when bottom drainage arrangements are used, the efficiency of water exchange and draining out of waste product is improved since water when existing from the bottom of the reservoir in the northern hemisphere has an anticlockwise vortex, whereas in the southern hemisphere has a clockwise vortex. Aerators are usually arranged parallel along the banks and about 5 to 10m distant from the dikes depending upon the pond size. Each paddle wheel should normally separated by a distance of 30 m and 50 m for optimum efficiency of the paddle wheel. Avoid relocating paddle wheels during culture.

Removal of Dissolved Metabolic Organics

One of the important stress factors is the increase of dissolved metabolic orgamcs ill culture water. It can increase ammonia and microorganisms.

Dissolved Metabolic Organics

1. Decaying excess high protein feed
2. Dead larvae, orgamsms, algae

Causes Stress

(a) Toxic ammonia nitrate

Animals stop eating, weakness and become susceptible to diseases.

(b) Microorganisms bacteria, fungi and protozoans

Lower water quality

3. Animals excreta, urine

(c) BOD (Biological oxygen demand)

This explains why water quality deterioration could quickly cause a high mortality rate. To prevent the buildup of dissolved organics, frequent partial to total water change is necessary; or the pollution could be reduced by the chemically removing the pollutants by adsorption using activated carbon.

The best way to facilitate the removal of metabolic wastes in a pond is by flushing out water from the bottom. Constantly maintaining high DO in the pond through supplemental aeration and water exchange, enhances nitrification. Nitrification is a major mechanism for ammonia removal in well-aerated ponds. Paddlewheel aerators are usually operated during dark (7 pm to 7 am) when oxygen depletion is likely to occur and at noon (12 noon to 2 pm) when temperature and oxygen stratification can become significant.

Phytoplankton Management

Phytoplankton play a significant role in stabilizing the whole pond ecosystem and in minimizing the fluctuations of water quality. A suitable phytoplankton population enriches the system with oxygen through photosynthesis during day light hours and lowers the levels of CO_2, NH_3, NO_2 and H_2S. A healthy phytoplankton bloom can reduce toxic substances since phytoplankton can consume Nl-I, and tie- up heavy metals. It can prevent the development of filamentous algae since phytoplankton can block light from reaching the bottom. A healthy bloom also provides proper turbidity and subsequently stabilizes shrimp and reduces cannibalism. It decreases temperature loss in winter and stabilizes water temperature.

Pond Bottom Treatment

For farms adopting advanced technology, it is necessary that pond bottom should be completely dried and aerated to get rid of toxic gases.

Many ponds in low-lying areas cannot be completely drained and dried. To overcome this, Aquafarmers apply waste digesters to the ponds. The digesters are harmless bacteria (probiotics) and enzymes that consume organic matter on the pond bottom.

After the application of digesters farmers apply a disinfectant, either organic silver or organic iodine. Copper sulphate is not used as a disinfectant nowadays as It IS not biodegradable and accumulates in the pond upto levels that are toxic to aquatic life. Organic silver is highly effective against bacteria and viruses and its toxicity to aquatic life is very low. Organic silver is applied at the rate of 18 litres (4 gallon) per hectare after lowering the water depth to 12 inches. Seven days after the application, this disinfectant disintegrates, so there is no need to flush the pond. Organic silver also prevents the development of algae that grows on shells.

Organic iodine, can cure gill or shell diseases, kills bacteria on contact and has low toxicity. Its effect can be noticed within 24 hours and the pond bottom can be disinfected without emptying the pond.

The suggested dosage is 5 ppm to 10 ppm. Its affectivity lasts for two to three days compared to about seven days in the case of organic silver.

Nitrogen Metabolites

Large quantities of organic matter ongmating from the heavy feed load and feacal matter accumulate in aquaculture ponds. These undergo oxidation-reduction reactions leading to decomposition, mainly through the action of bacteria. Different forms of inorganic nitrogen like ammonia, nitrite and nitrate are produced during decomposition.

Maintaining Water Quality and Preventing Diseases

Environmental conditions vary considerably at different times of the year and the bacterial and fungal, load of seawater also varies. During the dry moths; there is less dilution of organic and toxic pollutants from human and industrial wastes. During this time the absence of rains also reduces water exchange between clean seawater and polluted coastal water. The result is a rise of viral, bacteria, protozoa, fungi and toxic pollutants in the water. This is partially upset during the hot summer months by phytoplankton and zooplankton blooms, which assimilate some of the bacteria and toxic substances. Under such conditions, cultured animals become vulnerable to infection. They are stressed by the following:

- Overcrowding in captivity.
- Temperature fluctuation of water, especially during water change (A onedegree Celsius difference can cause stress).
- A temporary decline in dissolved oxygen level due to power failure.
- Increase of free-carbon dioxide, un- ionized ammonia and organics due to decaying excess feed and dead animals.
- Physical manhandling during water change.
- Poor nutrition - improperly fed fish and prawn.
- The high level of toxic pollutant in seawater that may contain heavy metals such as copper, zinc, lead, nickel, mercury and chemicals like poly-chlorinated biphenyl compounds, chlorinated hydrocarbons such as DDT and other pesticides.

While there is no known practical way to remove these pollutants. Effect should be made to limit these stress- inducing factors to keep the animals strong enough to fight infection. Healthy animals, do not easily succumb to diseases. Where adequate filtration is not possible, treatment of water is suggested to lower the bacteria and fungal load of the water.

Self-Pollution as a Possible Factor

When the accumulation of nutrients within ponds is high, self- pollution of the culture environment reduces production, frequently as a result of a severe disease-outbreaks. Although, in some cases, production losses can be linked back directly to diseaseoutbreaks, it is often difficult to separate the effect of disease and poor water quality. Disease-outbreaks occur when (1) a pathogen infects a population previously not exposed to the microorganism. or (2) poor culture conditions weaken resistance to pathogens permanently present in the culture environment. Outbreaks of new infectious diseases will be difficult to prevent as long as there are no strict regulations for transfer of culture stocks between regions. From a practical point of view, more attention should be paid to culture conditions, with special attention to water quality.

Some farms, experienced a collapse in production from 15-18 Mt ha^{-1} to 4-6 Mt ha', but were able to restore production levels of 10-12 Mt ha' ycar' on a continuous basis. These farmers concentrated on water quality management, introducing measures such as high levels of pond flushing (>30 per cent day") excavation and tilling of pond bottoms upon harvest, emergency aeration and the use of drugs, chemicals and biological agents to suppress disease-outbreaks. Farmers apply these measures empirically.

Water Quality Parameters for Shrimp Farming

Water Parameter	*Optimum Level*
Temperature	26-33 °C
Salinity	10-25 ppt
Dissolved Oxygen	>3.0 ppm
P^H	7.5-8.5
Total Ammonia Nitrogen	<1.0 ppm
Total Nitrate Nitrogen	<5.0 ppm
Nitrate Nitrogen	<0.01ppm
Sulphide	<0.03ppm
Biological Oxygen Demand	<10.0ppm
Chemical Oxygen Demand	< 70.0ppm
Sechidisc Visibility	25-45cm

Details of the analysis methods and the required equipment for the physico chemical parameters and microbiological aspects of water and sediment in aquaculture ponds

Sl.No.	Parameter	Methodology	Equipment I Chemicals
1	Temperature	Visible	Digital Thermometer
2	Salinity	Visible/titration	Refractometer
3	pH	Visible/titration	pH pen/meter
4	Alkalinity	Titrimetric method	Glassware and Chemicals
5	Transparency	Light intensity	Sacchi disc
6	Turbidity	Tendol effect principle	Tindalometer
7	Dissolved Oxygen	Visible/titrimetric method (Wrinklers)	YSI DOmeter
8	BOD	Visible/titrimetri method (Wrinklers)	YSI BOD probe
9	COD	Reflection method	Reflexor unit
10	Nitrate - N	Wood *et al.* (1967) method American Public Health Association (APHA)	Spectrophotometer meter (Chemicals and Glassware)
11	Nitrite - N	Wood *et al.* (1967) method (APHA)	Analysis kits (Merck, Loba, YSI, BDH *etc.*)
12	Ammonia-N	Soloranzo (1969) (APHA)	Analysis kits (Merck, Loba, YSI, BDH *etc.*)
13	Phosphate - P	Murphy and Riley, 1962 (APHA)	Analysis kits (Merck, Loba, YSI, BDH *etc.*)
14	Hydrogen Sulphide	p- Phenylene Diamine Hydro-chloride (APHA)	Analysis kits (Merck, Loba, YSI, BDH *etc.*)
15	Sediment Organic Matter	Gaudette *et al.,* method Titration	Glassware and Chemicals

Details of the Importance of Physico-chemical Parameters and Microbiological Aspects in Aquaculture

Sl. No.	Parameter	Purpose	Equipment Required	Chemicals and Glassware Required
I	**Water quality parameters**			
1	Temperature	Maintenance of optimal temperature, fluctuations at high level, leeds to severe effect on entire body of pond and leeds thermal stress on shrimp, algal crash *etc.* (28 - 32°C)	Mercuric thermometer/ Digital thermometer	—
2	Salinity	Eurihaline; tolerance capacity with broad range of salinity (10 - 25 ppt)	Clinometers (Refract meter)	Titrimetric method of knewdson's for standardization of salinometer
3	pH	Little bit basic conditions are favourable (7.4 - 8.5)	Digital pH meter/ pHpen	Standardized by Titrimetric method
4	Transparency	Optimal maintenance of plankton density (30 - 40 cm)	Sacchi disc	Primary productivity estimation by dark white bottle method by C^{14} method

Sl. No.	Parameter	Purpose	Equipment Required	Chemicals and Glassware Required
5	Dissolved Oxygen	Is the main requirement for physiological and biochemical activities of living things (> 3 mg/l)	D.O. Meter/ Wrinklers method	Chemical and Glassware required burette, pipette, conical flasks, D.O bottles (reagent bottles), beakers *etc.*
6	BOD	It is necessary to estimate the requisite of oxygen content by the enclosed body of water	D.O. method/Wrinklers method	Chemical and Glassware required burette, pipette, conical flasks, D.O bottles (reagent bottles), beakers
7	COD	The amount of oxygen required for the complete oxidation of all organic/ chemical components in the pond environment	Oxidized method	Chemical and Glassware required reflexor, hotplate *etc.*
8	Nitrate - N	The nutrient content is very high leads to heavy blooming of algae	Spectrophotometer meter/ Calorimeter/ kits (YSI, Merck, Loba)	Chemicals and Glassware
9	Nitrite - N	It is an toxic constituent	Spectrophotometer/ Calorimeter/kits (YSI, Merck, Loba *etc.*)	Chemicals and Glassware
10	Ammonia-N	It is an toxic constituent	Spectrophotometer/ Calorimeter/kits	Chemicals and Glassware
11	Phosphate-P	Nutrient	Spectrophotometer/ Calorimeter/kits	Chemicals and Glassware
12	Hydrogen sulphide	Toxic and makes anoxic conditions	Spectrophotometer/ Calorimeter/kits	Chemicals and Glassware
13	Sediment analysis			
	a. Sediment organic matter	It shows how much of organic matter and organic carbon produced in the pond bottom by the shrimp farming activity.	Titrimetric method	Chemicals and Glassware
	b. Sediment composition	It exhibits the ratio's of sediment components *i.e.*, sand, silt and clay	Pipette method	Chemicals and Glassware
II	**Biological**			
A.	**Bacteriological**			
1	TVC (Total Viable count of Bacteria)	It reveals the total bacterial forms that harbour the pond environment	Laminar flow chamber, oven, incubator, autoclave, petridishes, test tubes, conical flasks, test tube stand, micropipette, beakers, digital colony counter.	TGY media (Tryptone Glucose Yeast extract Agar)

Sl. No.	Parameter	Purpose	Equipment Required	Chemicals and Glassware Required
2	TVLO (Total Vibrio like organizers)	It reveals the total bacterial forms that harbour the pond environment	Laminar flow chamber, oven, incubator, autoclave, petridishes, test tubes, conical flasks, test tube stand, micropipette, beakers, digital colony counter.	TCBS media (Thiamine Citrate, Bile sucrose salt Agar)
3	Other pathogenic bacteria	It reveals the total bacterial forms that harbour the pond environment	-DO-	Different media for respective various bacteria
B.	**Virology**			
1	Dot Blot method	Viral diseases like WSV, MBV *etc.*	Kits	—
2	PCRmethod	Viral diseases like WSSV, MBV *etc.*	Primers, Thermal cyclar, UV laminar flow chamber, electrophoresis unit, UPS system *etc.*	—
3	Histopathology	Viral diseases like WSV, MBV *etc.*	Microtome, staining and distaining equipment and stains.	—
C.	**Ectocommensal**			
1	Parasites	Like Zoothamnium, Vorticella, Fungi *etc.*	Simple binocular microscope and Compound monocular microscope with CCTV.	—

Conclusion

In fish culture systems, increasing rearing density is one way of optimizing productivity. However, poor growth and increased incidence of disease have often been observed in fish reared at higher densities.

On the other hand, stocking density is one of the most important biotic factors in aquaculture because it directly influences survival, growth, behavior, health, feeding and production.

Higher densities may interfere with intra- population interactions and can consequently affect biomass gain. But lower density also is not profitable and sometimes causes several problems. So it is advisable to maintain the stocking density upto an optimum level.

Chapter 2
Environmental Hazards in Aquaculture

Introduction

In general environment is the physical surrounding of an organism.There is a direct relationship between environment and related organisms. One of the greatest problems that the world is facing today is that of environmental pollution. It is increasing with every passing year and causes grave and irreparable damage to the earth.Thus any problem in the environment is a matter of great concern. Aquaculture system is a part of the environment and two-way interactions are abundant.The success of aquaculture depends upon the adequate exchange of tidal water and a natural food supply.Major causes of impact are associated with feeding and nutritional wastes, the existence of diseases and the interbreeding of wild and selected strains.

Aquatic environments are complex ecosystems resulting from the interactions among physical, chemical, biological, and human-engineered factors. In essence, water, air, soil, and aquatic organisms can be viewed as overlapping compartments through their closely associated interfaces in the environment.Pollutants are defined as undesirable substances that produce injury in living organisms. Chemical pollutants generally act on the ecosystem by three major pathways: (a) settling on the substrate and smothering life there, (b) through acute toxicity leading to death of organisms, or (c) depleting oxygen values to a threshold level causing death of organisms. Therefore, pollution causes a departure from normal functioning of the system. As a consequence, chemical pollution of the water may profusely modify the biota and affect the ecosystem.

For a full understanding of the nature of the problem of aquatic environmental problems, it is necessary to examine the nature and extent of pollutants and the mechanisms of the actions of these pollutants on biota. This chapter highlights the knowledge regarding the problems in aquatic environments and their impact on fisheries impact and water resources.

Environmental Pollution is of Three Basic Types

The farming of aquatic animals produces waste in the form of fecal matter and unused feed. These largely nitrogen-based wastes can cause oxygen depletion in their aquatic environments and a net loss of productivity in certain areas. Additionally, the use of antibiotics, antifoulants, and pesticides are all problems that aquaculture can be introduced into the aquatic environment.

Pollutants enter aquatic environments and exert adverse effects on biota at the cellular, organism, population, and community levels, eventually altering the functioning of an aquatic ecosystem as a whole. Pollutants entering freshwaters and the oceans remain partly in solution and are partly adsorbed onto the surface of sedimentary matter.

1. Air Pollution
2. Water Pollution
3. Soil Pollution

1. Air Pollution

The emission of any impurity into the air, such as smoke, dust, cinders, solid particles, gases, mists, fumes, odours and radioactive substances. Air pollution is by far the most harmful form of pollution in our environment.Air pollution is caused by the injurious smoke emitted by cars, buses, trucks, trains, and factories, namely sulphur dioxide, carbon monoxide and nitrogen oxides. Even smoke from burning leaves and cigarettes are harmful to the environment causing a lot of damage to man and the atmosphere. Evidence of increasing air pollution is seen in lung cancer, asthma, allergies, and various breathing problems along with severe and irreparable damage to flora and fauna. Even the most natural phenomenon of migratory birds

has been hampered, with severe air pollution preventing them from reaching their seasonal metropolitan destinations of centuries.

2. Water Pollution

Water resources are an indispensable part of human society. Deterioration of the aquatic environment is associated with urbanization by human populations commencing with the industrial revolution in the eighteenth century. Of course water is very important for our society, and if we start saving it we will pay less because we consume less and of course we will keep our environment clean. After we use water, it is not the same as it was before. It is polluted. In many rural areas one can still find people bathing and cooking in the same water, making it incredibly filthy. In addition to these, thermal pollution and the depletion of dissolved oxygen aggravate the already worsened condition of the water bodies.

Water pollution can also indirectly occur as an offshoot of soil pollution – through surface runoff and leaching to groundwater.During the aquaculture process, irrespective of the species, water is the vehicle for both feed and waste.

3. Soil Pollution

Soil pollution, which can also be called soil contamination, is a result of acid rain, polluted water, fertilizers *etc.*, which leads to bad crops. Soil contamination occurs when chemicals are released by spill or underground storage tank leakage which releases heavy contaminants into the soil. These may include hydrocarbons, heavy metals, herbicides, pesticides and chlorinated hydrocarbons.

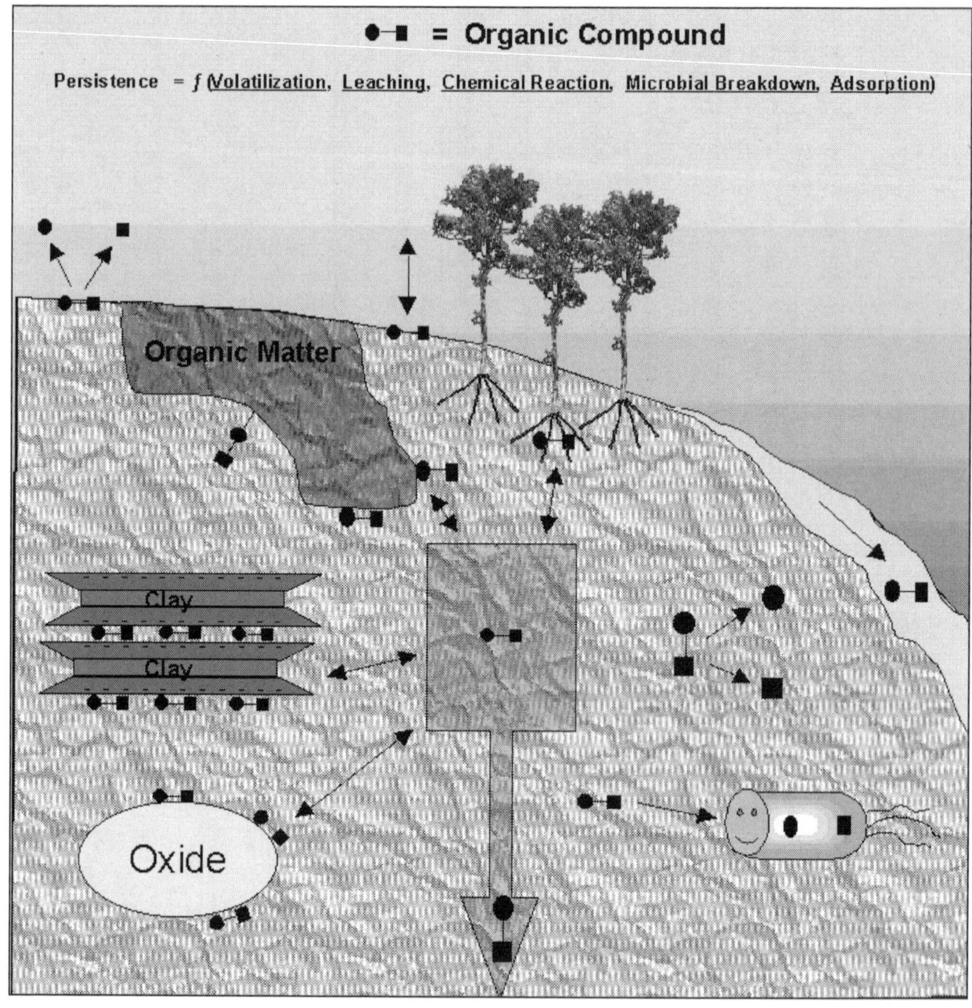

Interaction between Aquaculture and the Environment

The development and intensification of aquaculture has revealed a broad spectrum of associated environmental issues. The interactions between aquaculture and the environment are diverse and complex. They are directly related to sustainable development (IUCN, 2007). Domestication Domestication in aquaculture is the acclimatisation to captive conditions, the total control of the life cycle and the manipulation of breeding in captivity of aquatic organisms (Hassin *et al.*, 1997). The advantages of domesticating organisms are: to secure seed supply and improve production efficiency through the mastering of breeding and feeding to select organisms that can grow faster; to achieve better feed efficiency; and therefore to alleviate the pressure on fishes used as feed.

It will also minimise the potential negative impacts on wild stocks by attempting to make cultured organisms unable to live in wild ecosystems. However, domesticated animals become significantly different over time from their wild counterparts, both genetically and physically. The escape or release of strongly domesticated organisms into the environment can lead to unpredictable effects on the ecosystems, both on wild populations of the same species and/or other organisms.

In the case of aquaculture, the risk posed by the escape of domesticated organisms is far greater than that of terrestrial animals or plants in similar circumstances, because of their ability to disperse and the difficulty of recapture. As a principle for the sustainable development of aquaculture, as the domestication of species for aquaculture is necessary, the interaction of these domesticated organisms with their wild counterparts should not have negative effects. And therefore: selective breeding of aquacultured organisms should be designed to reduce their capacity to survive or reproduce in the wild; research for domestication should be encouraged and supported, and the creation of gene banks of wild species should be promoted as a reservoir of genetic sources.

The Major Environmental Factors that can Affect Fish Community

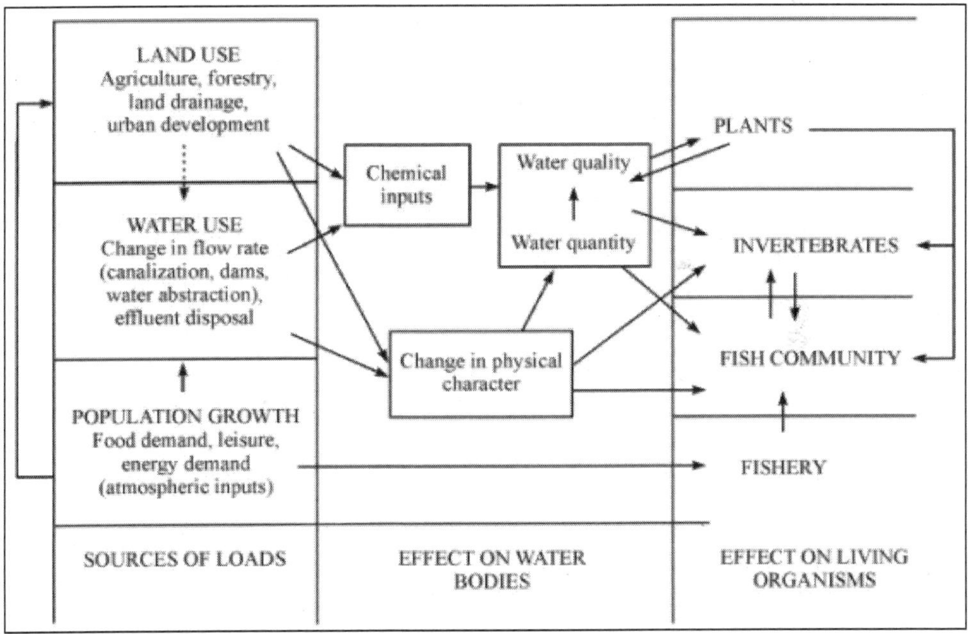

All attempts at increasing human food supplies inevitably produce environmental consequences elsewhere. With aquaculture, those consequences are only now starting to be understood. Reporters looking to cover the aquaculture sector should be well versed in the environmental factors of aquaculture. Fisheries are directly or indirectly affected by each type of pollution and global warming is an arising problem to it.Almost all pollution comes to an end in water and hence aggravates the problem.Some of the key environmental factors are as follows:

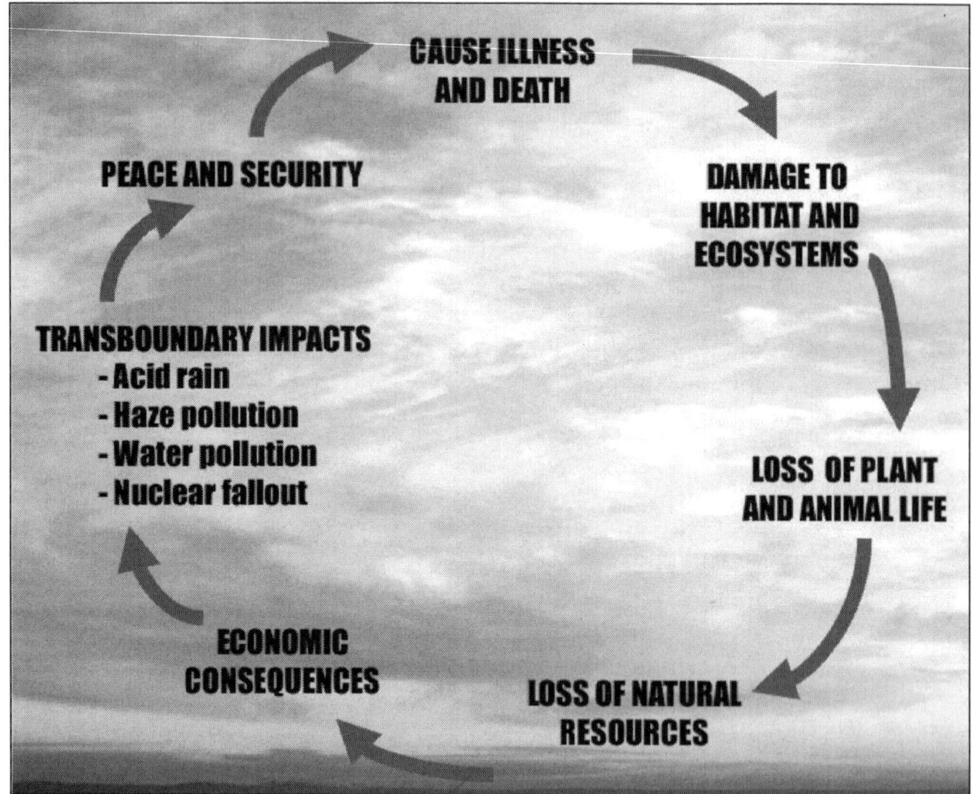

Pollution Caused by Aquaculture Wastewater

If continuously discharged wastewater without treatment, which contains high concentration of nitrogen and phosphorus nutrients, may result in a remarkably chronic elevation of the total organic matter contents, especially in badly managed or poorly located sites. Consequently, a series of negative ecological impacts may occur: (1) serious oxygen deficit caused by the decomposing of organic substances. (2) eutrophication or algae bloom caused by the accumulation of organic nutrients like nitrogen and phosphorus, which promotes a high biomass in the superficial water. Apart from increased phytoplankton production, eutrophication can cause many other effects which may be more sensitive and relevant indicators such as changes in: energy and nutrient fluxes, pelagic and benthic biomass and community structure, fish stocks, sedimentation, nutrient cycling, and oxygen depletion (Gregory and Zabel 1990, Fang *et al.*, 2004). (3) Water deterioration will bring about low productivity (4) Diseases may break out. Aside from this, inadequate handling of wastewater has serious consequences for human health, the environment and economic development (Enelld and Lof 1983). It contaminates water supply, increasing the risk of infectious disease and deteriorating groundwater and other local ecosystems, for instance after flooding.

However, all phases in the life of a mine can discharge metals to estuaries, rivers, streams, and lakes (Ripley and Redmann, 1978). Metals dissolve in water and are easily absorbed by fish and other aquatic organisms.

Metal toxicity produces adverse biological effects on an organism's survival, activity, growth, metabolism, or reproduction. Metals can be lethal or harm the organism without killing it directly. Adverse effects on an organism's activity, growth, metabolism, and reproduction are examples of sublethal effects.

Acid Rain

Acid rain results primarily from emission of sulphur dioxide (SO2 and nitrogen oxides (NOx) into the air Main sources: coal burning, and other fossil fuel use.A wide range of pesticides and herbicides, such as atrazine, dichlorobenyl, 2,4-D, and glyphosate, are widely used in modern agriculture, and this practice has become a potential threat to fish and fisheries. Many of these toxic substances are highly persistent and can bioaccumulate in the food chain of fishes at sufficient levels to threaten aquatic ecosystems and human health.

Eutrophication

Eutrophication is the leading cause of water quality impairment around the world. It is the overenrichment of water with nutrients such as nitrogen and phosphorus as a result of human activity. Eutrophication can be defined simply as the increase in the rate of production and accumulation of organic carbon in excess of what an ecosystem is normally adapted to processing (Nixon, 1995; Rabalais, 2004). Eutrophication can be harmful to both freshwater and marine ecosystems, and leads to a progression of symptoms that include:

- Excessive phytoplankton and macroalgal growth that is the source of organic carbon for accumulation. This can also reduce light penetration and lead to a loss of submerged aquatic vegetation (SAV).

- ☆ An imbalance of nutrient ratios that can lead to a shift in phytoplankton species composition and creating conditions that are favorable to nuisance and toxic algal blooms. Harmful algal blooms (HABS) can cause kills of living marine resources and shellfish poisoning in humans.
- ☆ Changes in species composition and biomass of the benthic (bottom-dwelling) community; eventually leading to reduced species diversity and increased dominance of gelatinous organisms such as jellyfish.

Low dissolved oxygen and formation of hypoxic or dead zones (oxygen-depleted waters). These oxygen-starved areas stress aquatic ecosystems, often leading to kills of living marine resources, altered ecosystem energy flows, and in severe cases ecosystem collapse.

The Environment

Like a giant aquarium, land-based fish farms live in tanks containing dirty water that must be changed. Depending on the system's set-up, this can result in the discharge of significant amounts of wastewater containing feces, nutrients and chemicals released into the environment. The release of this matter can result in algae blooms which eventually remove dissolved oxygen in the receiving waterway, or eutrophication. Zero oxygen content results in deadly fish kills.

Additionally, chemicals such as antibiotics and water treatment agents are commonly used in the aquaculture industry and aquaculture systems need to be closed, or wastewater treated prior to discharge.

Escapes

Aquaculture is one of the largest causes for the occurrence of foreign species introduced into new areas, which creates invasive species under the right conditions. Farmed fish can escape from their pens, damaging both the environment and threatening native fish populations.

As a result, escaped farm fish can compete for food and habitat, displace indigenous species and interfere with the life of wild species. They can also carry diseases and parasites that might kill native species. Additionally, escaped farm fish are able to breed with the wild stock which can dilute the natural gene pool and threaten the long-term survival and evolution of wild species.

Accordingly, about escapement management: aquaculture systems should be designed to effectively contain organisms and minimise the possibility of escape; contingency plans should be set up for the eventuality of escapes (additional preventative measures should be incorporated for high risk activities such as organisms transfer, grading and harvesting); and research on surveillance of escaped organisms should be encouraged.

Effects of Construction

Both land-based and aquatic wildlife can lose their habitats through the building of aquaculture facilities along coastal property, where clean and natural water can be accessed for its processes. For example, in Asia and Latin America, mangrove forests have been cleared to make space for shrimp farms.

Until recently, many aquaculture operations have failed to adequately address the issue of escapes of farmed fish. Farmed fish and shrimp are often genetically different from fish inhabiting the adjacent environment and ecologists have expressed concern that farmed escapes are diluting wild populations' genetics. This is particularly worrisome in salmon aquaculture where millions of fish escape every year from netcages suspended in the open ocean. Tilapia too, which are grown in freshwater ponds have a great proclivity to take over watersheds should they escape.

Diseases and their Control

Most of the diseases of aquatic animals are multifactorial and are initiated by complex interactions of host, pathogen and environment. Poor environmental conditions or bad management may increase the susceptibility of fish to a pathogen, although it is necessary for a specific pathogen to be present. Certain bacterial diseases, such as vibriosis, pseudotuberculosis, streptococcosis, and *Edwardsiella, Flexibacter columnaris* and *Aeromonas hydrophila* infections commonly occur among fish kept in fresh, brackish or sea water in Asia. Vibriosis is the most serious bacterial disease of penaeid shrimps in Asia. All of these diseases occur in ponds in a deteriorated environmental condition. Diseases of fish and prawns can be prevented by good management of the environment.

In.recent years various diseases of fish have occurred frequently under intensive management conditions in Asia. Therapeutic substances commonly used to control these diseases are antibiotics (chlortetracycline, oxytetracycline, chloramphenicol, neomycin), sulphonamides (sulfamonomethoxine, sulfadimethoxine, sulfadimidine) and furane derivatives (nifurpirinol, furazolidone).

Formaldehyde, trichlorfon and copper sulphate have been used against parasitoses. Frequent use of drugs has resulted in a high incidence of drug resistance among pathogenic bacteria isolated in Asia. Among strains of *Aeromonas hydrophi/a* isolated from eel ponds in Taiwan, 79070 were resistant to one or more antimicrobials, including chloramphenicol, chlortetracycline, furazolidone, kanamycin and sulfadimethoxine. Nearly all (98.5 per cent) of the strains isolated from eels were resistant to these antimicrobials.

The use of drugs in aquaculture may result not only in pollution and increased bacterial resistance, but also in a risk to human health from drug residues. To protect human health, drug residues in aquaculture products should be monitored. Certain drugs, such as oxolinic acid and oxytetracycline, have been recommended for bacterial diseases of cultured shrimps and fish, but observance of a withdrawal period is still needed to guarantee that the product is free from residues. The withdrawal periods for drugs vary with water temperature

Disease

Aquaculture operations can spread parasites and disease into the wild. Just like commercial chicken coops must be kept clean and are notorious for the spread of disease, farmed fish and shellfish are subject to the same circumstances.Also, farmed fish have an increased chance of getting parasites such as sea lice, as opposed to fish that live and breed in their natural environment.Farmed fish are also exposed to disease through the use of unprocessed fish used as a food source, as opposed to safer processed fish pellets.

Disease Transfer

The farming of species in wild environments can be a vector for disease proliferation in the wild environment. Disease transfer in salmon aquaculture is perhaps the most reported instance of this phenomenon. The disease infectious salmon anemia first appeared in Chile in the 1990s and has since been noted in other environments around the world. Poor biosecurity and global transfer of salmon larvae helped speed the transfer of the disease from country to country and even continent to continent. Sea lice primarily of the genuses *Lepeophtheirus* and *Caligus* are another frequently noted aquaculture side effect. Sea lice attach themselves to the skin of their intended victim and draw nutrients from the body of their host.

This is particularly damaging to salmon juveniles. Critics of aquaculture note that large farm operations established near the migration routes of wild salmon can cause sea lice to aggregate and jump from farmed animals to wild.

Oil Spill

Oil spills can cause damage to fishing and aquaculture resources by physical contamination, toxic effects and by disrupting business activity. The nature and extent of the impact of an oil spill on seafood production depends on

- The characteristics of the spilled oil,
- The circumstances of the incident and
- The type of fishing activity or businesses affected.

Commercially exploited animals and plants may be killed as a result of oil smothering and toxicity.The reluctance of consumers to purchase seafood products from an affected region can also result in a loss of market confidence.The greatest impact is likely to be found on shorelines where animals and plants may be physically coated and smothered by oil or exposed directly to toxic components in the oil.

Secondary Impacts

Because farmed fish need a food source, other wild species are at risk for being over-fished for the manufacture of fish food. Because most farmed fish are carnivorous, they are fed either whole fish or pellets made from fish. Species such as mackerel, herring and whiting are threatened because of the need to create food for farmed species.

Environmental Protection

It is now time to join hands across the globe to protect and preserve our environment.Thereby safeguard human health along with the natural environment:

air, water, and land. We need to preserve our wetlands, our rainforests, our farmlands and arctic zone,and save the endangered species.

The treatments of aquaculture waste and wastewater is still in its preliminary stages and there is no quantitative data to show how much of the polluted areas are treated or not. Additionally, the insufficient detection system of aquatic environment and wastewater discharge standards for aquaculture promote the pollution caused by the direct waste discharge.

It is, therefore, necessary that proper policies and practical management systems for regulating aquaculture waste be adopted. Different stages of waste management in aquaculture should be integrated so that the overall rate of pollutant removal and reuse is optimized.Such stages will comprise feed quality manipulation, feeding management, waste pre-treatment, primary separation, secondary thickening, sludge stabilization and sludge reuse or disposal. It would probably be more efficient and beneficial to combine two approaches more in the future. The traditional procedures for aquaculture waste treatment, mainly based on physical and chemical means, should be overcome by more site-specific approaches, taking into account the characteristics and resistibility of the aquatic environment. Further research needs to improve or optimize the current methods of wastewater treatment and reuse. Proposed new treatment technology should be evaluated for their feasibility at a larger scale for practical application.

- Small pressure groups should be formed to influence industries to install filters, electrostatic precipitators, scrubbers, *etc.*, to control atmospheric pollution.
- Education is the most important solution
- Plan campaigns in schools and colleges on simple issues such as the ill-effects of smoking.
- Substitutes for domestic fuel consumption such as smokeless chulhas, *etc.*
- Mobilize your immediate locality to protest against local polluting industries.
- Adopting sustainable aquaculture and agriculture.
- Following crop rotation.
- Use of biofertilizers and green manure.
- Promoting gene banks.

Plankton and the Environment

Maintenance of a proper phytoplankton bloom and growth of zooplankton is important for fish and shrimp culture. At the beginning of culture, the water can be stabilised by adding fertiliser to obtain a good phytoplankton bloom with a water transparency of 30-45 em.

The advantages of plankton (particularly phytoplankton) bloom in water are: enhancement of dissolved oxygen, stabilisation of water quality (with reduction in the toxicity of toxic pollutants), a source of feed, improvement of the conversion ratio of artificial feed, inhibition of plant growth in the pond, stabilisation of changes in water quality brought about by rapid temperature fluctuations, and diminished occurrence of the environmental diseases of fish and prawns.

It is important to avoid water that is too clean and unstable. However, excess phytoplankton bloom may promote environmental deterioration, resulting in heavy mortality of fish or shrimps.

Management

To ensure successful management of intensive prawn culture and to decrease the occurrence of bacterial infections, the following are recommended.

1. Selection of a Good Site

- ☆ Avoid eutrophic areas
- ☆ Avoid areas which produce hydrogen sulphide (especially in mangrove areas) avoid areas with soil containing peat, humus or pyrite
- ☆ Avoid areas polluted by domestic, industrial or agricultural sources (particularly
- ☆ Areas in which herbicides, insecticides and other pesticides are used)
- ☆ Avoid areas contaminated with toxic organisms
- ☆ Avoid areas which undergo abrupt changes in temperature or pH avoid areas with a low soil pH (4.0 and below)
- ☆ Ensure an adequate supply of clean water.

2. Good Pond Design

- ☆ Availability of sea or freshwater; the pumping system should take into account tidal variations
- ☆ The pumping system should have an adequate capacity
- ☆ A large reservoir (approx. 30070 of the cultured area) is desirable
- ☆ Proper design of gates for inlet and outlet of water
- ☆ Sufficient water depth (1.5.m or more)
- ☆ Construction and sloping of dike walls, strong enough to allow the use _ of vehicles and to prevent flooding by heavy rain.

3. Good Pond Preparation

- ☆ Complete drying and liming of sediments
- ☆ Avoid pollution by tilling and ploughing sediments up.to the dikes, instead of flushing into a waterway or the sea
- ☆ Avoid mixing inlet and outlet water.

4. Stock Larvae of Good Quality

Establish an **appropriate stocking density,** and **avoid overcrowding**

5. Maintain a Stable Phytoplankton Bloom

- ☆ Through fertilisation, proper feeding and water exchange
- ☆ Establish and maintain a plankton bloom, preferably greenish brown, to obtain a water transparency of 30-40 em
- ☆ Avoid clear water and an unstable aquatic environment

6. Management of Water Quality

- ☆ Ensure adequate water exchange and mechanical aeration and circulation (6-7 paddle wheels or air jets per ha)
- ☆ To ensure a stable environment, avoid adding large volumes of new water which may be of variable quality or may contain pollutants (allow new water to settle for a few days in the reservoir) .'
- ☆ Improve eutrophic water by using biological procedures
- ☆ Avoid pollution by appropriate treatment of, discharged water avoid using unsuitable drugs and chemicals.

7. Routine Monitoring of Water and the Pond

- ☆ Water colour, turbidity, dissolved oxygen, salinity, temperature, pH, NH_3, NO_2, H_2S redox potential of the bottom.

8. Proper Feeding Programme

- ☆ Use pelleted or natural feedof high quality to achieve a low feed conversion ratio, reduced 'feed wastage and improved productivity
- ☆ Avoid pollution by not overfeeding pellets or fresh feed, faecal strands
- ☆ Microscopic examination of fresh preparations of haemolymph, hepatopancreatic impression smears, gill mounts, presence of obvious lesions through microbiology, histopathology

9. Early and Accurate Diagnosis of Disease Problems

10. Place New Stock in Quarantine

11. Treatment with Drugs and Chemicals

- ☆ As a first step, correct any primary environmental problem or management fault accurate diagnosis before treating treat only when necessary, and early in the course of the disease
- ☆ Use drugs (particularly antibiotics) therapeutically instead of prophylactically at proper dosage and for a sufficient period to test bacterial isolates for sensitivity and resistance for ensuring an adequate withdrawal period.

13. Keep **records** and encourage **good communications** among the entire team.

14. Intensive systems for fish and shrimps are more easily controlled than semi-intensive and extensive systems, and the capacities of the environment and the organism can be enhanced by adopting the technology described above.
15. However, an increase in the carrying capacity of ponds will augment the amounts of organic and inorganic products. These substances can be removed by changes of water and self-purification through aerobic and anaerobic bacteria. To a certain extent the organic and inorganic substances can be discharged or mineralised with the aid of micro-organisms and additives. Beyond this extent the life of cultured animals may be endangered by the presence of H_2S, NH_3 and NO_2.
16. Adoption of a given technology may prove to be a limiting factor for subsequent increases in productivity. Overstocking may result in disease and mass mortality. When this occurs, the blame should be attached to damage to the environment of the pond, rather than to the occurrence of a pathogen. Attempts to exceed the limits of nature and to degrade the environment should be resisted. Successful results are achieved with balanced ecological systems.

Conclusion

Negative effects of waste from aquaculture to aquatic environment are increasingly recognized, though they were just a small proportion to land-based pollutants. Properly planned use of aquaculture waste alleviates water pollution problems and not only conserves valuable water resources but also takes advantage of the nutrients contained in effluent.

It is highly demanding to develop sustainable aquaculture which keeps stocking density and pollution loadings under environmental capacity. Saving the environment is not an issue anymore. It is a survival truth. Individuals, organizations and governments need to come together. And join hands to protect what is left of our planet so that the future is not wiped out before it's time for curtain call.

Aquatic ecosystems as life support, supplier of biomass and waste assimilator are a key factor to the sustainability of aquaculture. Aquaculture development raises many environmental issues involving the need for coordination of individual actions at a collective level. A framework has been proposed to build a grid of the main issues.The aquaculture industry is facing significant environmental challenges, among other things in the form of high figures for escaped fish, salmon lice and extensive losses due to disease. - The extent of these and other environmental challenges is so great that it is necessary to strengthen the management of aquaculture in order to ensure environmental sustainability and the possibility of future growth in the industry.

Chapter 3

Fish Poison: Its Prospects, Problems and Solutions

Introduction

Fish poison or **fish stupifying plants** have historically been used by many hunter gatherer cultures to *stun* fish, so that the fish become easy to collect by hand. Professionals today, to control fish populations or to eliminate alien or destructive species, use the plant toxins and some chemicals to poison the fish. The process of documenting many fish toxins and their use is ongoing, with interest in potential uses from medicine, agriculture, and industry. (Jeremy, 2002)Use of the herbal fish poisons has been documented in a number of sources involving catching fish from fresh and sea water. Herbal fish stupefying agents are proven means of fishing, which do not kill fish like chemical poisons. Many of the fish stupefying plants have been used for a long time by local people, and have been tested and found to have medicinal properties like *Carreya arborea*as a tested plant used as analgesic (Ahmed *et al.*, 2002) and anti diarrheal (Rahman *et al.*, 2003). In earlier studies, bark extracts of *Lannea coromandelica* caused lyses of cell membrane followed by fragmentation of cellular materials. These studies scientifically confirmed the toxic property of these plants.

Types of Fish Poisons

The poisons are of two basic types:

1. **Chemical Originated**

   a) Chlorinated hydrocarbon b) Organophosphates

2. **Plant Originated Fish Poisons**

   a) Saponin b) Rotenone

Chemical Originated

Chlorinated Hydrocarbons

Chlorinated hydrocarbons like Aldrin ($C_{12}H_8Cl_8$), Dialdrin ($C_{12}H_8Cl_8O$), Eldrin ($C_{12}H_8Cl_6O$) and Taffdrin-20 are used for the eradication of predatory and unwanted fishesSpecially the eldrin formulated fish poison chemically are very strong poisonous and a little high dose of the poison may affect.Its action takes 2-3 hours to affect IMC and 4-8 hours to hardy fishes like tilapia, singhi, magur,koi *etc.* Aldrin @ 0.2 ppm, endrin @ 0.01ppm and dialdrin @ 0.01 ppm are effective in killing fishes.

Organophosphate

The organophosphates used in india are DDVP, phosphamidon, Thiomatin *etc.* The dose phosphamidon and Thiomatin @ 3 to 30 ppm can be applied for complete eradication of unwanted fishes from the culture system.

Plant Originated Fish Poisons

There are quite good number of plants which could be used for fish poisoning. Various parts of plants such as stem, seeds, leaves, roots can be used as the materials for fish poisoning.

Some common plants are:

1. Barring-tonia acutangula
2. Crotton tiglium
3. Derris root powder
4. Mohua oil cake (Bassica latifolia)
5. Sugar cane Jaggery

Application Dose

Plant Effective	Dose
Barrring-tonia	10-20 ppm
Crotton tiglium	4-20 ppm
Derris root powder	4-20 ppm
Madhuca indica	200-250 ppm

Derris Plant

In Assam, *Milletia piscida* and *Milletia pachycapra* are used to eradicate unwanted fishes.Derris root powder: active ingredient is rotenone is found effective upto a depth of 1.5m on hot sunny days when temperature is more than 25°C. In deeper water, its effect reduces gradually and beyond 6 to 10 m,its effect is insignificant. The poisoning effect last for 4-12 days when applied at the range of 4-20 ppm.

Mohua Plant

Bassiaca latifolia or Madhuca indica is very commonly known as Mohua.Oil cakes of these plants contain an active ingredient is mowrin or saponin. Poisoning effect remains for 21 days from the day of application. MOC acts as a manure when its poisoning effect is lost.

Saponins

Saponins normally break down in the digestive system and must enter the bloodstream to be toxic, but fish take in saponins directly into their bloodstream through their gills. The toxin acts on the respiratory organs of the fish without affecting their edibility. Saponins also cause the breakdown of red blood cells that help the toxin to spread quickly. Even though the effects of the poison are powerful, they are not usually fatal. Fish that are washed away into untainted water revive, and can return to their pre-toxic condition. Because of this, the fishermen would have to gather the stunned fish quickly as they floated to the surface.

Saponins are one of a group of glucosides found in many plant species with known foaming properties when mixed with water. Saponins lower the surface tension of water allowing the formation of small stable bubbles. The amount of foam created by a crushed plant sample, shaken with water in a jar, is a good indication of the amount of saponins present.

Saponins have been used in modern times in the manufacture of fire extinguisher foam, toothpaste, shampoos, liquid soaps, and cosmetics and to increase the foaming of beer and soft drinks.

Plant Families that contain significant saponins are: Amaryllidaceae, Convolvulaceae, Dioscoreaceae, Lamiaceae, Lecythidaceae, Liliaceae, Loganiaceae, Meliaeae, Menispermacea, Papilionaceae, Solanaceae, Sapindaceae, Sapotaceae, Scrophulariaceae, Solanaceae, Verbenaceae

The saponins are naturally occurring surface-active glyco-sides. They are mainly produced by plants, but also by lower marine animals (star fish, sea cucumber, shark) and some bacteria (Riguera, 1997; Yoshiki *et al.*, 1998). They derive their name from their ability to form stable, soap-like foams in aqueous solutions. Saponins can be classified into the two groups: pentacyclic triterpenoid saponins and steroidal saponins.

The steroidal saponins are mainly found in monocotyledons (such as in the families Agavaceae, Dioscoreaceae and Liliaceae), while triterpenoid saponins mostly are present in dicotyledons (Fabaceae, Araliaceae and Caryophyllaceae.

Chemistry of Saponin

Saponins consist of a sugar moiety usually containing glucose, galactose, glucuronic acid, xylose, rhamnose or methylpentose, glycosidically linked to a hydrophobic aglycone (sapogenin) which may be tri-terpenoid.The great complexity of saponin structure arises from the variability of the aglycone structure, the nature of the side chains and the position of attachment of these moieties on the aglycone. According to the structure of the carbon skeleton of the aglycone, saponins are sometimes further classified into 12 main classes, namely the: dammaranes, tirucallanes, lupanes, hopanes, oleananes, 23-nor oleananes, taraxasteranes, ursanes, cycloartanes, lanostanes, cucurbitanes, and steroids (Vincken *et al.*, 2007).

Effect on Fish

Saponins have been reported to be highly toxic to fish because of their damaging effect on the respiratory epithelia (Roy *et al.*, 1990). Fish have also been shown to exhibit stress reactions to the presence of saponins in water. Permeabilisation and effects on other membrane properties. Saponins have long been known to have a lytic action on erythrocyte membranes and this property has been used for their detection. The haemolytic action of saponins is believed to be the result of the affinity of the aglycone moiety for mem-brane sterols, particularly cholesterol with which they form insoluble complexes.

Detoxification

Detoxification can be done by a heat treatment but the digestibility decreases significantly through the heat treatment.The levels of saponins can also be reduced by treatment with isopropanol.Detoxified seed meal appears to be a good source of protein for food and feed.

Rotenones

Plants containing rotenones are the second most utilized as a fish poison. Rotenone is an alkaloid toxin, in a group called flavonoids and stuns fish by impairing their oxygen consumption. The plant is toxic only to cold-blooded creatures and is found almost exclusively among the family comprised of legumes (Papilionaceae, Mimosaceae, Cesalpiniaceae). Rotenone is also used today as an insecticide. **Rotenone** is an odorless chemical used as a broad-spectrum insecticide, piscicide, andpesticide. It occurs naturally in the seeds and stems of several plants, such as the jicama vine plant. Emmanuel Geoffroy first isolated rotenone (which he called *nicouline*) from a specimen of *Robinia nicou*, now called *Lonchocarpus nicou*, while traveling in French Guiana.

STRUCTURE OF ROTENONE

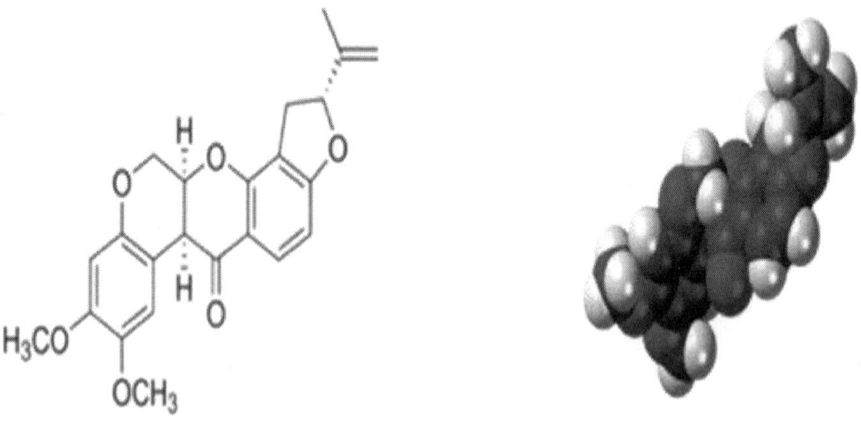

IUPAC name
(2*R*,6a*S*,12a*S*)-1,2,6,6a,12,12a-hexahydro-2-isopropenyl-8,9-dimethoxychromeno[3,4-*b*]furo(2,3-h)chromen-6-one

Presence in Plants

Rotenone is produced by extraction from the roots and stems of several tropical and subtropical plant species, especially those belonging to the genera *Lonchocarpus* and*Derris*.

Some of the plants containing rotenone:

- ☆ Hoary pea or goat's rue (*Tephrosia virginiana*) – North America
- ☆ Jícama (*Pachyrhizus erosus*) – North America
- ☆ Cubé plant or lancepod (*Lonchocarpus utilis*) – South America[7]
- ☆ Tuba plant (*Derris elliptica*) – southeast Asia and southwest Pacific islands
  - ❒ The root extract is referred to as derris or derris root
- ☆ Jewel vine (*Derris involuta*) – southeast Asia and southwest Pacific islands
  - ❒ The root extract is referred to as derris or derris root

Effect and Mehod of Action

When poisoned with rotenone, fish swim erratically and move to shallower water or come to the surface gasping for air. After that, their ventilation rate slows and they sink to the bottom where they remain until death. This happens not because rotenone removes oxygen from the water, but because it inhibits a process that occurs during cellular respiration, called oxidative phosphorylation. The specific site of action of rotenone is in the electron transport system where it blocks a mitochondrial enzyme called NADH ubiquinone reductase (or complex 1). This means that the blood oxygen content of the fish will increase because oxygen is now unavailable for respiration.

Toxicity

Rotenone is classified by the World Health Organization as moderately hazardous. It is mildly toxic to humans and other mammals, but extremely toxic to aquatic life, including fish.This higher toxicity in fish and insects is because the lipophilic rotenone is easily taken up through the gills or trachea, but not as easily through the skin or the gastrointestinal tract. Rotenone is toxic to erythrocytes in vitro. Thecompound decomposes when exposed to sunlight and usually has a lifetime of six days in the environment.In water, rotenone may last six months.

Resistant Fishes

Some fish are more resistant to rotenone than others and require a higher concentration to have the same effect. Resistant fish include bullheads, goldfish and bowfin. Rotenone treatments will usually kill fish within 24-36 hours.

Detoxification

Uses Rotenone poisoned fish can be revived by simply putting them into untreated freshwater. Rotenone can also be detoxified by solutions of potassium permanganate, chlorine or methylene blue. The use of methylene blue is not recommended because it increases bacterial infection on fish. Potassium permanganate is the most effective at detoxification and is used to reverse the toxicity in the affected fish or to accelerate the natural breakdown in water. It is often used on water in outlet streams flowing out of treated lakes so that rotenone will not kill desirable fish species that live in the stream. Rotenone has historically been used by indigenous peoples to catch fish. Typically, rotenone-containing plants in the *Fabaceae* family of legumes are crushed and introduced into a body of water, and as rotenone interferes with cellular respiration, the affected fish rise to the surface in an attempt to gulp air, where they are more easily caught.Small-scale sampling with rotenone is used by fish researchers studying the biodiversity of marine fishes to collect cryptic, or hidden, fishes, which represent an important component of shoreline fish communities.Rotenone is the most effective tool available because only small quantities are necessary. It has only minor and transient environmental side effects.

Other Poisonous Materials

Application of urea @100kg followed by the addition of bleaching powder @ 175 kg/hectare meter of water is also have good poisonous effect.This urea acts as fertiliser in the latter days.Sometimes bleaching powder alone@ 350 kg/ha meter of water is applied to kill unwanted fishes.Even torpentine @ 250 litre/ha is applied for eradicatioin. It is better to suggest plant originated fish poison because they affect on their respiratory system and the fishes die usually of Asphysia.As a result, the flesh of these fishes may be used for human consumption.Moreover, the toxicity effect of plant derived fish poisons does not last for long time as compared to chemical poison.Further toxicity effect of chlorinated chemicals is very much prolonged in to the water and as a result the treated water can not be used for a considerable period for fish culture.

Fishing with Poisons

Most fish poisons, also called icthyotoxins or piscicides, occur in several related plant species. A variety of chemicals found in these plants will stun fish when it passes through the gills or in some cases ingested. The fish then floats to the surface for easy capture. The two primary chemicals that occur in most plants used for stunning fish are saponin and rotenone. Poisoning was generally done in stagnant pools or slow-flowing streams and rivers, that allow the pounded bark, leaf, seed, root or fruit, to concentrate its power without being washed away or diluted by a strong current. Gathering the fish was usually done by hand, but baskets, spears and nets were sometimes employed.This primitive method is still in use in many parts of the world.

Most fish poisons, also called icthyotoxins or piscicides, occur in several related plant species. A variety of chemicals found in these plants will stun fish when it passes through the gills or in some cases ingested. The fish then floats to the surface for easy capture.

The active ingredient is released by mashing the appropriate plant parts, which are then introduced to the water environment. Poisoning was generally done in stagnant pools or slow-flowing streams and rivers, that allow the pounded bark, leaf, seed, root or fruit, to concentrate its power without being washed away or

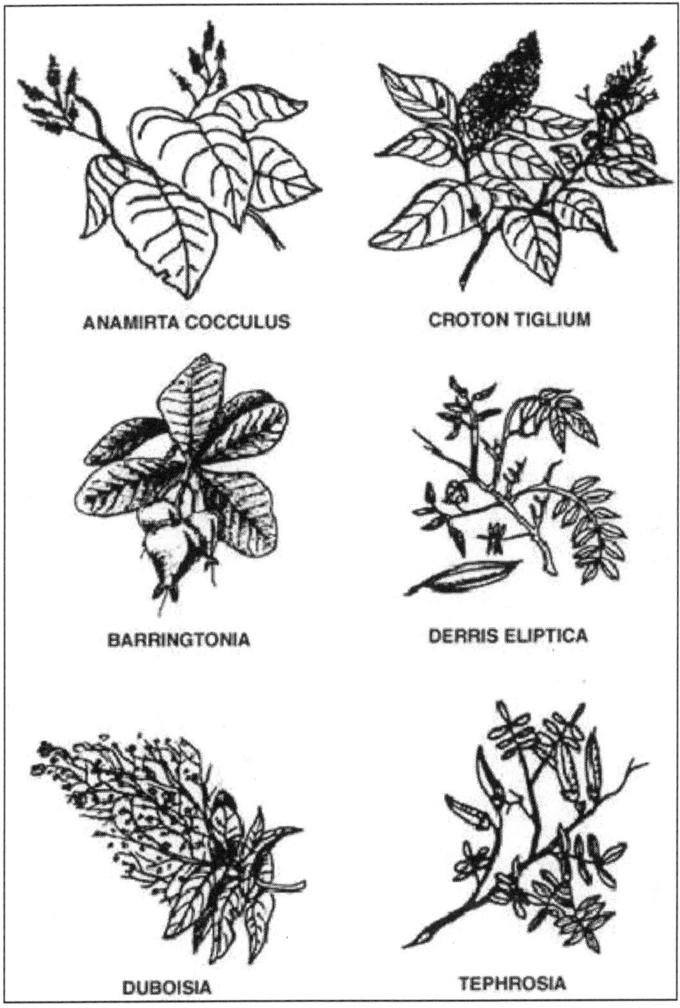

Fish Poison Plant Families of the World.

diluted by a strong current. Sometimes streams would be partly blocked to slow down the water flow. Gathering the fish was usually done by hand, but baskets, spears and nets were sometimes employed. Although primarily used in freshwater areas, Australian Aborigines and Californian Indians also used this technique in saltwater environments for octopus and low-tide shellfish fishing as well as for catching fish trapped in inter-tidal pools.This ethnological report from Cape York Peninsula, Queensland, Australia, shows that with some Native Peoples, the cultural and material world was not separated:

A secret, sacred song of the Pascoe River bora (initiation cult) was sung by 60-year-old George Morton accompanying himself with his own drum. The singer was born a Kandyu but married a Wutati woman who was the daughter of one of the great Wutati bora singers who handed down the entire repertory of ancient bora songs to him. This song tells of a turtle that used a medicinal vine as a poison to catch fish in a rock pool at low tide.

The use of plant poisons to catch fish is still used in many places in the world today. In Guyana, fishers pound the root of Lonchocarpus on logs fallen across a stream and allow the juices to drip down into the water. Brazilian gold miners, who probably learned the technique from the displaced Yanomami Indians of the Amazon, also toss pulped plant material into a very slow moving stream where the fish would surface down stream and be washed into a net set in place by the fishermen.

The two primary chemicals that occur in most plants used for stunning fish are saponin and rotenone.

Leaves are Used for Fish Poison

Fish Poison Wattle (*Acacia holosericea*)   Fish Poison Tree seedpod (*Acacia ditricha*)

Below is a short list of indigenous peoples and the plants they used to poison fish:

Location or Tribe/Common Name, (Latin Name)/Part Used

United States

Catawba, Cherokee, and Delaware/**Black Walnut**, (*Juglans nigra*)/**Bark and green nut husk**

Yuchi and Creek/**Devil's Shoestring**, (*Symphoricarpos orbiculatus*)/**Roots**

Horse Chestnut, (*Aesculus hippocastanum L*)/**Fruit, twigs and buds**

Cherokee/**Polk Sallet, Polkweed**, (*Phytolacca americana*)/**Berries**

Central and coastal California/**Turkey-Mullein**, (*Eremocarpus setigerus*)/**Leaves**

California Buckeye, (Aesculus California)/**Nut or fruit**

Soap plant, soap root, (*Chlorogalum pomeridianum*)/**Bulb**

Indian hemp, (*Apocynum cannabinum*)/**Stalk, leaves**

Pokeweed, Polk sallet, (*Phytolacca americana*)/**Leaves**

Indian Turnip, (*Arisaema triphyllum*)/**Leaves**

Wild cucumber, Manroot, (*Marah fabaceus*)/**Seeds**

South America

Mexico/**Lechuguilla**, (*Agave lechuguilla*)/**Leaves**

Venezuela/**Soapberry**, (*Sapindus drummondii*)/**Berries**

Mexican Buckeye, (*Ungnadia speciosa*)/**Nut or fruit**

Ecuador/Barbasco, (*Jacquinia sprucei*)/**Bark, roots**

Barbasco, (*Tephrosia toxicofera*)/**Bark**

Barbasco, (*Lonchocarpus nicou*)/**Roots**

Acariquara, (*Minquartia guianensis*)/**Bark**

Brazil/**Fish poison leaves**, (*Euforbia cotinifolia*)/**Leaves**

Pacific Islands

Rarotonga, Moorea/**Fish poison tree**, (*Barringtonia asiatica*)/**Seeds and leaves**

Hawai'i/**'Auhuhu**, (*Tephrosia Purpurea*)/**Roots and bark**

'Äkia, The fish poison plant, (*Wikstroemia uva-ursi*)/**Roots, bark and leaves**

Australia

Pituri, (*Duboisia hopwoodii*)/**Cured leaves**

Austral Indigo, (*Indigofera australis*)/***Leaves and fruits***

Fish Killer Tree, (*Barringtonia asiatica*)/**Seeds and leaves**

Fish Poison Tree, (*Acacia ditricha*)/**Leaves**

Fish Poison Tree, (*Barringtonia racemosa*)/**Seeds and leaves**

Fish Poison-wood, (*Barringtonia vitiflora*)/**Seeds and leaves**

Fish Poison Wattle

Soapy Wattle, (*Acacia holosericea*)/**Leaves**

India

Pongam, Indian Beech, Derris, (*Pongamia pinnata*)/**Seeds**

Fish Berries, (*Anamirta cocculus*)/**Seeds**

Bloodflower, Curassavian

Swallowwort, (*Asclepias curassavica*)/**Roots**

Africa

Pencil tree, Milk bush, (*Euphorbia tirucalli*)/**Leaves, sap**

Guele, Ironwood, (*Prosopis africana*)/**Dry fruits**

Common Mullein (*Verbascum thapsus*)

Turkey Mullein or Dove Weed (*Eremocarpus setigerus*)

Ecological Responsibility

Professionals today, to control fish populations or to eliminate alien or destructive species, use the same plant toxins: saponin and rotenone. Practicing primitives may be eager to experiment with the techniques listed above, but great care must be used, as the toxins are not selective and will eliminate all fish in the water where it is introduced. Keep in mind what is down stream and may be affected by these poisons. These chemicals will generally break down in sunlight. If you choose to use this technique, be aware that fishing with poison (even natural poison) is illegal in most states. Check your local laws.

Conclusions

A review of the available scientific and popular literature reveals that the ancient practice of poisoning fish was an important method of securing food, and continues to flourish in many cultures today. A modern medical perspective on these plants may uncover additional uses for these plants. Unfortunately, the poisoning of streams to capture fish has had ecological consequences. The efficiency of this method has ensured that its increasing use as populations increase may result in the extirpation of susceptible fish species from their native streams. Despite illegalization of this method in countries such as Guyana, the practice continues and many of the plants are grown commercially, both for personal use, for sale to one's neighbours, and for production of industrial products. It is with caution, then, that we should explore these uses, both from a standpoint of caution when using plants to poison fishes, but also (and perhaps more importantly), from the standpoint of exercising caution when infiltrating these cultures for industrial, pharmaceutical, or technical purposes.

Chapter 4
Fish Pond Management

Introduction

Ponds are very small and shallow of quiet standing water with slight wave action and may be naturally created or manmade. Naturally created ponds show the same geological history as lakes. However, unless polluted their small volume and shallowness results in very rapid progress to the point where autochthonous production dominates filling up the basin. Man-Made or artificial ponds are created to serve diverse purpose,Fish Farming, Duck Rearing, Ornamental, Cultivation of any aquatic weeds. The ponds used exclusively for fish culture purposes are designed and managed scientifically these are better called as fish ponds.

Scientifically, pond management is developing or improving impounded aquatic habitat and water quality for the better production of fish for domestic use or recreation. This is done mainly to improve or maintain a desired species composition and ratio, favourable water habitat for fish and other aquatic organisms which help to sustain the fish population, supplementing natural food supplies, and reducing competition from unwanted plants and animals. Proper planning and construction are the keys in building a whether they are primarily recreational, aesthetic, or agricultural.

Though a good pond management thought to be highly beneficial, but it is no wonder that landowners who place high value on their ponds are upset when some problems occur. A large number of ponds either fail to hold water, or contain muddy waters, or are choked with waterweeds, or do not produce good fish population. So it's necessary to know about the problems commonly arise in fish pond management, with their solution.

Types of Fish Ponds

Within the fish farming pond system there can be different pond components, namely nursery, rearing, production, segregation and breeding or spawning ponds.

The area percentage of these ponds in a fish-farming complex can be:

- ✰ Nursery pond - 3 per cent
- ✰ Rearing pond - 11 per cent
- ✰ Production pond - 60 per cent
- ✰ Segregation pond - 1 per cent
- ✰ Breeding pond - 25 per cent

Problem in Site Selection

Careful selection of the pond site is one of the most important steps in avoiding farm pond problems.

A good pond site contains:

1. A topography ("lay of the land") that provides for economical pond construction.
2. A water supply that provides adequate but not excessive amounts of water and
3 .A soil type with sufficient clay content to hold water.

Failure to select a good site will result in serious pond problems (flooding, leakage, pollution) that are very expensive to correct.

1. Topography

The topography of the pond site should be carefully evaluated to minimize costly soil removal. Ideally, the topography of the land should be above the flood plain, relatively flat, permitting construction of a pond of regular size, shape, and depth. Sometime flooding causes serious problems in managing fish ponds. Incorporating the topography of the land with the design of the pond dam will permit natural drainage and avoid expensive drain systems or costly pumping equipment.

2. Water Supply

A continuous supply of high-quality water should be available throughout the year to replenish losses due to evaporation, seepage, or partial drainage of the pond for irrigation, weed control, or fish management purposes.

The water supply should be sufficient to provide enough water to fill the pond within a reasonable period of time (less than 15 days) and maintain a relatively constant water level (fluctuations less than 2 feet year-round).Excessive inflows flush essential nutrients downstream and may allow valuable fishes to escape.

Problem Caused by the different Water Source

- ☆ Flowing water system (river, canal, stream): High turbidity of settleble solid. Undesired organism can get entry into the pond. Solids will settle in the bottom, increasing bottom thickness.
- ☆ Beel, reservoir: Very rare to construct pond near it.
- ☆ Underground water: Devoid of dissolved oxygen. Loaded with iron.
- ☆ Natural source: High load of bacteria, parasites (due to surface runoff). Physical turbidity.

3. Soil Type

Clay and silty clays are excellent soils for holding water because they stop water from steeping through. Since the pond is simply a container for holding water, its dam and bottom must be composed of a soil that minimizes seepage. If a pond's water level drops much more than 6 to 12 inches in one month, it may have a serious leak. Soils with a high clay content (Bentonite clay @1-3pounds per square foot of pond basin) are preferred because clay particles tend to swell when wet and, thereby, help to seal the pond bottom.

Problem in Pond Construction

Ideally, the pond owner should have complete control of the water entering and leaving the pond, so it requires an inlet outlet system. Since it is often necessary to lower water levels to control flooding, erosion, waterweeds, and fish populations, an "easily operated" drain outlet of "adequate size" (sufficient to drain the pond in 5 days or less) is essential. Similarly, since it is often necessary to prevent waters

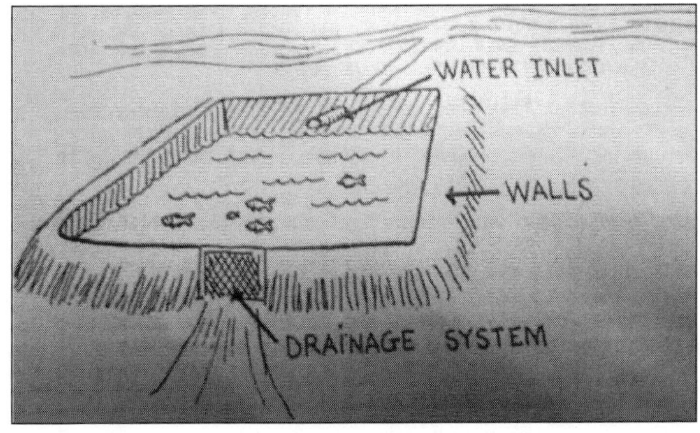

containing chemical contaminants, heavy silt loads, or unwanted fish from entering the pond, an inflow control system is important. When streams are used as a water supply, the stream should be diverted around the pond and an inlet pipe, which can be screened or closed as needed, should be installed. When surface runoff waters are used as a water supply, an embankment or diversion ditch with an inlet pipe, which can be screened or closed, is recommended.

Problem in Stocking

Fish selection and Stocking density of fish in pond is another problematic thing in pond management. A farmer should not stock the fishes with much more size differences otherwise predation will occur. A farmer will always try to make profit by overcrowding in pond, but this little mistake can cause a big problem in fish pond management. Overcrowding of fish will increase their stress, thus leading to various diseases and pathogen attack.More stocking density will also increase the competition of food, oxygen and space.

Problem of Poor Water quality

Temperature

Unfavorable temperature may cause lethal conditions to aquatic organisms, their populations, community structure, or ecosystem. Temperature often causes physical, biological, chemical stress on fish.

- Physical: High temperature decrease dissolve oxygen in water causing hypoxia.
- Chemical: Higher temperatures can alter concentrations of substances in water, which can have an impact on the ability of fish to withstand chemical exposure.
- Biological: High temperature increases metabolic rate, disrupt their life cycle, decrease diseases resistance capacity.

Water temperature can be adjusted to optimal levels in controlled condition like hatchery, but cannot be maintain in large water body. Aerator during calm and warm afternoon helps break thermal stratification by mixing warm surface water with cool subsurface water. Tree plantations also give shade on pond banks. Temperature should be 25°C-27°C.

Dissolved Oxygen

Dissolved oxygen is one of the most important water quality factors to consider in managing fish ponds. Without an adequate supply of dissolved oxygen, fish cannot survive. Aerator is commonly used to increase the Oxygen concentration in pond.

pH

The pH of pond water is a chemical measurement indicating how acidic or alkaline (basic) the water is on a standard scale ranging from 0 to 14. Pure water - >7, acidic water - >0-6.9, alkaline water - >7.1-14

The Effects of sub optimal pH leads to

(1) Increase of mucus on gill surface, (2) Damage to eye lens and cornea, (3) Poor phytoplankton and zooplankton growth, (4) Stress, (5) Increase susceptibility to disease, (6) Poor growth *etc.*

Problem of Fish Mortality

Oxygen depletion is the leading cause of fish kills in ponds. Because low oxygen kills are usually sudden and massive, many pond owners mistakenly jump to the conclusion that a pesticide must be responsible. The following signs point to low oxygen as the cause of a fish kill:

- Fish gulping at the water's surface.
- Large fish die before small fish of the same species.
- Pond water changes color.
- Light scum or a film is visible on the water.

Most fish kills can be attributed to one of three major causes:

1. Fish suffocation due to lack of oxygen,
2. Toxic Pollutant Exposure
3. Disease outbreak.

Death Due to Suffocation

A good supply of dissolved oxygen is necessary to support fish and other aquatic animals to live. But if there is lack of oxygen in the pond, mass mortality can occur. Over fertilization, algal bloom die off, winter kill,summer kill are the causes behind the fish death due to depletion of oxygen.

Aquatic Vegetation or Algae Die-off

Sometimes a heavy phytoplankton bloom or infestation of aquatic weeds in a pond will die off naturally or as a response to herbicidal treatment. As the plants decompose, oxygen is consumed and DO levels may fall to a level lethal to fish.

Winter kill

During severe winters, ice can forms on ponds and creates a seal between the water and the atmosphere. This prevents a pond from obtaining dissolved oxygen from the atmosphere and also retards the sunlight penetration, causing death of aquatic plants. If this continues for an extended period of time, fish will eventually suffocate.

Summer kill

Summer kill is most common in shallow ponds that are heavily vegetated and have high accumulations of decomposing organic matter. Cloudy weather during the heat of the summer can cause gradual reductions in the amount of dissolved oxygen in a pond. If this condition is present for several days it may cause mass mortality.

Problems of Unwanted Animals

Snail

Over population of snail:

1. Build up the organic load and detoriates the water quality.
2. Compete with fish for dissolved oxygen, food, other useful nutrients.
3. Transmit disease by acting as intermediate host to various parasites.

Control

1. Introduce snail eating fishes (Cichlids, Gouramies, puffer fish).
2. Manual removing.
3. Use of chemicals (Bug-geta, Natria *etc.*).

Predatory Fish

The presence of predatory fish may results in big problem in fish pond management. The carnivorous predators like *Wallagoattu, Clariousbatracus, Anabus testudinius* generally eat the other culturable small fish population and leads to economic loss.The predatory fishes not only complete with cultured carps for food and space but also directly prey upon them. Some successfully breed in confinement quite earlier to be spawning of carps. Therefore they are large enough till the carp spawn is released in 'bundhs' or spawning ponds, to take a heavy toll of the spawn. The majority of common weed-fished are also active competition with the major carps for the available food in pond. They have high fecundity and ripen in summer months and can breed even in the absence of rains. Young weed fishes directly feed on carp hatchlings and the spawn.

Water Bird

Fish are major food items for several water birds such as kingfishers, egrets and herons. However, their consumption is so low that there is no danger to pond population balance.

Livestock

Cattle and other livestock that have free access to a fish pond may cause damage in several ways. Their wading stirs up sediments, causing muddy water as well as bank damage. Wading will also hamper fish spawning in the spring and summer by damaging eggs and nests and by frightening fish from their nests. If livestock must be given access to the fish pond, they should be restricted into a small portion of the pond. This can be accomplished by the appropriate use of fencing.

Maintenance and Management of Fish Farm

Maintenance and judicious management of aquafarms, keeps the culture enterprise profitable and save the farmers from any unwanted loses. Feed and water quality are the two major factors governing the productivity of the fish culture pond. Besides, seed quality, stocking and other management measures also determine the extent of fish production.

An aquaculturist should be trained enough to manage and maintain the fish farm on the following lines:

- ✰ Fertilization of ponds
- ✰ Restorations of improvement of the pond bottom of liming
- ✰ Control of aquatic vegetation
- ✰ Control of weed fishes and predators.
- ✰ Prevention of fish disease
- ✰ Monitoring physicochemical characteristics of water and soil.

Fertilization of the Ponds

The agricultural soil is found most suitable for the construction of fish farms owing mainly to the fact that they are sufficient provided with nutrients. The natural productivity of the pound can greatly be enhanced by the use of fertilizers (both organic and inorganic).Fertilization can improve fish production by increasing the production of tiny plants and animals at the bottom of the food chain, the phytoplankton and zooplankton. However, negative impacts from fertilization can also result if the added nutrients stimulate growth of undesirable types of aquatic vegetation and algae. These produce oxygen on sunny days, but very heavy blooms will utilize nearly all of that oxygen during the night, leaving very little for fish respiration.

Improvement of the Pond Bottom

To retain the pond productivity of pond it is essential that the ponds are drained dry at least one in a year. This exercise is necessary to clean the ponds and repair the bottom. By doing so following objectives can be achieved.

- ✰ Fish parasites disease producing germ and various stages of life cycle of harmful insects are destroyed.
- ✰ Mineralization of soils is reduced particularly when there is great accumulation of submerged vegetation.

- ✰ Excessive deposition of muds makes the ponds shallower and hence the same may be removed by dredging at the bottom.

After the ponds are dried and necessary cleaning is done, the pond bottoms are subjected to 'liming' *i.e.* application of lime.

Feeding

Feeding is the most direct and reliable method to increase production. Apart from natural food, most cultured species take artificial feeds. In general artificial feed should contain 30-40 per cent protein, 5-10 per cent fat, 50-60 per cent carbohydrate, less than 5 per cent cellulose, 10 per cent water, vitamins and minerals. Animal and vegetable ingredients can be used in formulating feed pellets. Fishmeal, prawn meal, soybean meal, silkworm pupa, wheat, tapioca, slaughter house wastes, rice bran, aquatic weeds, oilcakes *etc.* are good feed ingredients. Usually Indian farmers give rice bran and oil cakes in powder form to major carps. Carnivorous fishes such as murrels and catfishes may be provided with trash fishes. For adult fish, daily supplementary feeding can be at 2 per cent of its body weight.

Control of Aquatic Vegetation

For sustained primary productivity the growth for producers in the form algae and aquatic macrophytes is essential but unwanted and undesirable growth of algae or plats place them in the category of menacing weeds. Generally weeds can be classified into four categories they are:

Floating weeds: *e.g.,* Pistia, Azolla, Lemna, Eichornia *etc.*

Rooted weeds with floating leaves: *e.g.,* Nyphaea, nelumbo, nymphoides *etc.*

Rooted emergent weeds: *e.g.,* Polygonum, *Typhacyperus etc.*

Submerged weeds: *e.g.,* Chara, hydrilla, Naja *etc.*

Control Measures

Manual, mechanical chemical and biological control measures have been suggested to eradicate the managing weeds the selection of the method has to be based on the type of the infection, nature of the farm and specie that is cultured.

Pond plants are generally beneficial, providing feeding areas for fish, refuge for small fish from bass, and protecting shorelines from wave erosion. However, when plant growth becomes a nuisance, several approaches can be used to manage the problem. Although limited numbers of aquatic plants growing in ponds are beneficial, dense growths of algae and other waterweeds can seriously create undesirable changes that threaten pond life. They not only retard light penetration in the pond, but also decomposition of waterweeds depletes dissolved oxygen, causing mass mortality.

Prevention of Fish Disease

Strictly adhering to the principles of biosecurity and good husbandry management will prevent many disease outbreaks. However, problems that may lead to disease will arise even in the best-managed systems and a scientific and methodical approach will help determine the best course of action, reduce losses, and get the system and fish back to normal. Record-keeping is an important tool in this approach. Problems may or may not result in diseased fish.

The key to preventing disease and reducing losses is early detection of any abnormalities in the system or the fish. Early detection relies on daily monitoring of water quality, general system checks, and observing fish behavior, fish appearance, and mortalities. Working with an aquaculture or recirculating system specialist will help solve problems.

Water Quality

A periodic monitoring of physio-chemical characteristic of water very essential water management is very important aspect of pond management. Exchange of water is best remedy prevention of fish disease. If the organic load is increased in the pond immediately aeration will be provided trait netting is very essential to check the growth and hygienic conditions of the fish.

The addition of pesticides, nutrients, or supplemental foods to enhance the fishery or the resultant by-products of feeding fish will affect the quality of both surface and groundwater.

The addition of any substance used to enhance the fishery, which dissolves in the water, will increase the movement of these substances into the groundwater. Nutrients added to the fishery will increase the biological productivity of food chains associated with all water related wildlife habitats.

Maintainance of Water Parameters

- ✰ Chlorine can be instantly neutralized by applying Aqua Med Safe and Care to the water, or by using a Tap Water Dechlorinator.
- ✰ Ammonia/Ammonium and Nitrite should be removed by a suitable Pond Filter which can be assisted with Starter Bacteria or Zeolite.
- ✰ Nitrate can be reduced by changing some pond water or by introducing Viresco Aqua.

- ☆ Phosphate can be binded and reduced using Phosless.
- ☆ Hardness can be raised by using Opti Pond.

Sanitation and Disinfection Protocols

Good sanitation and disinfection procedures reduce the numbers of disease-causing organisms present within a given system and prevent or reduce the spread of disease-causing organisms from one system to another.

In addition to those quarantine protocols listed in the previous section, some practical methods that will help prevent or reduce the pathogen load in a system include:

Proper attention to food sources

1. Be careful with live foods: although live or fresh foods can be a good source of nutrients, these may also be a source of pathogens
2. Ensure proper storage(in a cool, dry location) and usage (follow manufacturers recommendations/expiration date) of manufactured feeds to prevent loss of nutrients and build up of pathogens or toxins.

Good overall system maintenance and cleanliness to reduce environments that will favor parasites

1. Good husbandry (nutrition, water quality, social structure, habitat)
2. Regular siphoning of excess organic matter from sumps and tanks
3. Backwash of filters as needed to reduce organic loading
4. Check air and water pumps and lines
5. Flush sediment out of water lines as needed
6. Maintain UV sterilizers (regular bulb changes, as per manufacturers recommendations; cleaning quartz sleeves) and ozone units (check generator and other components, monitor for levels in the system)
7. Keep nets and other equipment off the floor

Conclusion

Fish farming has taken up great attention now-a-days for adopting scientific pond management methods in fish culture practices. Analyzing water parameters, replenishment of water, aeration, regular feeding, observation for mortality and disease symptoms should be routine checks in the management of aquaculture ponds.

Improved management practices have contributed the increased fish production. But with it, several problems sometimes led a heavy economic loss to the farmer. Good sanitation and disinfection procedures reduce the numbers of disease-causing organisms present within a given system and prevent or reduce the spread of disease-causing organisms from one system to another.

Chapter 5
Leech Problems in Fish Ponds

Introduction

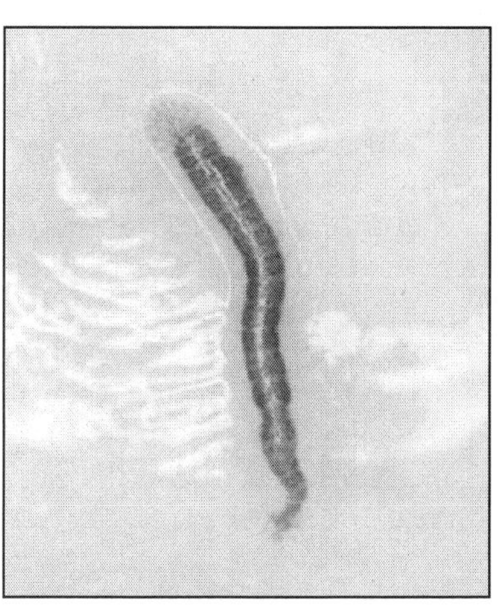

Leeches are external parasites and affix themselves on the body, fins, or gills of the fish. Usually they appear as heart shaped worms (they are just curled up) attached to the fish. They are usually introduced to the pond via plants or snails. Since leeches are sucking and borrowing into the surface of the fish, removal with forceps can cause great damage, if not death, to the fish. Leeches are segmented worms with long, slender, flexible bodies. They are usually 15-30mm in length and can easily be seen. They are brown with suckers at each end of the body. These suckers are used to attach to the fish and they enable the leech to move freely over the body surface. The leech usually attaches to the flanks or belly of a fish. They can also be found in sheltered areas such as on the gills and in the mouth. They have a simple life cycle. Leeches are common parasites in fisheries.

The Body Parts of a Leech

Physical Appearance

The body of a leech is divided into 33 segments that are then divided into three parts: head, body and rear. The shape of the body varies, but is generally flattened and cylindrical. Leeches come in many colors and patterns dependent upon geographic location. Leeches have eyes, but use them for detecting light and motion rather than seeing. There are two openings: mouth and anus.

Suckers

Leeches have two main suckers, one on each end, which are used for movement and feeding. The head sucker houses the jaw. The outer ring of each sucker is very muscular and clamps onto the victim after creating an opening with the jaw.

Reproductive Organs

Leeches are hermaphrodites, meaning that they have both male and female reproductive organs. They reproduce by copulation. Each leech deposits packages of sperm into the body of the other, which move to the ovaries for fertilization. The eggs are then wrapped in a gelatinous coating and left to hatch.

How do Leeches Attach to their Host?

Leeches are parasites that feed on the blood of fish, frogs, turtles and mammals. Leeches are freshwater carnivores found in lakes and ponds. Humans are scared of leeches because of their unattractive appearance and the fact that leeches can feed on human blood. Leeches are worms that detect their prey through vibrations and swim very fast to attach to their hosts and feed. Leeches can attach themselves to prey using their mouth and tail.

What do Leeches do in Fish Body?

When feeding, leeches insert a slender, tubular organ (called a proboscis) into the skin. This allows blood to flow from the fish to the parasite. The feeding action and attachment can result in damage to the fish, causing lesions and bleeding on the skin.

Infections Caused by Leeches

Leeches are external parasites with striated bodies, muscular body wall and two suckers used for feeding and movement. The parasite has been recorded in *Epinephelus bleekeri, E. coioides, E. fuscoguttatus, E. lanceolatus, E. malabaricus*and *Cromileptes altivelis* in Malaysia, the Philippines and Thailand.

Causative Agent

Zeylanicobdella arugamensis has been reported in *E. coioides* in the Philippines.

Stages Affected

Leeches are reported to affect nursery, grow-out and broodstock stages.

Gross Clinical Signs

The brownish-black parasites are attached in patches in affected areas such as the body surface, fins, eyes, brachial and mouth cavities. The fins of affected fish are frayed and the attachment and feeding sites are hemorrhagic and swollen. Affected fish lose their appetite, show sluggish movement, swimming at water surface.

Effects on Host

The affected fish have anemia and may show secondary bacterial infections. These leeches are known to act as vectors of viruses, bacteria and protozoan blood parasites. Mortality may occur in heavily infected fish.

Transmission

Transmission is horizontal. Mature leech leaves the host fish to deposit cocoons on hard substrate such as rocks, shells or vegetation. A cocoon contains a single egg which hatches onto a young piscicolid leech. The young leech then attaches to a host where it matures. Leeches usually die after deposition of cocoon.

Diagnosis

The leech has an elongated and cylindrical body narrowing at both ends containing the suckers. The characteristic oral (anterior) and caudal (posterior) suckers of the leech can be seen by gross and microscopic examinations. These suckers are alternately, strongly attached on the host. Mature leech can reach 15mm in length.

Minimising the Problems with Leeches

- There are no effective treatments for leeches once they have become established.
- Prevention of exposure and sensible fishery management (such as reduced stock densities and lower angling pressure) are the most effective methods for controlling leeches.
- Leeches rarely cause problems to whole populations of fish. When this does occur however, leeches are seen to be indicators of other problems within the fishery.
- A healthy fish can avoid these parasites, whereas sick fish cannot. If a fish is sick and resting on the bottom of a fishery, it is easy for leeches to attach to it. This is also the case during the colder months when fish are less active.
- For more information about managing stock density, good habitat and monitoring water quality, see the relevant fact sheets in this series.

Preventive and Control Methods

- Leeches may be removed from culture water by filtration.
- Manual removal using wet cloth has been effective in removing large patches of the parasite.
- A bath treatment of 200-250 ppm formalin for 1 hour, provided with strong aeration will detach most of the parasite.
- After treatment, treated fish should be transferred to clean, parasite-free facility.

☆ Culture facilities must be cleaned with detergent, disinfected with chlorine and exposed to intense sunlight for several weeks prior to use to eliminate cocoons of the parasite.

Controlling Measure

1. A measure which can be successful in controlling leeches is bait trapping. A metal can with a lid (a one pound tobacco or coffee can is ideal) drilled with small holes (depending on the size of the nuisance species) and baited with raw meat may trap large numbers of leeches from a heavily infested area. After feeding, the leeches will have difficulty leaving the can. Destruction of the contents of the can will help in reducing the size of the leech population.
2. Leeches are important in the food web of a lake. As such, many fish species will eat leeches. A healthy pond fish population will generally keep leeches under control, and redear sunfish are especially good leach eaters.
3. Ducks have also been used to control leech populations, but an over abundance of ducks can create other more serious problems. In order to keep ducks near your area for leech control they will also have to be fed. Duck feed is high in phosphorus and travels rapidly through the duck's digestion system. A readily available form of phosphorus is then added to the lakes. If the lake is small and the ducks numerous, enough phosphorus can be added to the lake to cause algae to grow. Ducks are also an intermediate host for a number of parasites that may not be desirable in a swimming area.

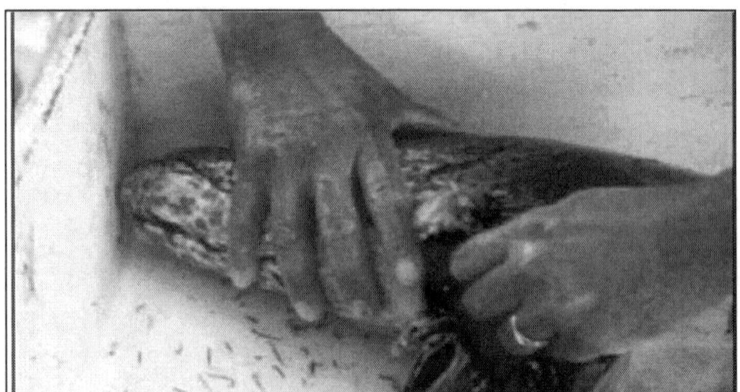

4. Water level manipulation can be used to control the bloodsuckers on small private ponds in areas where temperatures fall below freezing in the winter. Dropping the level of the pond at least four feet after ice has started to form on the pond will freeze the leeches in the mud. This measure is effective but it will also kill other forms of aquatic life that burrow in the mud.
5. The control methods mentioned in the attached sheet require draining the pond.

6. Salt solutions of 3 ppt are used by fish farmers in hauling water to reduce stress in fish. Such a treatment would require 2.5 lbs of rock salt for every 100 gallons of water.

Pathology and Epizootiology

Heavy leech infections have variable effects on fish hosts. Leeches were attached to the external perimeters of the mouth region and pathological changes were restricted to bite signs and mild tissue changes even in the most heavily infected individuals harbouring over 100 leeches. More severe damage was evident in piscicolid infection inside the mouth (mainly on the palate and the corners of the jaws) of grey mullets. Infections occur only in mature fish (over 200 mm in length). The number of leeches per fish was variable and could reach up to 67. Lesions only partially corresponded to observed infection loads and developed in the mouth in the area with the highest predilection for leech attachment, *i.e.* in the anterior part of the mouth roof and and the articulation zone of the lower and upper jaw. In heavy infections leeches were also attached in the craters of the deeper lesions. In some heavy infections the gular membrane also became perforated.

Fish confined to ponds seem to be more vulnerable to leech attack. Heavy leech infection of eyes and nostrils of cultured carp, which resulted in mortalities, have been reported from Ghana (Ghana Ministry of Agriculture, 1965). An increase in *B. tricarinata* was noted in pond cultivated *Clarias lazera* in Israel, but fish were removed before leech populations reached high levels. Leech attacks were reported from *Clarias* spp. and tilapia farmed in Southeast Asia. The leeches involved were unidentified local piscicolids or Glossiphoniids which induced pathology similar to that reported above (Kabata, 1985).

Leeches are vectors of haemoprotozoans. *Piscicola geometra* was shown to transmit SVC virus to Carp (Ahne, 1985). Feeding wounds may become contaminated by opportunistic bacteria and fungi (Kabata, 1985).

Treatment

Species of *Piscicola* were eradicated from pond fish by application of 0.5 ppm Dipterex (Dylox) or 0.25–0.8 ppm (Al) Masoten (Neguvon). Analogous organophosphates such as Malathion are effective too. In higher temperatures the higher dose is recommended. Removal of leeches from fish was also achieved by a 15 min. flush treatment with 5 ppm $CuCl_2$; 0.5 ppm copper sulphate for 5–6 hours or 25 pptNaCl for one hour (Prost, 1974; Kabata, 1985) [not effective for euryhaline leeches of grey mullets]. It was also recommended that ponds should be treated with unslaked lime (2500–3000 kg per ha) before stocking.If the fish is bathed in a 2.5 percent solution of salt for 15 minutes, most of the leeches should just fall off. Those that do not will be affected enough to remove with forceps with minimal damage. Another treatment is to add Trichlorofon at 0.25 mg/l.

Conclusion

An overpopulation of leeches in a pond can cause harm to fish and other aquatic life living in the pond. Adding chemicals to the pond water will kill leeches but will cause fish and other life more damage than the problem leeches. Using natural means to remove leeches from the pond will keep your pond healthy while reducing the threat of leeches. Removing leeches naturally from a pond requires patience and persistence to ensure all harmful leeches are removed.

Chapter 6
Problems Caused by Snails and its Solutions

Introduction

Freshwater snails are gastropods that live in a watery non-marine (freshwater) habitat. The majority of freshwater gastropods have a shell, with very few exceptions. Some groups of snails that live in freshwater respire using gills. Others need to surface to breathe air. Most snails act as a "clean up crew" in the pond, eating excess food, decaying plant or fish matter, and moving the pond bottom but the problem with snail is that they reproduce quickly creating over population and more serious penchant for disease transmission.

Problems Caused by Snail

- ☆ Snails help in cleaning up the pond bottom by feeding upon the decayed plants and animals and also on the unused feed materials.
- ☆ But Aquaculture producers have experienced significant monetary losses due to over population of snail in the pond because they create problems in the pond. The problems are:
  - ❒ Overpopulation by rapid reproduction.
  - ❒ Disease transmission by acting as intermediate host to various parasites

Over Population of Snail

The snail population should stay relatively constant and an excessive growth in the snail population indicates possible problems in the balance in the food chain in the pond. Overpopulation of snail builds up the organic load in the pond which detoriates the water quality. Snails may compete with the fish for dissolve oxygen and other useful nutrients.

Disease Transmission

Many species of freshwater snail are intermediate hosts of highly infective fluke (trematode) larvae of the genus *Schistosoma* which cause schistosomiasis.The snails are considered to be intermediate hosts because humans harbour the sexual stages of the parasites and the snails harbour the asexual stages.Freshwater snails are also intermediate hosts of foodborne fluke infections like Fascioliasis, caused by *Fasciola hepatica* affecting the liver, lungs and intestines of humans or animals. Besides the snails also carries various pathogens which may infect the cultured fish and proved to be economically detrimental for aqua culturists.

Disease Transmission Cycle Involving Snail as an Intermediate Host

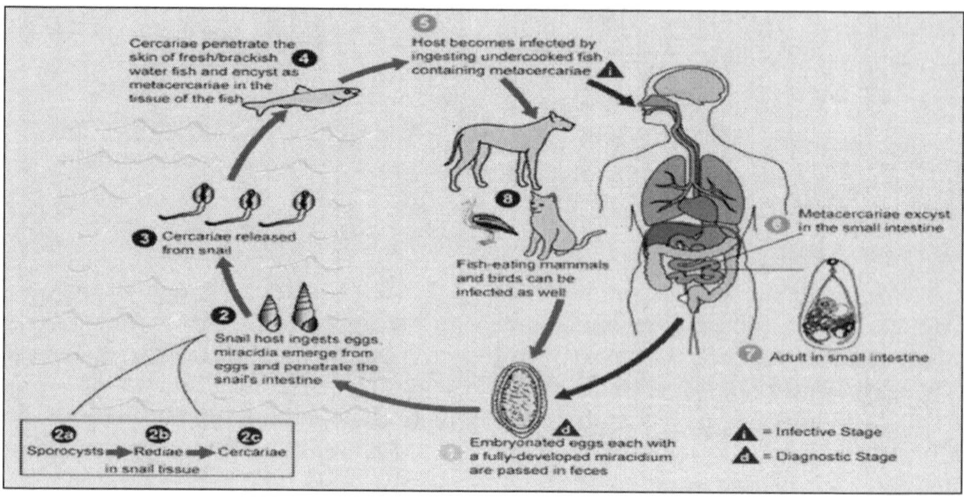

Control Measures of Snail

☆ As the snails create problems in the aquaculture pond,Control efforts to date have largely focused on reducing snail populations in aquaculture ponds through biological, chemical, physical means.

Biological Means; Introducing Snail Eating Fish

The introduction of snail eating fish can help to control the snail population. Not all snail eating fish are equally good in removing snails and some fish are great snails eaters, but aren't well suited for most pond. An example of the latter are puffer, which primary feed on snails, but also tend to get aggressive towards other fish. Common snail eating fish include; Black carp, Malawi Cichlids, varieties of Gourami, varieties of Bettas, and most Botia species. Ducks also can be introduced in the pond to control the snails.

Mylopharyngodon piceus **(Black Carp)**

Cichlids

Botia **sp.**

Manual Removal

- ☆ Hand picking won't eliminate all snails, but it's a good option to control population excesses.
- ☆ To facilitate the hand picking a bait can be used. Some lettuce or potato pieces are put in the pond. Many snails are attracted by the bait and will collect on it, which makes it easier to remove the snails.
- ☆ Some times branches of trees are put in the pond and after few days the snail settle on. Then the branches with the snail are removed from the pond.

1. Drain Out Pond Water

Physical alterations of culture ponds by draining or scraping can be an effective snail control mechanism. Although winter draining often eliminates pond snail populations completely, many production ponds cannot be drained or scraped due to year-round production. Negative impact is that scraped suspended particles have on pond nutrient dynamics, fish feeding, and gill obstruction.

2. Complete Cleaning and Reduction of Snail Habitat

- ☆ As a last resort, and the only 100 per cent effective method to remove snails is to clean the pond completely.
- ☆ Snails need vegetation for food, shelter and a substrate for their eggs.
- ☆ The removal of vegetation in ponds reduces the number of snails.
- ☆ But this method is very expensive and is not practicable in most cases.

Chemical Means

Snail Killing Chemicals

- ☆ To date, copper sulfate appear to be the best means to control snail populations in ponds.
- ☆ Treatment with lime or lime slurry along pond banks has also been used to control snails.
- ☆ Sodium chloride concentrations of 2.5 ppt have also been used to control snail populations in aquaculture ponds.

- ☆ Application of chemicals is the least desirable method of snail eradication, being too toxic and dangerous.
- ☆ They are usually effective, anything designed to kill snails is probably not healthy for other plants or fish.
- ☆ Even if the chemicals are harmless to the fish, the massive die off of snails will leave decaying snail bodies in the pond, which can affect the water quality.

Water Quality of Snail Infection

Snail population increase in a pond due to high pond water turbudity. If the organic load of a pond become high then snail population will increase. Because snails feed on the organic materials of the pond. Water temperature also an important factor which affect the snail population.

Different Species of Snails

Neritina natalensis

Pomacea bridgesii

Viviparus viviparous

Conclusion

Presence of snail may help algal control, and facilitate nutrient cycling. But snail does have definite downsides. They are known disease carriers for fishes, humans, and plants. So for fish farmers point of view snails are not desirable. To eradicate snail researchers should work to bring out easy and economical methods which the fish farmers can afford with and which will not harm the cultured fish or other biota of the pond.

Chapter 7
Use of Aqua Drugs: A Boon or Bane in India

Introduction

Aquaculture is an established and growing industry, and an increasingly important supplier of foods.The industry also produces baitfish for sport-fishing and ornamental fish for the pet trade.All aquaculture operations will have a demand for drugs, biologics, and other chemicals, collectively referred to as "regulated products". This may include: 1) disinfectants as part of biosecurity protocols, 2) herbicides and pesticides used in pond maintenance, 3) spawning aids, 4) vaccines used in disease prevention, or 5) marking agents used in resource management.The chemicals are used in the aquaculture mainly to treat diseased fish and, to a lesser degree, to improve water quality in culture facilities.

Drugs

- DRUG is a chemical substance used in the treatment, cure, prevention, or diagnosis of disease or used to otherwise enhance physical or mental well-being.
- Broadly speaking, DRUG is any substance that, when absorbed into the body of a living organism, alters normal bodily function.
- In aquaculture, this includes compounds that one would typically think of as drugs–antibiotics and other therapeutic compounds, fish sedatives and anesthetics, gender manipulators and spawning aids, *etc.*
- However, it's important to remember that innocuous, common household compounds – hydrogen peroxide, salt, and ice – are also considered drugs.

Drugs as a Chemical

Drugs are usually distinguished from endogenous biochemical's by being introduced from outside the organism.

For example:-Insulin is a hormone that is synthesized in the body; it is called a hormone when it is synthesized by the pancreas inside the body, but if it is introduced into the body from outside, it is called a DRUG.

Use of Drugs and Chemicals

- Soil and water quality of the pond is improved by using different types of chemicals and drugs, as a result fish culture and fish production increase.
- With the application of antibacterial drugs fish diseases can be controlled.
- Phytoplankton production may be increased by using different types of chemicals which increase the oxygen concentration in the pond and fishes get sufficient nutrition.
- Nutritional chemical compounds are supplied through feed to the fish, so as to increase the growth of the fish.
- Some chemicals help for colouration of fish.
- During transportation different types of anesthetic are used.
- By using different types of pesticides and herbicides, undesirable aquatic weeds and harmful insects are controlled.

Why Chemicals Used in Aquaculture?

- Chemicals have many uses in aquaculture, the types of chemicals used depending of the nature of the culture system and the species being cultured.
- They are essential components in:
- Pond and tank construction,
- Soil and water management,
- Enhancement of natural aquatic productivity,
- Transportation of live organisms,
- Feed formulation,
- Manipulation and enhancement of reproduction,
- Growth promotion,
- Health management and processing and value enhancement of the final product.

Concerns Regarding Chemical Usage

There are several important concerns with regard to the use of chemicals in aquaculture. These include:

- **Human health** concerns related to the use of feed additives, therapeutants, hormones, disinfectants and vaccines.
- **Product quality** concerns related to such issues as the occurrence of chemical residues in aquaculture products, their use in the enhancement of product quality and in the preparation of value-added products.
- **Environmental concerns**, such as the effects of aquaculture chemicals on water and sediment quality (nutrient enrichment, loading with organic matter, *etc.*), natural aquatic communities (toxicity, disturbance of community structure and resultant impacts on biodiversity), and effects on microorganisms (alteration of microbial communities and the generation of drugresistant strains of bacteria).
- **The general lack of knowledge** concerning the effects and fates of chemicals and their residues in cultured organisms and within the aquaculture system itself.

Some Drugs Commonly Used which Act as Boon in Indian Aquaculture

For Soil and Water Treatment

1. Lime

- Lime is a major chemical used for soil and water treatment in aquaculture. It is used to correct pond bottom and stabilize water pH. It is also reported to ensure a healthy plankton bloom.
- There are at least four types of lime use:
  - Agricultural lime/lime stone or crushed shell ($CaCO_3$)
  - Hydrate lime or slake lime($Ca(OH)_2$)
  - Quicklime/burnt lime or burnt shell lime (CaO)
  - Dolomite or dolomite lime ($CaMg(CO_3)_2$)
- **Use:** It is used @ 3kg/bigha – 150kg/bigha.

2. Iodine

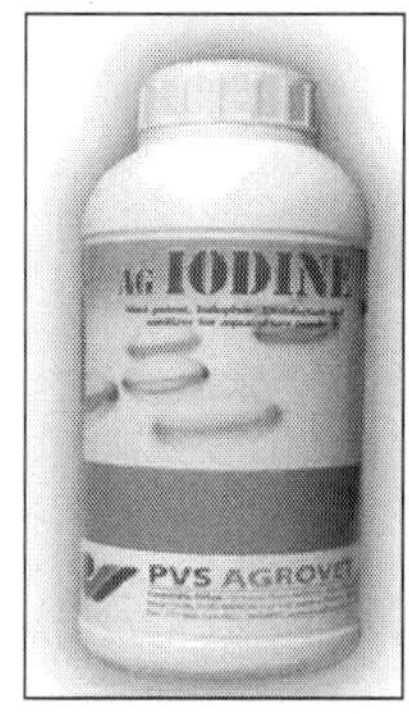

- The commercial solution of iodine, as used for disinfection of wounds, contains 10 per cent iodine, dissolved in alcohol. The solution has a dark brown Colour.
- **Method of use**: One part by volume of commercial I2 is mixed with nine part by volume of water. This solution is used directly for touching sore spot, wounds or fungus. This is done by brush or a tuft cotton wools, dipped into the solution. I2 is never used for preparing baths.

3. Formalin

- ☆ A weak formaldehyde solution may be used as a permanent bath for fish infected by skin or gill flukes.
- ☆ The water temperature should be not less than 18°c.
- ☆ **Method of use***:* 6-7 cu.cm. of stock solution of formalin is mixed with 1litre of water and spread throughout the pond. A different treatment is used to cure cases of sliminess of skin, caused by *Costia necatrix*. For this purpose a stock solution is spread by mixing 10 parts by volume of 32 per cent formaldehyde solution with 90 parts by volume of water. This fishes are bath in this solution 15-30 min. only.

4. Malachite Green

- ☆ Malachite green is a synthetic organic dye, consisting of green crystals with metallic luster.
- ☆ It is very soluble in water.
- ☆ It is recommended for treatment of fungus, especially fungus developed on eggs.
- ☆ It is very effective to treat "Ich" in both aquarium and food fish.
- ☆ **Method of use***:* Stock solution is made by mixing 1gm of MG in 450cu. cm. of distilled water. The lower conc. Is used for treatment of eggs.

5. Copper Sulphate

- ☆ Copper sulphate is vigorously used in aquaculture sector in W.B.
- ☆ It has been reported very useful against bacterial diseases. Cryptocaryon infections in marine aquaria where the target species tolerance is much bertter than in freshwater.

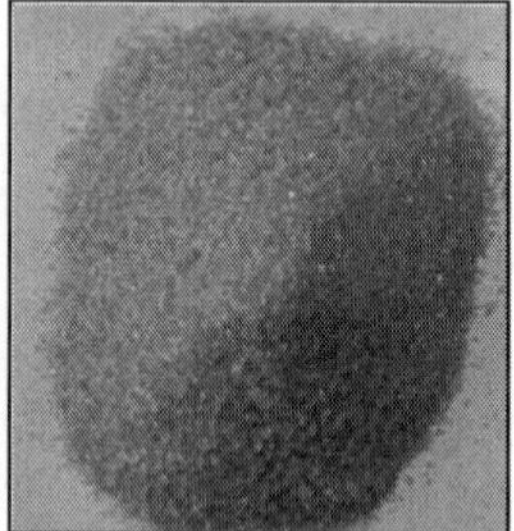

- ☆ It is also toxic to fish which have been exposed to a reduced osmotic pressure without time to acclimate; the loss of sodium ions exacerbates copper toxicity.
- ☆ **Used:** It may be used in case of ulcer disease by dip treatment at a conc. Of 500ppm for 1min.

6. Hydrogen Peroxide

- ✰ Hydrogen peroxide is used for treatment of **cotton wool disease**.
- ✰ Its effect on bacteria due to its decomposition to hydroxy radicals and on protozoa, and possibly on flukes, due to intracellular oxidation.
- ✰ **Method of use**: Hydrogen peroxide may be used only in fresh tap water, in containers in which no plants, sands or other materials are present,

  3 per cent commercial solution of hydrogen peroxide is added per 17.5cu.cm. of freshwater. In this solution fishes are bathed for 15-20 min. Treatment must be repeated for several times.

7. All Cure

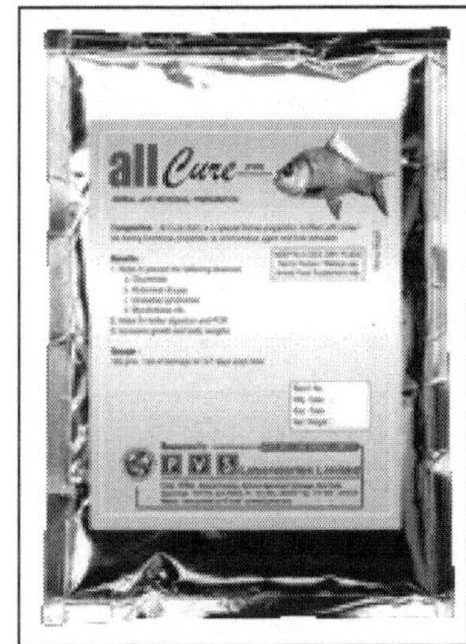

- ✰ **HERBAL ANTI MICROBIAL PREPARATION**
- ✰ **Main Benefits:** Helps to prevent the following diseases:
  - ❒ Cloumnaris
  - ❒ Abdominal dropsy
  - ❒ Ulcerative syndromes
  - ❒ Myxoboluses *etc.*
  - ❒ Helps for better digestion and FCR
  - ❒ Increases growth and body weights
- ✰ **Dosage:** 100 gms./ton of biomass for 5-7 days each time.
- ✰ **Presentation:** 1 kg. pouch

8. CIPRO – M

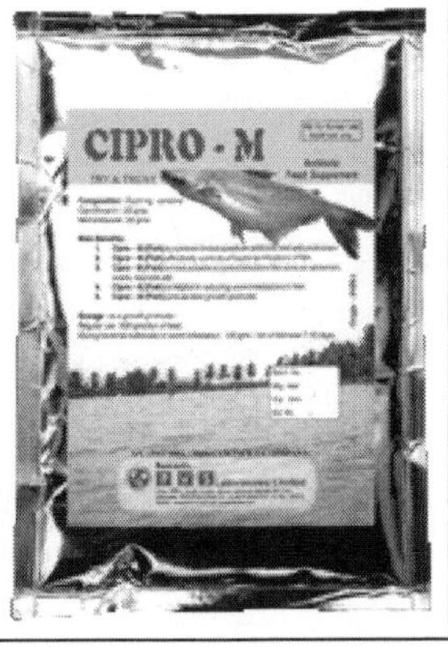

- ✰ ANTIBIOTIC FEED SUPPLEMENT
- ✰ **Main Benefits:**
  - ❒ Proven broad spectrum antibiotic and anti protozoan.
  - ❒ Effectively controls all bacterial infections of fish.
  - ❒ Suitable to control infections like columnaris, red sores on abdomen, ulcers, necrosis *etc.* and helpful in reducing worm infestations in fish.

- ❐ Best growth promoter.
- ❐ Very safe, does not interfere with growth of the animals.

☆ **Dosage:** At culture: 5 gms.,/kg. of feed with a suitable binder (Bind Gel) At Feed Plant: 500 gms./ton of feed.

☆ **Presentation:** 1 kg. pouch

9. Fish Clean

☆ DISINFECTANT and SANITIZER

☆ **Main Benefits:**

- ❐ Helps to control the lice infestations like Argulus, Lernaea *etc.*, and worm infestations like Dactylogyrus, Paradactylogyrus *etc.*, in fish.
- ❐ Controls microbial loads like virus, bacteria, fungus and protozoan organisms in aquaculture ponds.
- ❐ Controls external wounds, red blotches on stomach, infections of gills, tail *etc.* of fish.
- ❐ Prevents the chances of diseases on periodical application of Fish Clean.

☆ **Dosage**: Fish tanks: 1mtr. Water depth. 1.5 to 2.5ltrs./hectare depending upon the need. Double the dose if the problem in severe.

☆ **Presentation**: 1ltr. plastic container

10. Oxy Fish

☆ **OXYGEN GENERATOR IN AQUACULTURE PONDS**

☆ **Main Benefits:**

- ❐ Releases dissolved oxygen content.
- ❐ DO levels maintains quickly for longer periods.
- ❐ Helps for expulsion of toxic gases like ammonia, hydrogen sulphide.
- ❐ Relieves the animals from stress.
- ❐ Oxy fish functions well in variant salinities, pH fluctuations, in presence of heavy organic loads *etc.*
- ❐ Makes the animals to take feed well and to move well and to grow well.

☆ **Dosage:** Normal conditions: 1kg/hectare

☆ Stress/depletion of oxygen conditions: 2.5 kg/hectare

- ☆ In serious conditions: Repeat the next dose after 48 hours.
- ☆ **Presentation:** 1 kg. pouch

11. Fish Plankt

- ☆ PLANKTON DEVELOPER THROUGH BIOLOGICAL PROCESS
- ☆ Phosphate solubilizing bacteria plays major role in production of phytoplankton in fish ponds.
- ☆ FISH PLANKT comprises of two specific bacteria can perform both in low and high pH conditions. These bacteria take care the solubilization of phosphates in the water column as well as at the pond bottom. This consortium produces phosphorus pentoxide (P_2O_5) in the column and phosphoric acid ($H_3 PO_4$) at the bottom of the pond which results in development of blooms of phyto plankton.

12. Bottom-Lact

- ☆ A POWERFUL BIO-DEGREDANT BACTERIA
- ☆ **Benefits:**
  - ❒ Clears pond bottom sediments.
  - ❒ Prevents production of toxic gases.
  - ❒ Arrests growth of bacterial pathogens.
  - ❒ Makes pond water, clean, clear and visible.
- ☆ **Dosage:** 1mtr. Water depth.

  10 kg. per hectare once in 15 days or as advised by an aqua consultant.
- ☆ **Presentation:** 5kg. pouch, 10 kg. bag, 50kg. plastic drum

13. Micro Feed

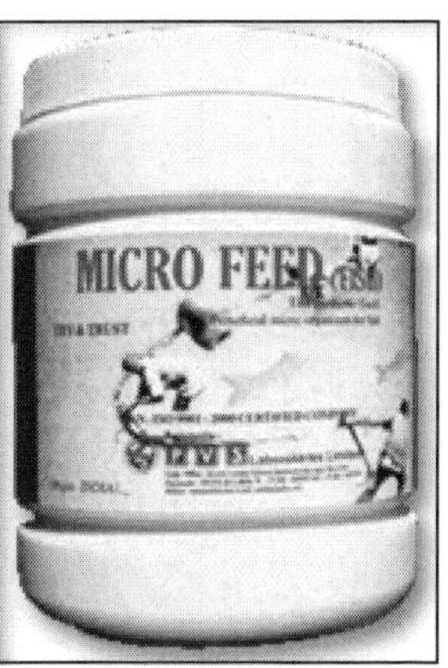

- ☆ FEED PROBIOTIC
- ☆ **Benefits**
  - ❒ Arrests the growth of pathogenic bacteria in the fish gut.
  - ❒ Ensures healthy balance of the digestive flora of fish.
  - ❒ Minimizes the usage of antibiotics.
  - ❒ Increases growth, body weights and disease resistance.
  - ❒ Reduces production cost.

- **Dosage:**

  At culture: 3-5 gms./kg. of feed.

  At Feed Plant: 1kg./ton of feed – regularly (or) 7 days in every month.
- **Presentation:** 1kg. tin and pouch

14. Ecto Mectin

- FOR CONTROL OF EXTERNAL PARASITES
- **Composition:**
  - Ivermectin: 1 per cent w/w
  - Herbal carrier having anti-microbial properties: Q.S.
- **Benefits**
  - Kills effectively external parasites like argulus, ergalasis, lernia *etc.*
  - Allows the fish for uninterrupted growth.
  - Prevents all sub-clinical stage of bacterial infections.
  - Very safe, no side effects and economical.
  - Control of gill flukes and other external parasites also seen.
- **Dosage:** 250 gms/ton of feed for 5 days continuously (or) 12 gms. per every ton of bio mass of fish 5 days continuously.
- **Direction for use:** Mix recommended quantity of Ecotomectin in 10 kg. of feed or rice bran and make the premix. Mix the premix thoroughly in the whole feed. Repeat the next dose after 90 days to prevent re-occurrence of parasites.
- **Presentation:** 1kg. tin/pouch

15. Cleantox+

- Adsorbs obnoxious gases
- Composition: HSCAS (Zeolite) with YUCCA
- **Benefits**
  - Cleantox+ adsorbs toxic gases like H_2S, NH_3 *etc.*, effectively
  - Cleantox+ improves water quality, transparency and alkalinity effectively
  - Cleantox+ consists High CEC, which increases adsorption of noxious gases.
  - Cleartox+ reduces turbidity of water and maintain water pH levels.
  - Cleantox+ enhances DO content because of specific action of silica and aluminum.

- Cleantox+ increase cation exchange capacity of soil resulting in less fertilizer requirement.
- Cleantox+ binds suspended organic matter and slime quickly.

☆ **Dosage and Usage**

Pond preparation: 50 Kgs/acre with a depth of 1 mtr

Regular usage: 10 – 20 Kgs/acre with a depth of 1 mtr for every 15 days

Note: Mix Cleantox+ with water or sand and spread all over the pond.

Dosage needs to be varied on the advice of Aqua consultant based on the environmental, pond conditions and stocking density *etc.*

16. Enrichmin

☆ All in one, essential feed additive

☆ **Composition**: Enrichmin is a unique proprietary blend of fat and non-fat soluble vitamins, Macro and Micro minerals, naturally sourced amino acids, predigested protein, instant energy source, herbal extracts, blend of probiotics and natural extracts rich in anti-toxins, liver stimulants, stress relieving and immune stimulating agents.

☆ **Benefits:**

- Enrichmin keeps fish/shrimp away from stress and improves the resistance.
- Enrichmin ensures prevention of Hepato pancreatic necrosis.
- Enrichmin is very effective in cramping disease
- Enrichmin improves digestion capacity and immunity.
- Enrichmin helps toabsorb and metabolize all kinds of nutrients
- Enrichmin reduces risk of infection and antioxidant profile.
- Enrichmin improves livability by reducing Early mortality.
- Enrichmin detoxifies and stimulates pancreatic function.
- Enrichmin prevents and control loose shell.
- Enrichmin improves FCR, growth and weight gain at a faster pace.

☆ **Dose advice:**

Powder: 10 – 20 grams per Kg of feed.

Liquid: 10 – 20 ml per Kg of feed.

(Or)

As suggested by Aqua consultant based on the environmental, pond conditions and stocking density *etc.*

17. Aquabloom

☆ "Ultimate phytoplankton booster"

☆ The growth rate and improved survival rate of cultured fish and shrimp mainly depends upon nutritious value of phytoplankton

- ☆ Aquabloom is specially designed to improve healthy and stable beneficial phytoplankton for aquaculture ponds as it contains necessary elements required to encourage growth of healthy and highly nutritional phytoplankton.
- ☆ **Composition:** Combination of naturally sourced 19 quality amino acids mix, derived from milk protein and 22 Micro and Macro Minerals along with phytochemicals and polyphenols of organic origin
- ☆ **Benefits:**
  - ❒ Aquabloom nurtures the existing planktons and boosts the growth of sustainable planktons in aquaculture ponds.
  - ❒ Aquabloom ensures pond ecosystem through optimum combination of different phytoplankton's ensuring a natural sustainable food chain for fish and shrimp.
  - ❒ Aquabloom mostly contain amphoteric molecules ensuring proper attachment of various objects.
  - ❒ Aquabloom ensures sustainable phytoplankton growth.
  - ❒ Aquabloom enriches the soil condition.
  - ❒ Aquabloom helps to maintain water pH.
  - ❒ Aquabloom maintains pond ecosystem.
  - ❒ Aquabloom improves feed efficacy, growth and survivability
  - ❒ Aquabloom is safe, reliable and effective
- ☆ **Dosage:**

  500ml - 1 Litre per acre

  (Or)

  As suggested by Aqua consultant.

18. Aquagrow

- ☆ "Efficient growth and production only matters in Aquaculture"
- ☆ It is essential to convert the genetic potential of aquatic animals to practical productivity.
- ☆ Aquagrow plays it role in growth modulation, enhancement of digestion and nutrient absorption through application. Aquagrow naturally induces and modulates important biological processes and nutrient sensing receptors in aquatic animals resulting in the optimization of nutrient absorption including vitamins, minerals, amino acids and protein hydrolysate and reducing the feed conversion ratio.
- ☆ **Benefits**
  - ❒ Aquagrow is readily digestible and absorbable with a high profile of essential and required amino acids for good growth of fish and shrimp.

- Aquagrow cannot be a substitute for any amount of plant proteins as it contains protein hydrolysate for fish.
- Aquagrow improves molting and shell formation in shrimp.
- Aquagrow patches up the protein deficiency and improves feed efficiency of the fish and shrimp by increasing positive nitrogen balance, amino acid anabolism and protein synthesis.
- Aquagrow is result oriented as it improves the overall pictures of the species like shinning eyes and skin, growth activity and FCR *etc.*
- Aquagrow helps faster and healthy growth of fish and shrimp with glossy look.
- Aquagrow contains 98 per cent predigested protein hydrolysate hence there will be increased muscle synthesis in fish.
- Aquagrow reduces FCR and improve growth rate due to increase in nutrient digestibility and absorption.
- Aquagrow consists herbal appetizers which enhances appetite resulting in higher feed intake and pepsin, amylase improves digestion and absorption of nutrients from gut.

☆ **Dosage**:

15 – 20 ml per Kg feed

Or

As advised by the Aqua Consultant

19. Y30

☆ "An efficient ammonia eliminator"

☆ **Composition**: Yucca Schidigera plant extract with Synergistic blend of probiotics and proven bioremediation agents.

Guaranteed Analysis: B50 ....Max 10 mg.

The B50 value indicates the milligrams of Yucca schidigera extract necessary to reduce 50 per cent of ammonia in aqueous solution. It contains B50 value less than 4.0 mg.

☆ **Benefits**

- Alleviates the ammonia stress
- Reduces the dissolved levels of obnoxious gasses, restores the levels of DO and reduces the BOD
- Neutralizes the off-flavor and creates eco-friendly environment.
- Actives including Sarsasapogenin, Furostanol, and Spirostanol are delivered in a protected form to retain their bioactivities for a longer period.
- Bioactive saponins and sapogenins in Y30 acts a s natural surfactants and supports the growth of micro biota at the pond bottom, helps effectively in bioremediation process.

- ❒ Improves steroid hormone synthesis as the composition is rich in Trigogenin.
- ❒ Improves gut micro flora and aids in better utilization of nutrients
- ❒ Improves and purifies overall water quality
- ❒ Promotes healthy growth of fish and shrimps. Increases the overall harvest yield.

☆ **Recommended Usage**

Powder/Liquid: Apply 80 – 140gm/70 – 80 ml per acre by thoroughly mixing with sand and sprinkling water evenly.

Application can be repeated once in 15 days up to 90 days of culture and Once in 10 days after 90 days of culture. Or As advised by Aqua consultant.

20. Complimin Aqua

☆ Micro fined trace mineral mixture

☆ **Composition:** Manganese, Zinc, Copper, Cobalt, Iron, Selenium, Iodine and Chromium

☆ **Benefits:**

- ❒ Complimin Aqua is specially blended, micro fined trace mineral concentrate for aquaculture.
- ❒ Ingredients have high biological value, which ensures even distribution in the feed.
- ❒ Prevents and defends against cramp disease and loose shell.
- ❒ Improves natural feed and stabilizes water color.
- ❒ Minimizes stress due to fluctuating environmental conditions
- ❒ Plays a major role in the metabolism of shrimp and fish
- ❒ Quickly form shell after molting
- ❒ Stimulates muscle mass growth
- ❒ Improves immunity and growth.

☆ **Dosage**

For water treatment

- ❒ For pond preparation: 10 – 15 Kg/acre
- ❒ Regular Use: 3 – 5 Kg/acre

Mix required quantity of Complimin Aqua with water, then spread all over the pond. Use in the morning for the best performance.

For Feed addition

- ❒ Daily 5 – 10 g/kg of feed, once daily increasing frequency during the molting period

  (Or) As advised by the Aqua consultant

21. Minano Forte

Organic Nano Minerals

Minano consists Ionized nanometer sized colloidal particles of minerals embedded in amino acids and then bio encapsulated for use in aquaculture nutrition.

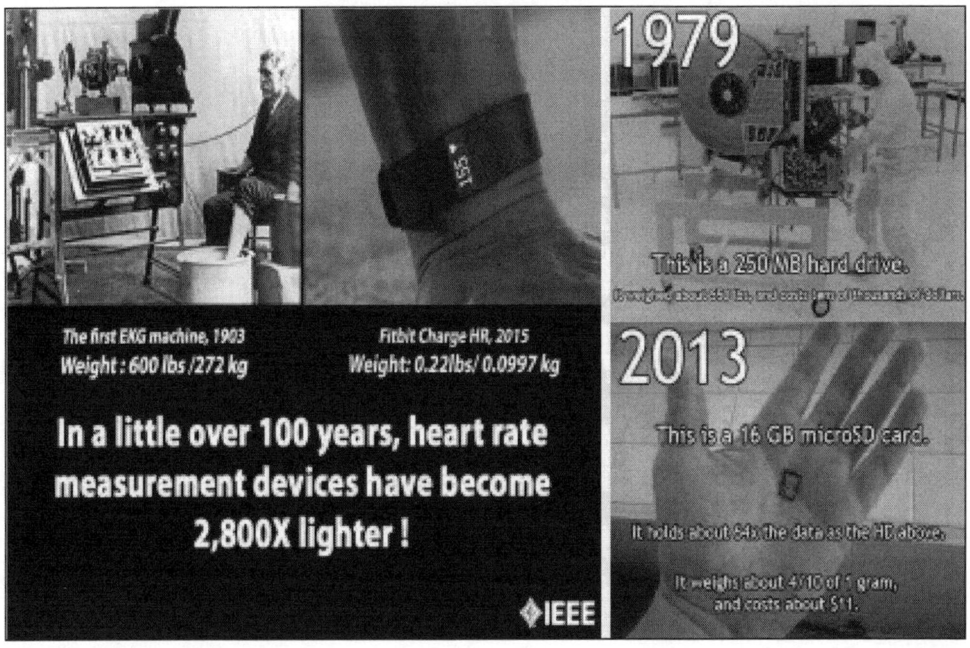

- ☆ **Technology**
  - ❒ Colloidal particles of ionized minerals, having nanometer size, embedded in matrices of amino acids and then encapsulated using a biopolymer for use in nutrition
  - ❒ Colloids occupy an intermediate place between (particulate) suspensions and solutions, both in terms of their observable properties and particle size.
- ☆ **Benefits**
  - ❒ Colloidal Nano products are Organic feed supplements which supports
  - ❒ High bioavailability
  - ❒ Cellular function
  - ❒ Improves immunity
  - ❒ Lower stress levels
  - ❒ Muscle function
  - ❒ Antioxidant protection
  - ❒ Growth

- ❒ Doesn't settle the ingredients at pond bottom
- ❒ Controls pH of water and maintain water quality
- ❒ Increases FCR and Weight gain.
- ❒ Overall improves productivity and performance.
- ❒ Cost and Convenience
- ❒ Easy to use, 10 Kg of mineral mixture was replaced by 1 lt of Minano.
- ❒ Reduction in labour and transportation cost.
- ❒ No wastage of active ingredients
- ❒ Improves ROI.

☆ **Dosage**: 500ml of anion+250 ml of cation per acre mix both with water and then spread all over the pond.Use in the morning hours for the best performance.

22. Aquahum

☆ Ultimate growth and nutrients utilization

☆ **Composition:** Aquahum is composed of concentrated humic acid and hydrolyzed fruit pulp containing reducting and non-reducing sugars.

☆ **Benefits:**

- ❒ Aquahum improves the growth and food utilization
- ❒ Aquahum detoxifies the harmful metals and chemicals in water
- ❒ Aquahum strengthens the immune system
- ❒ Improves plankton growth and pond soil quality
- ❒ Having the ability to neutralize the negative effects of pesticide residuals and toxins in the pond.

☆ **Dose advise**: 300 gm/acre (Or) As advised by the aqua consultant.

23. Trueyeast

☆ Excellent Yeast – Excellent Fish

☆ **Composition:** Natural Yeast Saccharomyces Cerevisiae

☆ **Benefits:**

- ❒ Release of nutrients content to microflora.
- ❒ Increases number of digestive bacteria population
- ❒ Improves FCR

☆ **Dosage:** For Fermentation in Freshwater – 500gm/acre and Brackish water – 2 Kg/acre.

For Feed use – 10 gm/Kg of feed (Or) As advised by aqua consultant

24. Anta Mix ME

☆ Synergistic functionality for extra performance

- **Composition**: Anta Mix ME is an acidifier consist of organic acids and their salts, phosphoric acid, and protected with micro encapsulated organic acids and essential oils.
- **Benefits:**
  - Anta Mix ME – Gut control through continuous release of organic acids which
    - limits E. Coli colonization
    - limits development of gram –ve bacteria (+++) and Clostridia (+)
    - reduces risk of wet litter
    - improved protein digestion
    - fast energy source
    - reduction of salmonella
  - Mode of action of encapsulated organic acid and essential oils release
    - increase of fat capsule digestion through lipase
    - all microencapsulated acids are strong organic acids (antimicrobial activity)
    - improving feed intake
    - stimulating saliva secretion and immune responses
    - improving enzyme activities
    - acting as antioxidant
    - synergistic antimicrobial activity with organic acids
    - direct antifungal, antiprotozoal and antibacterial activity
  - Better feed intake, improved digestion and nutrient absorption
  - Ca-content and P-content with a high bio-availability versus mineral deficits
  - Reduced buffering capacity (BC-Value) in feed
  - Additional source of energy in times of stress
  - Better health, reduced mortality, improved performance
  - Feed utilization (Egg production, FCR, DWG, ROI)
- **Dosage:**

  Along with AGP:

  Breeders – 1 kg/ton of feed

  Layers/Broilers – 500 gm/ton of feed

  Along without AGP

  Breeders – 2.5 to 3 kg/ton of feed

  Broilers – 2 to 2.5 kg/ton of feed

  Layers – 1.5 – 2 kg/ton of feed

  (Or)

  As suggested by Poultry Consultant

25. IM 5

- ☆ Powerful antioxidant and Immunostimulant
- ☆ **Composition:** Vitamin E, Selenium, Yeast, Vitamin B12 and Protein Hydrolysate
- ☆ **Benefits:**
  - ❒ IM5 is a powerful antioxidant and important for normal neurological functions
  - ❒ Plays a major role in the removal of free radicals.
  - ❒ IM5may protect cell membranes against lipid oxidation and exert synergistic effect in hydro peroxide detoxification.
  - ❒ IM5 required to prevent muscular dystrophy
  - ❒ IM5 improves growth and immunity.
  - ❒ Protects cells and membranes against peroxide danger.
  - ❒ Ensures fastest recovery from the diseases
  - ❒ Improved resistance to infections and under stressful conditions
- ☆ **Dose advise:** 5 -10 gm per kg of feed for 5-7 days

26. TH4+

- ☆ Disinfectant: Quaternary ammonium and glutaraldehyde
- ☆ A benchmark international disinfectant based on 4 quaternary ammonium compounds, glutaraldehyde, pine oil and terpene derivatives offering powerful bactericidal, virucidal and fungicidal action as well as acting as a deodorant.
- ☆ French and European Standards
- ☆ 22 approved uses including thermonebulisation
- ☆ Broad spectrum action (H5N1, Gumboro, Circovirus, salmonella, *etc.*)
- ☆ Effective in hard water or in the presence of organic matter
- ☆ Effective at high and low temperatures
- ☆ Non-corrosive and biodegradable
- ☆ Contains no phenol or formaldehyde

> Formats: 1L – 5L – 10L -25L – 60L – 200L – 1000L

Use biocides safely. Always read the label and product information before use. Only for professional use. Safety Data Sheets available on request.

27. Oxy Pond

- ☆ Improves DO
- ☆ Dissolved oxygen is the most important and critical parameter that decides the success in aquaculture. Lower dissolved oxygen (below 3 ppm) level is harmful to the growth and survival of prawn/fish and mortality rates can reach up to 50 per cent or higher, Usage of Oxy Pond is the most cost

effective and natural alternative to increase the Dissolved oxygen levels in the water. Oxy Pond enhances the oxygen levels in the pond water.

- **Composition:** Sodium Perborate with adsorbents and de-odorizes in suitable concentration.
- **Benefits:**
  - Releases required amount of oxygen and helps in restoring the pond bottom.
  - Fast recovery of oxygen level during severe DO drops
  - Reduction of BOD levels
  - Regulates pH fluctuations
  - Aquatic animal ensures complete relaxation from stress
  - Improvement in feed conversion
- **Dosage and Usage:**

  500 gm per acre or As directed by the aqua consultant.
  - Use Oxygen tab preferably during late evenings to compensate dissolved oxygen depletion that generally occurs at early morning hours.
  - Use for once a week whenever pond bottom and water quality deteriorated or according to water quality or as advised by Aquaculture consultant.
  - Application procedure: Broadcast all the suggested Oxy Pond to spread uniformly throughout the aquaculture pond bottom.

28. LIQPRO – PS

- Unique liquid probiotic
- The great challenge facing by Aqua Farmer in aquaculture is to maintain the sustainable pond water quality. The organic load, fecal matter, dead algae, molted exo-skeleton, Un-utilized feed, excessive feed, which is rich in protein and carbohydrates, settles at the pond bottom and produces toxic gases, ammonia/hydrogen sulfide. This deteriorates the healthy environment and causes damage to the shrimp/fish culture.
- LIQPRO-PS is the one stop solution for stink, thickness, slime, pollution and toxic algae pond water
- **Benefits:**
  - Liqpro-PS plays a major role in water purification by organic matter degradation (sludge).
  - Liqpro-PS reduces toxic gases (like ammonia/nitrite/H2S) build-up in the pond.
  - Liqpro-PS promotes growth of beneficial phytoplankton and stabilizes the bloom and pH.
  - Liqpro-PS suppresses the growth of pathogenic bacteria especially VIBRIO in the pond

- Liqpro-PS quickly reduces the water thickness causes by pollution.
- Liqpro-PS reduces foul smell and very effective in a wide range of pH and Salinity.
- Liqpro-PS improves pond environment, water color and algal profile of the pond.
- Liqpro-PS increases DO, stimulates the natural food production (BIOFLOC).
- Liqpro-PS stimulates floc granule formation-a natural food for shrimp.

☆ **Composition**: *Rhodobacter* sp., *Rhodococcus* sp. with growth factor aqueous media

Potency: 4 x 10^9 cfu/ml

Dosage and Usage: For 1-meter depth of water

Species	*Prevention*	*Treatment*
Shrimp	1 – 1.5 liters	2 – 3 liters/acre
Fish	1 liter/acre	1.5 – 2 liter/acre

☆ **During Pond Preparation (1 acre)**

Before creating the "water color", spread 15 – 20 liter of Liqpro-PS into shrimp pond for utilizing the residue of organic matter in pond bottom and for establishing good bacteria groups on pond bottom

- Early stage, 0.5 ppm/week
- Middle stage, 1ppm/week
- Last stage, 1-2 ppm/week

☆ **Caution**

- Apply between 08.00 - 11.00 AM (sunny day)
- Paddle wheel should be turn on during the application
- Store at room temperature 27 °C

29 YUPRO R

☆ Unique blend of yucca and probiotics

☆ Dead Plankton, fecal material and other decaying matter are toxic resulting gases like ammonia, nitrate and nitrate. Absence of beneficial microbes causes increase in pollution levels, depletes DO levels and promotes the growth of pathogens and harmful algae.

☆ **Composition**: Concentrated Yucca Schidigera (30 per cent) extract with Blend of beneficial probiotic species (Bacterial sps)

Potency: 10 x 10^9 cfu/g

☆ **Benefits:**

- Yupro R speeds up the natural process of organic load/sludge degradation.

- Yupro R probiotic organisms can consume a broad spectrum of pollutants, including surfactants, fats, sugars, starch etc. while eliminating the formation of hydrogen sulphide.
- Yupro R neutralizes the off-flavor and creates eco-friendly environment.
- Yupro R creates probiotic farming environment.
- Shrimp quickly recover from stress and has long lasting effect.
- Yupro R improves the water quality, helps environment to get stable and balanced.
- Yupro R is effective in high range of pH and salinity

☆ **Dosage and Usage**: For one-meter depth of water

Species	*Prevention*	*Treatment*
Shrimp	250 gm/acre	400 gm/acre
Fish	150 gm/acre	250 gm/acre

☆ In feed: 1 Kg/ton of feed (Or) As advised by Aqua Consultant.

☆ **Directions of use**: Dissolve required quantity of Yupro R with wet sand and apply all over the pond.

30. NUTRIPRO-R

☆ Unique combination of Probiotics, Protein, Vitamins and Minerals

☆ Nutripro-R is a unique combination of beneficial probiotic species, Protein, Vitamins and essential growth factors that will help in:

- reduced effects of mycotoxins,
- increase in beneficial bacteria and
- impair the proliferation of bad bacteria in the gut,
- improved intestinal integrity,
- improved immune response,
- increased palatability of feeds and
- better feed efficiency.

☆ Nutripro-R provided with yeast cultures have important benefits for Shrimp/Fish as source of nucleotides and beta-glucans, and as potential growth promoters.

☆ **Benefits**

- Best source of stable protein with highest nutritional value
- Decreases the effects of mycotoxins and increases in beneficial bacteria
- Improves healthy micro gut flora helps in inhibiting the disease causing pathogens

- ❒ Improves immune system and helps in promoting excellent growth
- ❒ Helps in regulation of acid-base equilibrium and metabolic activity
- ❒ Enhances feed efficiency and Survival rate

☆ **Composition:** Blend of beneficial probiotic – 1000 x 109 cfu/kg and Yeast/Protein – 40 per cent and unique proprietary blend of fat and non-fat soluble vitamins, Macro and Micro minerals, naturally sourced amino acids, predigested protein, instant energy source, herbal extracts, blend of probiotics and natural extracts rich in anti-toxins, liver stimulants, stress relieving and immune stimulating agents.

☆ **Dosage:** Shrimp/Fish: 2 – 3 gm/Kg of feed

31. MAXPRO-PS

☆ Protect with Probiotics

☆ Excessive Organic load is an indicator that a pond is out of balance.Ponds with low oxygen levels and high bottom sludge (nutrient) accumulations with Un-utilized feed, molted exo-skeleton, dead algae, fecal matter and excessive feed are tending to go anaerobic which produces toxic gases, ammonia/hydrogen sulphide. This deteriorates the healthy environment and causes damage to the shrimp/fish culture.

☆ **Composition**: Unique blend of beneficial probiotic species (Bacterial sps), Rhodobacter sp, Rhodococcus sp, Nitrosomonas sp, Nitrobacter sp, and Aspergillus Oryzaewith growth factor aqueous media

☆ **Potency**: >50 x 10^9 cfu/lt

☆ **Benefits:**

- ❒ Maxpro-PS produces variety of antibacterial, antifungal substanceswhich suppresses the growth of pathogenic bacteria like Salmonella, Aeromonas, E. coli and VIBRIO in the pondenvironment.
- ❒ Maxpro-PS is water and soil probiotic for shrimp, fish which produces large amount of waste degrading enzymes like amylase, protease, lipase and cellulose, it improves pond environment by converting complex compounds to simple form for easy bio-degradation.
- ❒ Maxpro-PS oxidizes ammonia to form nitrites and nitrates, enhances the natural nitrification and denitrification cycles, to convert nitrate in to nitric oxide, and harmless nitrogen to create and maintain a high water quality and an ideal environment for shrimp/fish growth.
- ❒ Stabilizes pH, alkalinity and algal blooms and effective in high range of salinity (0-50ppt) and pH (4.5 – 10)
- ❒ Eliminates malodorous in the pond as well as geosmin, which causes off flavor for fish/shrimp
- ❒ Decomposes organic wastes and sludge
- ❒ Stabilizes soil pH

- ❒ Helps in balancing the ecology in the pond
- ❒ Improves water quality and reduces turbidity.

☆ **Dosage and Usage**

For Shrimp – 5 Lt/acre

For Fish – 3 Lt/acre.

☆ Take require quantity of product and mix with pond water and apply over the pond. Suitable time for application is 2 – 5 PM (Or) As suggested by the Aqua Consultant.

32. Nuculture

☆ Clean – Care - Protect

☆ Sludge and water management is the key factors for shrimp production. Sludge deposits of the previous culture and polluted water from canals and creeks are concern areas, which effects the survivality and growth of the shrimps. When the right strains are chosen they can even degrade toxic compounds that are man-made or those that are present in nature

☆ **Composition**: Unique blend of *Bacillus* spp. – 6, *Aspergillus* spp. – 2, *Acetobacter* spp.- 2, *Lactobacillus* spp. – 2, *Bifidobacter* spp. – 2 with active functional prebiotics as base.

☆ **Potency**: 10 x 10^9 billion cfu/gm

☆ **Benefits:**

- ❒ Control vibriosis by producing bacitracin, gramicidin, polymyxin, tyrotricidin, and competitive exclusion.
- ❒ Compete with other bacteria in ponding and clearing the organic matter
- ❒ Better yield and good control of disease and immunity enhancement in Litopenaeus vannamei
- ❒ Protection against Vibrio harveyi by stimulation of cellular and humoral immune defenses
- ❒ Better water quality parameters, increased length and weight of the animal
- ❒ Enhance humoral immune response
- ❒ Decomposes organic wastes and sludge
- ❒ Stabilizes soil pH
- ❒ Acts as a feed supplement by promoting the growth of healthy phyto/ zoo plankton growth.

☆ **Dosage and Usage:** Initial dose 1.5 kg/acre and repeat the dose 300 – 500 gm/acre in one-meter water depth every 12 – 15 days depending on the pond condition during culture period. In Vibrio endemic areas and in ponds with high salinity, the frequency of application may be reduced to 7 days. Specific usage decisions need to be taken considering other factors like soil nature, water quality, stocking density *etc.*, in consultation with aquaculture consultant.

33. NULIFE

☆ **Introduction:** Sludge and water management are the factors for better fish and shrimp production. Sludge and silt deposits of the previous culture and polluted water from the canals and creeks are the main areas of concern, which affect the survivability and growth of the organisms. When the right biological strains are chosen they can even degrade toxic compounds that are man made or those that are present in nature. Cotrolling pathogens will require a combination of environmental manipulation techniques that allow the balance to be restored, changing the production paradigm. The causative agent of white faecal matter are the Pir toxins encoded from a plasmid,which are highly mobile genetic elements. The intestinal bacterias like *Vibrio alginolyticus, V. Parahamolyticus, V. vulnificus* can be reduced by application of probiotics as well as disposal of sludge and water exchange. The disease occur in different soil conditions and it results in detoriated water quality. Peak mortality rates are seen in the face of extremely low oxygen. Also the parasitic protozoans known as gregarines have been related with disease. So the present herbal preparation have depressant effect on causative organisms and effective against controlling the disease.

☆ **Benefits:**

❒ Tonic to improve the immune System

❒ Reduces the Infection rate and stress

❒ Antiviral and antibacterial capability

❒ Highly efficient to control white gut and white fecal matter

❒ Effective against loose shell syndrome

❒ Cost effective and ecofriendly

❒ Control vibriosis by producing bacitracin, gramicidin, polymyxin, tyrotricidin, and competitive exclusion.

❒ Compete with other bacteria in pond and clearing the organic matter

❒ Better yield and good control of disease and immunity enhancement in Litopenaeus vannamei

❒ Protection against Vibrio harveyi by stimulation of cellular and humeral immune defences

❒ Better water quality parameters, increased length and weight of the animal

☆ **Dosage and Usage**: 5 ml per Kg feed for 7 days in all meals or as directed by the aqua consultants.

Presentation: 1 litre.

Contents: Scientific blend of time tested herbal ingredients:

Aegle marmelos, Allum sativum, Aristolachia indica, Azardirachata indica,

Cassia fistula, Catharanthus roseus, Curcuma lango, Cynodon dactylon, Lantana camara, Melea azedarah Mimosa pudica Momordica charantia, Morus alba Ocimum americanum Phyllanthus emblica Psidium guajava, Salanum nigrum Tridax procumban Tylophora indica

Common Pesticides and Antibacterials Used in Aquaculture and their Manufacturers

Pesticide/Chemical/Antibiotic	*Brand Name*	*Manufacture*	*Rate (Rs)/kg*
Dichlorvos	Nuvan	Hindustan Ciba-Geigy Ltd.	225
Malathion	Cythion	Cyanamid India Ltd.	75
Quinophos	Ecalux	Sandoz India	150
Benzene hexachloride	Gammexene	Local companies	40
Copper sulphate		Local companies	20
Sodium chloride	Common salt	Local companies	0.50
Calcium oxide	Lime	Local companies	1.25
Oxytetracycline	TM-50	Pfizer Ltd.	120
Chlortetracycline	Aurofac-20	Cyanamid India Ltd.	68
Doxycycline	AFS Forte	Vesper	90
Nitrofurazone +Furazolidone	Bifuran	Eskaylab Ltd.	120
Furazolidone	Groviron	Glaxo Laboratories	120
Sulphadiazine + Trimethoprim	Bactrisol Neochlor Forte	Alved Pharma Pvt. Ltd. Vetcare Division of Tetragen Chemie Ltd.	290
Chloramphenicol Neomycin+ Doxycycline	Neodox Forte	Vetcare Division of Tetragen Chemie Ltd.	NA

Chemical Drugs-Bane in India

Excessive use chemicals and drugs brings adverse effect to the fish.

- The use of synthetic drugs in hatcheries may adversely affect larval growth and inhibit defense mechanisms.
- They may reduce the activity of acid or alkaline phosphate acid, which are considered to play a major role in the protein metabolism of shrimp.
- There are considerable constraints to the promotion of safe and effective use of drugs and chemicals in aquaculture in INDIA. They include:
  a) Lack of trained manpower and related capacity-building schemes and support services to disseminate information on fish health management;
  b) The misapplication of some chemicals (*e.g.* the excessive prophylactic use of antibacterials);
  c) Insufficient understanding of mode of action and efficiency of certain chemicals, especially under tropical aquaculture conditions.

Conclusion

- ✰ Clearly, chemicals and drugs play an important role in present aquaculture systems, whether they are intensive, semi-intensive, semi-closed or closed systems.
- ✰ Certain types of chemicals, if used inappropriately, can cause damage to animals and the environment. Overuse of chemicals, especially antibiotics, not only increases production costs but also intensifies adverse consequences.
- ✰ The adverse impacts of chemical use in aquaculture may be outweighed by their advantages; however, there has been increasing concern about their uses, and they must be used with great caution.
- ✰ Actual and scientific application of drugs may control the disease, otherwise it will be toxic to fish.

Chapter 8

Problems of Predatory Fishes in Culture Units

Introduction

Predators are a major source of stress to fish and can also cause significant losses. In modern days wild fishes like weed and predatory fishes are most problematic in culturing edible freshwater fishes as they prey up on the larval forms (spawn, fry and fingerlings of carps) of cultured fishes. These fishes causing adverse effect on culture fishes, as a result the yield is decreasing and leads to economic loss. Weed and predatory fishes carrying so many pathogens like viruses, bacteria and some external, internal parasites cause serious diseases to the cultured fishes. These fishes compete for food, space and oxygen with the carps that are cultivated.

Predators

1. Consume the fish in the pond,
2. Consume the fish's feed,
3. May transmit parasites and other infections to fish,
4. Scare the fish when they are chasing them up, and
5. Cause physical injury to several fish in the process of hunting.
6. May translocate fish to a different pond.

For example, Large African catfish can predate on smaller ones; so, maintaining uniform sizes within ponds is critical. And the wounded fish left out in the ponds consequently cannot get to the feed as well as the other normal fish. This is because, for example, their eyes might be injured or their open wounds might get infected, *etc.* Consequently, their growth rate slows and chance of survival drops. Controlling predators is therefore important in commercial production.

Common Predatory Fishes

Chitala (*Notopterus chitala*)

The adult Chitala is a predatory fish which subsists mainly on small fish. In the early fry stage, it feeds voraciously on carp fry and aquatic insect larvae. The fish breeds in confined waters, attaching the eggs to submerged objects. The species is neither cultivated at present nor is anything known about its cultural possibilities.

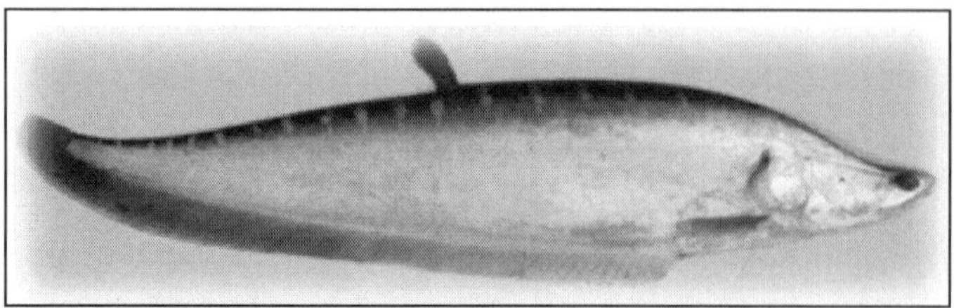

Freshwater Shark (*Wallago attu*)

It is highly predatory in nature and is extremely destructive to carps. Its large mouth, well-armed with teeth, enables it to successfully attack and destroy even medium size carps. On account of its predacious habits, this species should be carefully removed from carp ponds. No attempts appear to have been made to culture this quick-growing species in ponds on scientific lines.

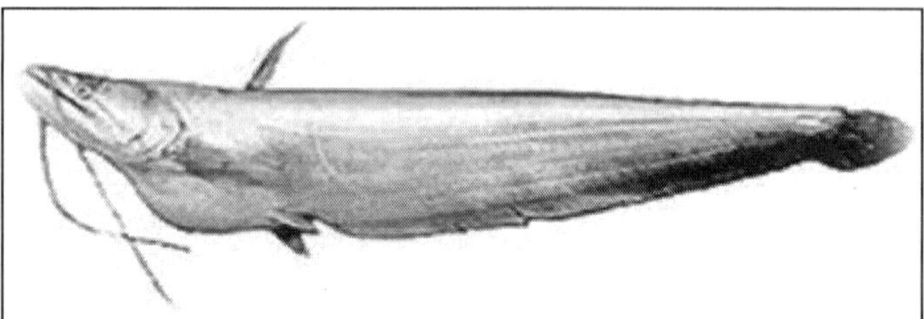

Catfish (*Mystus seenghala*)

The fish is a predator with marked piscivorous tendencies, *Barbusfilamentosus* and *Cirrhinareba*having being found most frequently in its stomach in the Cauvery river. Insects and their larvae, crustaceans, gastropods and very rarely aquatic weeds are also consumed. Very little is known about the development and growth of the species.

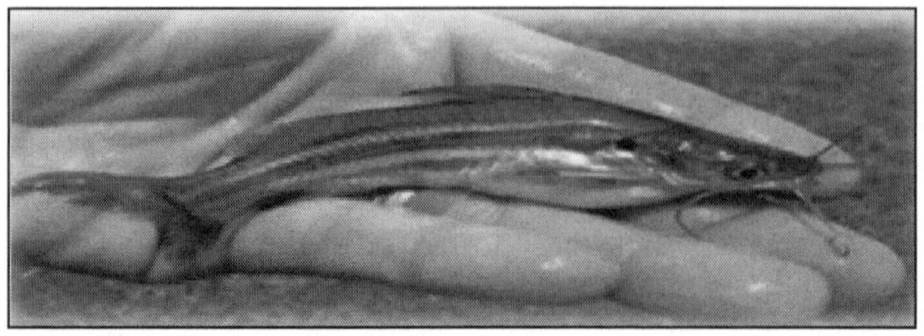

Pangas (*Pangasius pangasius*)

This is yet another of the common major catfishes attaining a length of about four feet and found all over India. It is a predatory fish in its natural haunts, but unlike *Wallago* and *Mystus*, it seems to live largely on gastropods whose shell remains appear to constitute the largest single item amongst its stomach contents.

Feeding Habits of Predatory Fishes

In predatory fishes, the mouth is generally large and provided with a number of teeth, which, however, are not so much for chewing the prey as for preventing its easy escape. With its terminal mouth, the predator can easily snap at and catch relatively larger victims. Unlike the non-predators, the intestine is very short with a distinct bag- like stomach. The Large Goby (*Glossogobiusgiuris*), the Murrel (*Ophicephalus* spp.) and the freshwater shark (*Wallagoallu*) are familiar instance.

Common Predatory Fishes

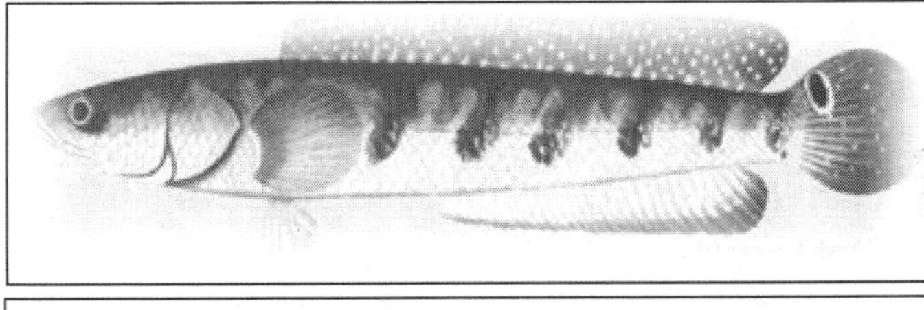

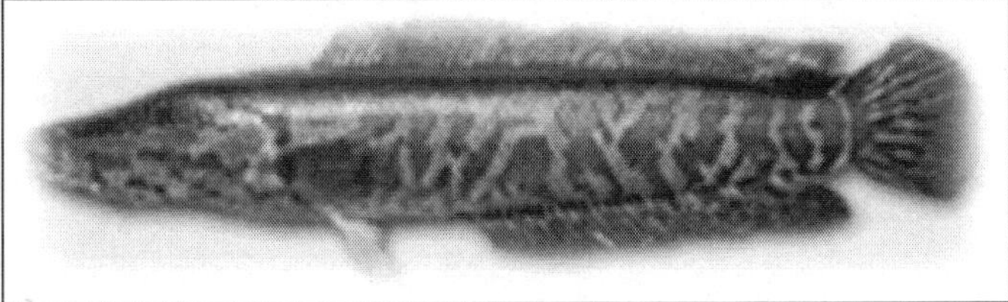

***Channa* sp.**

***Clarias batrachus* (Magur)**

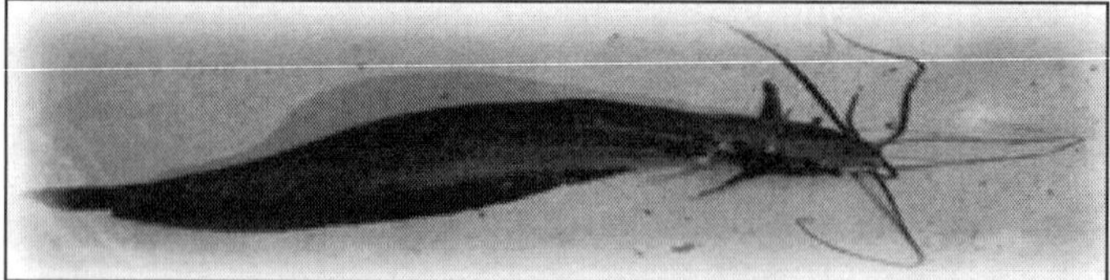

Heteropneustes fossilis

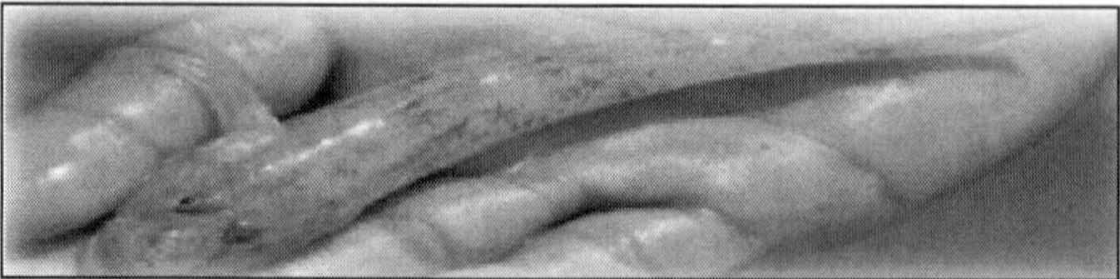

Glossogobius guiris

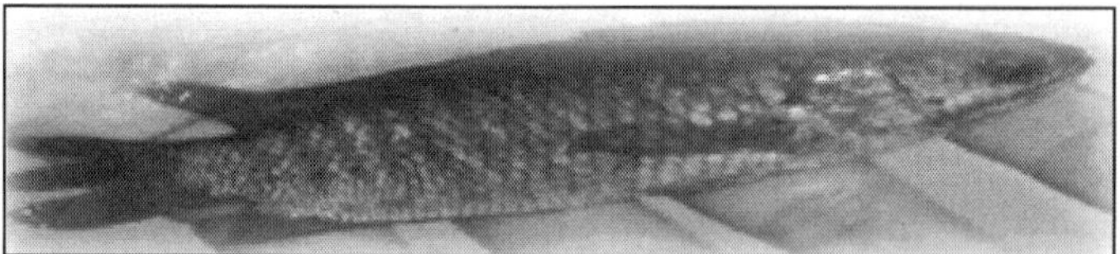

Anabus testudineus

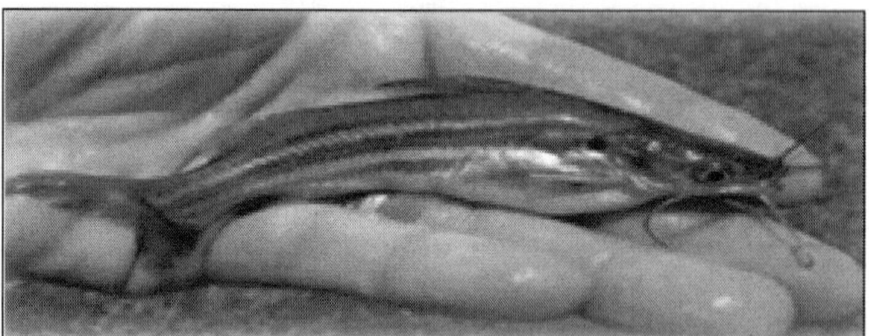

***Mystus* sp.**

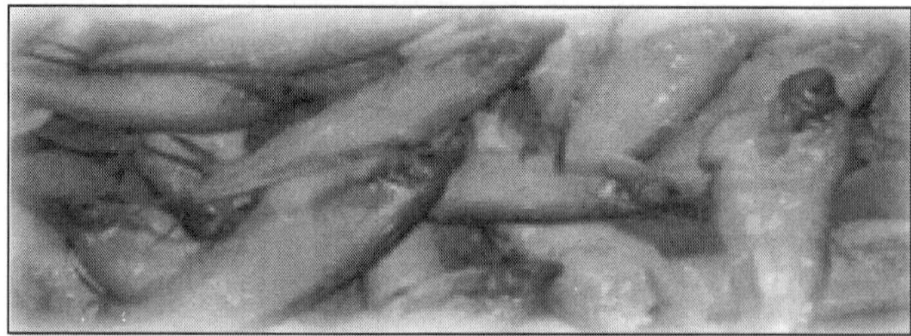

***Ompok* sp.**

Predatory fishes are almost invariably present in ponds, though in varying number. Even in ponds completely isolated from other sources of water, some of the predatory fishes like the *Murrels, Anbas,* and others, crawl in over wet ground from neighboring waters. And once they get in to the pond it is extremely difficult to eradicate them by the usual netting operations. All the predators feed on carps, adult specimens preying upon fingerlings and yearlings, and the young ones destroying the carp fry. Some of the predatory fishes like the *Murrels, Anabas, Nandus, Heteropnuestes, Clarias*and *Notopterus* breed in ponds a little prior to or at the commencement of the monsoon. Their young ones, feeding on the plankton in the nursery ponds, grow so fast that when the carp fry are ready to be introduced into the pond, they are large enough to destroy them in large numbers.

Clarias batrachus **(Magur)**

Heteropneustes fossilis

Adverse Effect of Weed and Predatory Fishes

1. Compete with the cultured species for food and spawning habitat
2. Shortage of natural food in cultured ponds.
3. Predatory fishes prey up on the larval forms (spawn, fry and fingerlings of carps) of cultured fishes; this causes economic loss to the farmer.
4. Overpopulate the pond.
5. Contribute to water quality deterioration.
6. Interfere with seining and grading.
7. Make harvest more difficult because fish must be sorted.
8. Cause weigh-backs or dockages when fish are sold to processing plants.

   Transmit disease to cultured species.

Solution

- The presence of carnivorous fishes in the fish pond especially in nursery and rearing ponds causes serious and big loss of fries and fingerlings.
- Therefore, in managing a fish pond, firstly facilities and secondly the management need to be improved so that the wild fish do not get chance to enter into the fish pond. Predatory and weed fish should be completely eradicated from nursery, rearing and stocking ponds before these ponds are stocked.

Prevention Methods

All ponds, except for those filled directly by a spring or by rainwater, need water inlets. The water inlet must be constructed with fine screen so that it supplies adequate quantities and quality of water, and so that it does not allow unwanted fish or other materials to enter the pond.

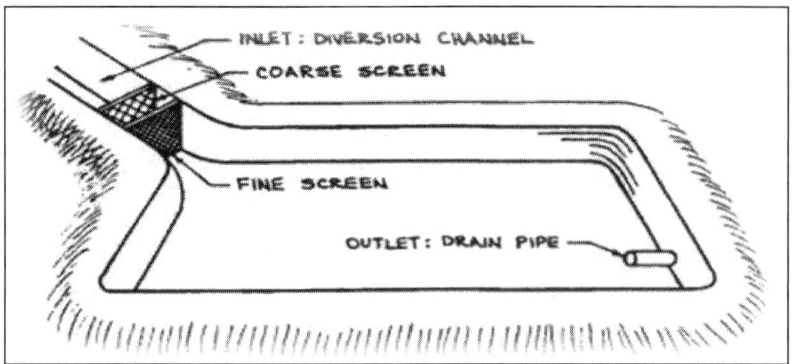

First and foremost thing to prevent weed and predatory fishes use vertical, horizontal screensat the inlet. The horizontal screen at the left is very effective. Here the screen is placed so that the water passes through it and falls in to the pond. This screen is just out from the wall at the inlet. In the version below the horizontal screen has a vertical screen wall attached to it. This screen prevents the entering of unwanted, weed and predatory fishes.

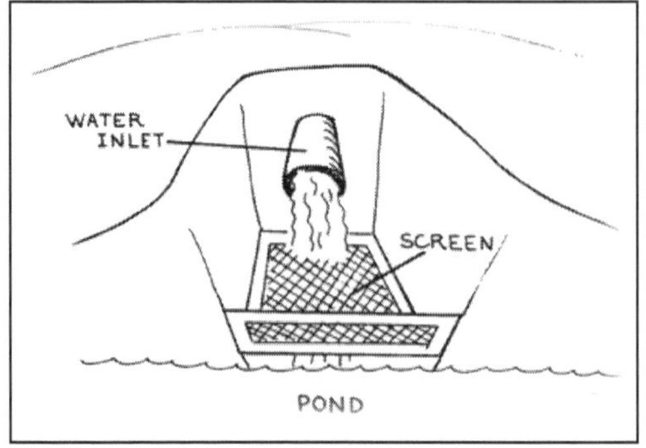

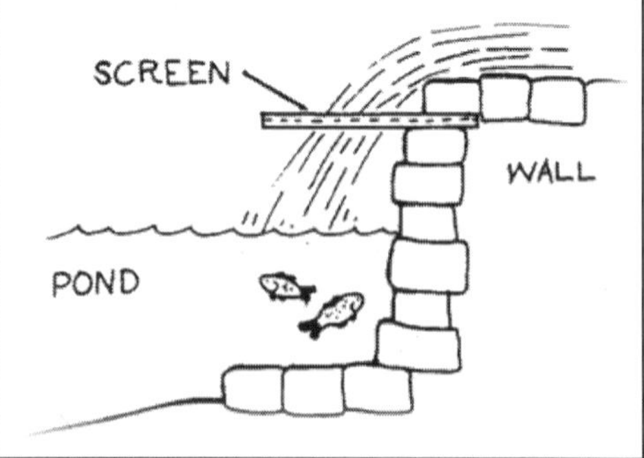

In second method, using nylon bag nets at the inlet pipes, these nylon bag nets prevent the entering of unwanted, weed and predatory fishes. A nylon mesh bag makes a good filter, as long as it is partially submerged in the pond so that it does not tear as the bag catches fish or other material from the water source. Check it periodically.

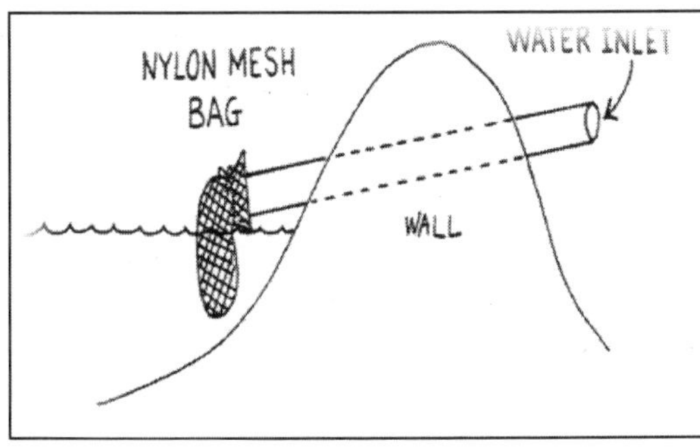

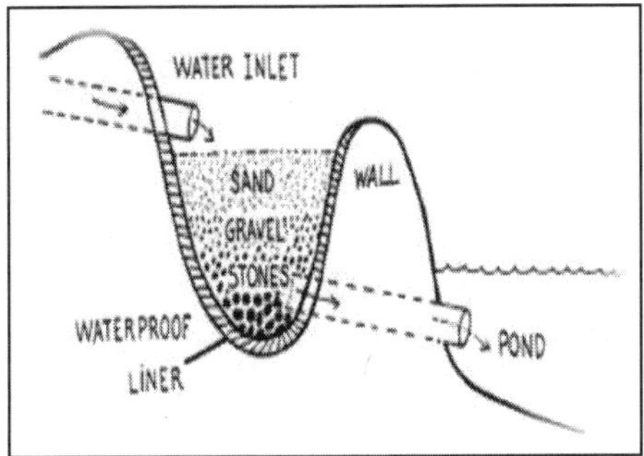

In the third method, using a saran fibre filter is basically like a wire screen that is placed horizontally underneath the water inlet. However, it is placed in a box standing in the water and uses saran fiber material instead of wire. These filters all have good and bad points. All must be cleaned often to remove debris that collects in them from the water source. The best filters are the sand and gravel filter, and the saran filter, but these are more costly than the others.

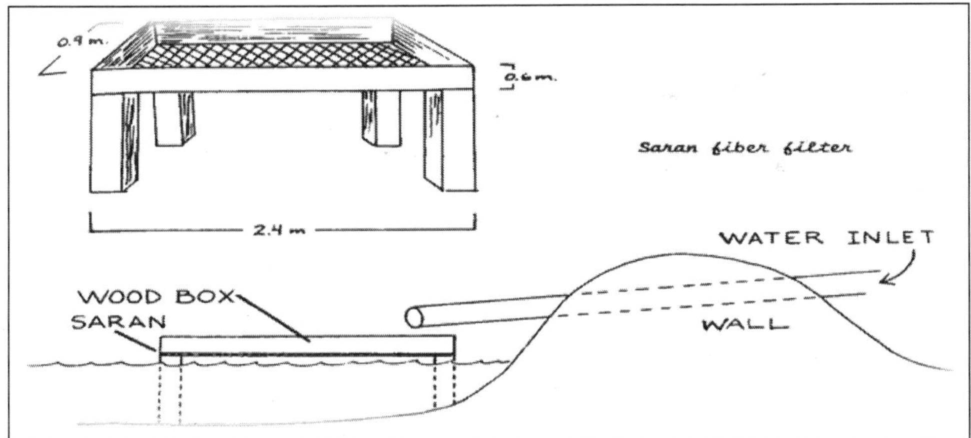

Control Methods

There are three methods to control the all unwanted, weed and predatory fishes.

1. Chemical method
2. Biological method
3. Physical method

1. Chemical Method

- In this method chemical toxins are used to control the unwanted, weed and predatory fishes.
- Economic point of view the poisoning should be done during pre-monsoon season when the water level is usually low, requiring the minimum quantity of poison material.
- Poisoned fish should be safe for human consumption, least adverse effect on the pond biota, toxicity period should be of short duration, should not have residual effect, easy commercial availability, simplicity of application, cost considerations.
- So many fish toxins are used such as Endosulphan, Taxaphene, TFM, PCIP (polychlorpinene), Endrin, Copper sulphate *etc.* Bleaching powder and Ammonia are considered suitable.

Application of Toxicants in Ponds

Bleaching Powder

- Bleaching powder or Calcium hypochlorite ($CaOCl_2$) is another practical and safe fish toxicant.
- It kills all the predatory and weed fish of the pond when applied at the rate of 25–30 ppm (Tripathy *et al.*, 1980). Within 1–3 hours we can find fish mortality and the toxicity lasts for 3–5 days.
- Plankton and benthic fauna start developing from the 7th or 8th day after treatment.
- Chlorine content of the bleaching powder thoroughly disinfects the pond which is essential in undrainable ponds where disinfection by sun drying is not at all possible. Chlorine killed fish are safe for human consumption.

Ammonia

Anhydrous ammonia when applied at the rate of 20–25 ppm kills the predatory and weed fishes. Besides, it also controls the aquatic weeds and later acts as nitrogenous fertilizer. Toxicity of ammonia lasts for 4–6 weeks.

Calucation of Dose

The required quantity of poison can be calculated using the following formulae.

For Rectangular Ponds

$$\text{Required amount of poison in kg} = \frac{\text{Length (m)} \times \text{Width (m)} \times \text{Average depth} \times \text{Dose (ppm)}}{1000}$$

For Circular Ponds

$$\text{Required amount of poison in kg} = \frac{3.143 \times \text{Square of pond radius (m)} \times \text{Average depth} \times \text{Dose (ppm)}}{1000}$$

2. Biological Method (Crude extracts of different plant parts)

Most of the indigenous cultures across the globe have used poisonous plants (ichthyotoxicants or pesticides) to catch fish by making use of the various chemicals present in plants to stun or stupefy them. In addition to their use as traditional piscicidal agents for catching fish, plant derived fish toxicants are widely used in aquaculture management also for eradicating unwanted/wild species of fishes including the predatory ones.

Recommended Doses of Fish Poison

Poison	*Dose (kg/ha/m)*
Mohua oil cake (*Bassialatifolia*)	2 500
Powdered seed of *Croton tiqlium*	30 – 50
Root powder of *Milletiapachycarpa*	40 – 50
Seed powder of *Milletiapiecidia*	40 – 50
Seed powder of *Barrinqtoniaacutanqula*	150
Seed meal of tamarind (*Tamarindusindica*)	1 750 –2 000
Tea seed cake (*Camellia sinensis*)*	750

Mohua Oilcake

This is of plant origin; it kills all the fish species within a few hours when applied at the rate of 250 ppm (CIFRI, 1968). It contains about 4–6 per cent of active ingredient, which is saponin, which on dissolving in water haemolyses the red blood cells and thus kills the fish (Bhatia, 1970). The toxicity of doses up to 250 ppm lasts for about 96 hours (Jhingran and Pullin, 1985) and subsequently it serves as organic manure in the pond. It should be applied at least two weeks before stocking the ponds.

Rotenone

Rotenone began to supplant copper sulphate as a fish toxicant in 1934. Rotenone kills fish by inhibiting cellular respiration and the ability to use dissolved oxygen, leads to suffocation of fish. Fish exposed to lethal rotenone concentrations move to the surface and gulping for oxygen as if the water was oxygen-depleted. However, rotenone does not reduce the amount of oxygen dissolved in the pond water.

Approved Fish Toxicants

- ☆ Chlorine, rotenone and antimycin A are the only fish toxicants currently approved by the U.S. Environmental Protection Agency (EPA).
- ☆ Chlorine is a non-restricted, general use pesticide that is used as a fish toxicant and algaecide.
- ☆ The Food and Drug Administration (FDA) has not approved the use of rotenone in ponds that contain fish destined for human or livestock consumption.

3. Physical Methods

- ☆ If ponds are used for other purposes where a farmer cannot use toxicants to remove fish, the only option is repeated netting and angling which may reduce the unwanted, weed and predatory fish population.
- ☆ There are physical methods of removing unwanted fish, such as water level draw-downs, seining and electro fishing.
- ☆ Lowering the water level increases predation on small fish and limits the spawning habitat of the unwanted species.
- ☆ Combining water level draw-down and seining may be effective, particularly in small ponds, but usually a pond must be completely drained to remove all fish.

Ethics of Predator Management

Aquaculturists are in the business of raising livestock, and accordingly, they have a high awareness and regard for the value of animal life. In general, farmers have a legal right to protect their livestock and other property from predation. While the legal right to destroy a predator may exist, most farmers adhere to established ethical standards, respecting the value of all animal life. Most fish farmers try to use preventative methods, such as net covers or other barriers, to reduce predation impacts.

Conclusion

Farmers should have awareness about the control methods of these wild fishes because they cause severe economic loss. Even though some of the wild fishes (*e.g.* Murrels) can fetch more price than cultivable species, they will compete for food and space in the pond environment. Natural plants extract as pesticides is better than synthetic pesticides (chemical method) to control the weed and predatory fishes. Prevention is better than control for wild fishes like weed and predatory fishes.

In summary, predator control is a management precaution necessary to ensure the health and safety of captive livestock and to protect the economic interests of the farmer. Not unexpectedly, there are emerging ethical issues surrounding the appropriateness and humaneness of some of the control methods currently employed. Society must constantly weigh the benefits of predator management to farmers and the animals in their care, againstthe potential negative impacts to those wild animals which maybe affected. In this context, the aquaculture industry enthusiastically supports innovations in non-traumatic predator control technologies, and utilizes strategies which prevent interactions between.

Chapter 9
Risks in Aquaculture

Introduction

Aquaculture is the practice of cultivating aquatic animals and plants in managed aquatic environments. Not too long ago, aquaculture was perceived as a cure for hunger and dwindling wild fish supplies. At the end of the twentieth century, given the rapid growth of the aquaculture industry, critics began questioning the real social and environmental impacts of aquaculture. The social impacts are generally felt more acutely in poorer countries. For example, people have been displaced from their homes and jobs by aquaculture operations, sometimes by operations that pollute land and water previously used by local residents.

Aquaculture, like any farming activity, produces effects on the environment. Aquaculture uses energy and creates wastes. As aquaculture replaces wild habitat, changes to the ecosystem inevitably occur. Even where aquaculture operations are placed in non-pristine areas, potential exists for exotic (non native and genetically altered) aquaculture species to escape and adversely affect native species by competing for food and space, interbreeding and hybridizing native species, and spreading disease. Collecting wild larvae for rearing in aquaculture units can decimate native populations of fish, and can affect biodiversity. Toxic and bio-accumulative compounds can be harmful to people, including fish farmers themselves, and to plants and animals. Excessive discharge of organic wastes causes pollution.

The environmental impacts of effluents depend on the type of aquaculture practiced, and on farm management. Aquaculture can, in local situations, improve the environment or be environmentally benign. If ponds are properly managed, nutrient-rich discharges (soil and water) can be dredged for use in crop production, thereby reducing the need for soil amendments such as inorganic fertilizers.

Fish ponds can increase bird populations, which are pleasing to birdwatchers, but are disdained by fish farmers. While poorly managed fish ponds can serve as breeding grounds for vectors of animal and human disease, well-managed fish ponds can be used to control these vectors. Thus, fish farm management geared at minimizing negative environmental effects can be critical for balancing the farm's impact on the environment, and for its own long-term success.

Natural Resources Prone to Risk

- Four of the most important resources to aquaculture, outside human and technological resources, are:
- Land
- Water
- Seed
- Feed
- Efficient uses of these resources are necessary to guarantee optimum production from aquaculture. A number of regional and international events highlighted some of the most pressing issues concerning seed in global aquaculture development.

What is Risk?

A function of the probability of an adverse health effect and the severity of that effect consequential to a hazard(s) in food.

Risk has two parts:

1. The likelihood that a hazard will affect us
2. The severity of its consequences if it does

What is Hazard?

A biological, chemical or physical agent in, or condition of, food with the potential to cause an adverse health effect.

Common Risks Include:

- Establishment of infectious diseases and pathogens and non-target organisms through the discharge of transport water, use of live feeds, inappropriate disposal of mortalities or on equipment and personnel are considered to be high risk.
- Release of chemicals, pollutants, or domesticated endemic, non-endemic or introduced species during the translocation are considered moderate risks.

Environmental Impact of Aquaculture

- The environment exerts a major influence on aquaculture. Interaction between aquaculture and changes in the environment induced by human activities is becoming increasingly important.

- Examples of the impact of the environment on aquaculture are the various types of pollution from urbanization, industrialization and intensification in aquaculture which cause stress in fish breeding, reduced growth, disease, mortality and contamination of surviving stock.
- The situation in Dhanmondi Lake in Dhaka city is a typical example of pollution due to urbanization. Aquaculture in Bangladesh faces high risks as the country suffers perennially from floods and droughts. It is thus, important to offer pond preparation and production models which are well adapted to flooding and dry season risks.

Important Burning Issues of Environmental Adverse Effect on Aquaculture

- Excess eutrophication of water and sediments from discharge of aquaculture farm effluents, particularly from integrated aquaculture systems
- Misuse of chemicals in disease treatment, with potentially adverse effects on fish production, product residues and farm worker health;
- Reduction of biodiversity due to introduction of exotic species and use of toxicants to kill the predators or so-called wild fish; and
- Social conflicts due to water becoming unusable for various other purposes because of the use of manures and fertilizers, and non-availability of pond water for other purposes, such as agriculture, are also increasing.
- While poorly managed fish ponds can serve as breeding grounds for vectors of animal and human disease.

Recent Threats

- When acid rain falls to the earth, it undergoes a new round of physical and chemical changes. These changes the chemical characteristics of the water reaching the streams and lakes thus affecting aquaculture production.

- Higher amounts of acids can mobilize other toxic metals from the insoluble to the soluble ion forms. The toxic metals include lead, mercury, zinc, copper, cadmium, chromium, manganese and vanadium.
- Tannery wastes disposal also creates a huge problem due to heavy metal cadmium (Cd) resulting in reduced growth of fish.
- Lot of fish seeds are being wasted during by catch which changes the whole ecosystem's biodiversity.

Chemical Pollution

- Use of chemicals in ponds and laboratories constitutes considerable risk to the environment. These chemicals can become disruptive and when they find their way into natural aquatic systems they can cause irreparable damage to the ecosystem.

- Chemicals such as fertilizers, pesticides, antifoulants (for cages), chemotherapeutants are all considered risk factors in the environment.

Common Chemicals

Metomidate, Etomidate, Propoxate, Ketamine Hydrochloride, quinaldine Sulfate-are generally used to prevent diseases. Improper dosage of these following chemicals leads to severe fatal cases in fishes or to aquatic animals such as:

- Benzalkonium Chloride (BKC)
- Cutrine Or Chelated Copper Compound
- Furazolidone
- Chloromycetin (Chloramphenicol)
- Calcium Hypochlorite, *etc.*

Water Quality Analysis

- Water quality and quantity vary from place to place. On a whole, groundwater is considered more desirable for aquaculture because it has more consistent water quality than surface water and is less likely to be contaminated by pathogens.

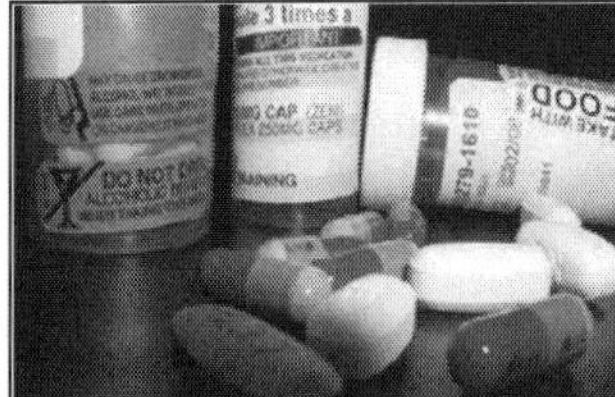

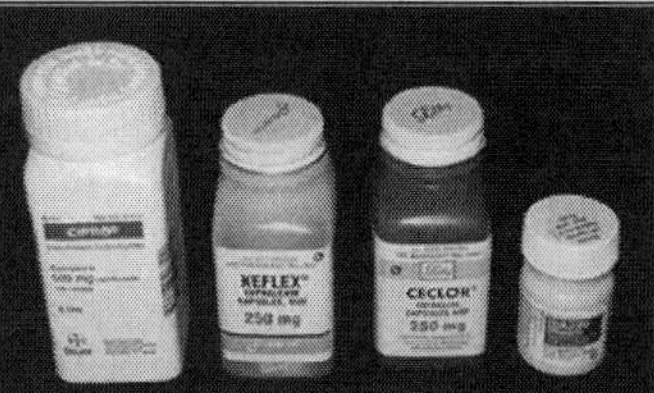

- ✰ The different parameters for analyzing water quality are mentioned in the successive slides.
- ✰ So to maintain a sound aqua farming system water quality analysis is compulsory.

Physico-Chemical Risk Factors

1. Dissolved Oxygen

- ✰ Dissolved oxygen is probably the most critical water quality variable in freshwater aquaculture ponds. Oxygen levels in ponds systems depend on water temperatures, stocking rates of aquaculture species, salinity, and the amount of aquatic vegetation and number of aquatic animals in the ponds.
- ✰ Dissolved oxygen concentrations will vary throughout the day. Dissolved oxygen in the water is obtained through diffusion from air into water, mechanical aeration by wind or aeration systems, and via photosynthesis by aquatic plants.
- ✰ Low dissolved oxygen can be lethal to our aquaculture species. Some effects include stress, increased susceptibility to disease, poor feed conversion efficiency, poor growth and even death.
- ✰ Use of air-lift pumps, paddle wheels helps to increase the D.O. level.

Paddle wheels

Air lift pumps

2. Water Temperature

- ☆ Fish and crayfish are ectotherms as heat is obtained from their external environment thus temperature therefore has a direct effect on important factors such as growth, oxygen demand, food requirements and food conversion efficiency.
- ☆ The higher the temperature, the greater the requirement for oxygen and food and the faster the growth rate.
- ☆ It is important to select a culture species that is best suited to the climate in that native area for pond culture, or in the case of tank culture, be prepared to heat or cool the water supply.

3. pH Levels

- ☆ The pH is the measure of the hydrogen ion (H^+) concentration in soil or water. The fluctuation of pH levels will depend on algae levels within the pond.
- ☆ Sub-optimal pH has a number of adverse affects on culture animals. It can cause stress, increase susceptibility to disease, low production levels and poor growth. Signs of sub-optimal pH include increase mucus on the gill surfaces of fish, damage to the eye lens, abnormal swimming behaviour, fin fray, poor phytoplankton and zooplankton growth and can even cause death.
- ☆ In the case of freshwater crayfish low pH levels will cause the shell to become soft. This is due to the shell of the crayfish being composed of calcium carbonate which reacts with acid.

4. Turbidity

- ☆ Water turbidity in freshwater ponds is caused by phytoplankton and zooplankton and suspended solids such as clay and silt particles in the water column. Water turbidity is important as it determines the amount of light penetration that occurs in the water column of a pond.
- ☆ A combination of less sunlight and lower temperatures will result in a decreased amount of vegetation present with in the ponds which depend on sunlight and warmth to grow. A low turbid pond will of course have the opposite affect (sluggish movement of the fish).
- ☆ Turbidity is measured in centimetres using a sechii disk which consists of a round plate divided into alternate black and white "pie" sections.

5. Salinity

- ☆ The term salinity refers to the total concentration of all dissolved ions in the water, it is not, as many people think, the concentration of sodium chloride in the water.
- ☆ If salinity is too high, the fish will start to lose water to the environment. As freshwater fish are not physiologically adapted to osmo-regulate within

a saline water source, decreased growth and survival can occur under these conditions.

- The salinity of the water source that is to be used for aquaculture should be tested before a project commences. Salinity tolerances will vary amongst species therefore it is important to choose an aquaculture species that is best suited to the salinity of the water source.

6. Alkalinity

- Alkalinity is a measurement of carbonate and bicarbonate ions (ions are atoms or groups of atoms with a negative or positive charge) dissolved in the water expressed in ppm. As the amount of carbon dioxide fluctuates, the pH of water changes.
- A suitable range of alkalinity is 20 to 300 ppm.
- Alkalinity in excess of 300 ppm does not adversely affect fish, but it does interfere with action of certain commonly used chemicals (*e.g.*, copper sulfate). Alkalinity remains relatively constant in ponds, but decreases steadily in non supplemented, recirculating systems. Alkalinity can be increased by adding agricultural limestone to ponds or sodium bicarbonate to recirculating systems.

7. Nitrogenous Wastes

- Most fish and freshwater invertebrates excrete ammonia as their principle nitrogenous waste. Analytical methods are used to determine total ammonia-nitrogen (TAN). As pH and temperature increase, the amount of TAN in the toxic un-ionized form increases. Fish continuously exposed to more than 0.02 ppm of the un-ionized form may exhibit reduced growth and increased susceptibility to disease.
- When fish are cultured intensively and fed protein-rich feeds they can produce high concentrations of ammonia in the water. Ammonia and other metabolic wastes are gradually removed by natural processes in ponds or through the use of biological filters in recirculating and reuse systems. Nitrite is toxic to fish and causes "brown blood" disease. Concentrations of 0.5 ppm have reduced growth and adversely affected fish.

Brown Blood Disease

8. Hardness

- ☆ Calcium and magnesium ions comprise hardness. Test procedures usually determine both ions as "total hardness," expressed as ppm calcium carbonate.
- ☆ In most waters the concentrations of alkalinity and hardness are similar, but they can differ vastly as alkalinity measures negative ions (carbonate, bicarbonate) and hardness measures positive ions (calcium, magnesium).
- ☆ Hardness is important, especially in the culture of several commercial species such as striped bass and crawfish.
- ☆ If hardness is deficient, these species do not grow well. Hardness should be above 50 ppm; low hardness can be adjusted by the addition of lime or calcium chloride.

9. Iron

- ☆ Many groundwater's contain elevated levels of dissolved iron.
- ☆ When exposed to the air, this iron interacts with oxygen, becomes insoluble, and forms a red deposit.
- ☆ Small clumps of iron are produced that can settle on fish gills, causing irritation and stress.
- ☆ Problems can be avoided if the iron-bearing water is exposed to air and the resultant clumps of iron removed by settling or filtration before the water enters the culture system.

Pond with Iron Deposition

10 Chlorine

- ☆ To control bacteria, municipal water supplies are typically treated with chlorine at 1.0 ppm. If municipal waters are used to culture fish, residual chlorine must be removed by aeration, with chemicals such as sodium thiosulfate, or filtration through activated charcoal. Chlorine levels as low as 0.02 ppm can stress fish.

11. Carbon Dioxide

- ☆ Only when using groundwater, transporting fish at high densities, or in recirculating systems are carbon dioxide problems likely to develop.

- At high concentrations, carbon dioxide causes fish to lose equilibrium, become disoriented and possibly die.
- Testing groundwater before use and aerating it, if necessary, will reduce carbon dioxide to acceptable levels.
- Careful planning, aeration or oxygenation, and buffering of water will keep carbon dioxide at acceptable levels when large numbers of fish are hauled extended distances or cultured in recirculating systems.

12. Hydrogen Sulfide

- Ponds with oxygen-poor bottoms and accumulated organic material can release hydrogen sulfide when seined or disturbed. Substratum beneath heavily fed cages/net pens can accumulate wastes (*e.g.*, uneaten food, faeces) and produce hydrogen sulfide gas if oxygen becomes deficient.
- Hydrogen sulfide gas has a rotten egg odour and is extremely toxic to fish. Any detectable odours or levels should be avoided and extreme care should be taken when handling fish in an afflicted pond.
- Ponds can be drained, exposed to air or excavated to mitigate the problem.

Soil Quality Analysis

- Where land-based aquaculture is concerned, earthen ponds are considered the most cost-effective way to retain water for fish and shrimp farming operations.
- Comprehensive soil horizon analysis must be studied.
- The best types of soil suited for aquacultural purposes includessilty clay, silty clayey loam and clayey loam.
- Soils suitable for earthen pond construction must contain a minimum of 20-30 per cent clay and preferably not more than 30 per cent sand.
- Acidic soil in certain areas and sandy soils in other areas of the country are also a constraint to aquaculture development.
- The use of pesticides and fertilizers in rice fields is increasing daily, where 4000-5000 tons and 242 different types of pesticides are used annually.
- The use of pesticides is important for crop production, but improper doses or unapproved chemicals are harmful to fish and other aquatic fauna.
- In a survey conducted in Bangladesh, it was revealed that the killing of fish with pesticides occurs mainly as a result of;
- The use of pesticides in improper doses;
- The use of banned chemicals; and
- Aerial spraying of chemicals for mosquito control.

Risk in Seed Resources

- These include inadequate and unreliable supply of quality seed, genetic quality, inadequate hatchery technology and facilities for rearing fry/

fingerlings, distribution mechanisms, impacts of releases of cultured seed stocks, the need for more hatcheries with business orientation and others.

- In order to secure stable seed supply for major freshwater aquaculture species, factors affecting seed availability, seed quality, seed production technologies and support services, seed distribution networks, breeding technologies, genetic improvement and domestication need to be understood well if resources are best to be targeted and policy decisions on future investment and management options improved.
- Recognizing these issues, a project Study and Analysis of Seed Production in Small-scale Rural Aquaculture was undertaken by the Aquaculture Management and Conservation Service of FAO's Department of Fisheries and Aquaculture.

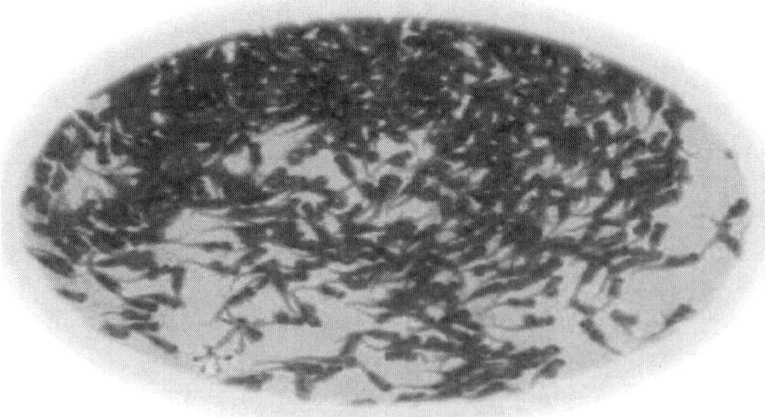

View of Fish Seeds

Feed and Fertiliezr Problem

- Besides fish seed, the two most important inputs required for more intensified aquaculture are:
    1. Feed
    2. Fertilizer.
- Feed and fertilizers are both required to intensify the system in order to increase profitability.
- Feed ingredients that are predominantly used are rice bran, wheat bran, mustard oilcake and sesame oilcake, although as many as 83 different types of ingredients, of both plant and animal origin, could be listed as potential feed ingredients.
- Farmers need to be trained in the preparation of farmer-made aquafeeds, as the use of commercial feeds will be uneconomical for freshwater aquaculture farmers. They need to be made aware of locally available ingredients and simple but effective feed preparation processes.

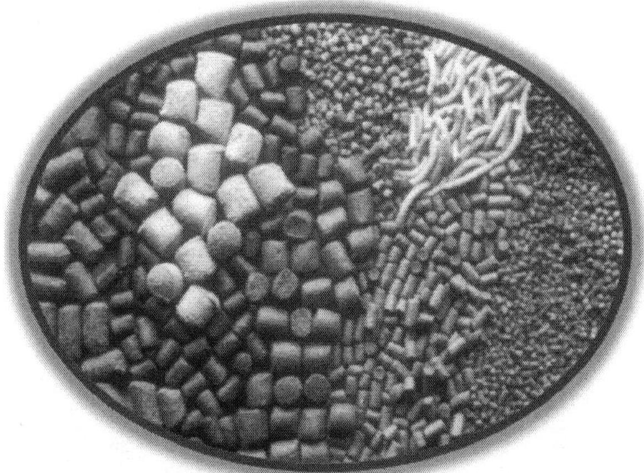

Fish Feed

Organic and Inorganic Fertilizer

Inorganic fertilizer use mainly follows the N-P-K regime, for which Urea, TSP and MP are used. Cow dung and to a lesser extent, poultry manure are used as organic fertilizers.

- ☆ If available manure alone can supplement inorganic fertilizers, that will be an added benefit. However, since manure has other important uses in the country, a balanced use of both organic and inorganic fertilizers, along with supplementary feed, seems to be the best option.
- ☆ Another recent problem is that of the presence of SSP (single super phosphate) in the market. Either the import of SSP should be banned or farmers, especially aquaculture farmers, should be educated about the negative impacts of using SSP in place of TSP.

Introduction of Diseases

- ☆ Disease in freshwater aquaculture mainly results from two causes:
  - ❒ Degradation of natural balance of the environment; and
  - ❒ Intensification.

- The introduction of non-endemic species [European carp and mosquito fish (*i.e. Gambusia* sp.)] into natural water arising from their inadvertent release from aquaculture facilities is considered a serious environmental threat.
- The introduced species may have the tendency of out-competing the native stock partly because they do not have natural predators, parasites and pathogens in their new environment. This trend creates biodiversity loss in natural waters. They may also carry diseases and parasites alien to the native with disastrous Consequences.

Gambusia* sp.** ***Cyprinus carpio

Non-endemic Species

- The introduction of genetically modified organisms such as transgenic fish is considered hazardous to the environment. It is found that that the ecological hazards would include the possibility of the transgenic fish being a voracious predator or competitor thereby impacting negatively on key ecological processes.
- Inter breeding of introduced or transgenic fish with the native stock could cause dilution of the gene pool.
- Transmission of diseases and parasites to native stocks from cage and pen facilities is a major problem. In many countries disease testing and certification programs for animals are not implemented with the result that native stocks are exposed to non-endemic parasites and diseases from aquaculture facilities.

Disease Management

- Practicing good husbandry is the best way to keep away most diseases experienced in the field of aquaculture.
- However, once disease strikes, it has been observed that it becomes very hard to find a suitable 'doctor'. The mechanism of developing disease experts and proper diagnosis systems is yet to be effectively developed.
- With the growing inclination of all concerned towards greater intensification of aquaculture systems, both frequency and intensity of

disease occurrence are bound to considerably increase. Therefore, it is important to take necessary administrative measures to strengthen the field of combating disease.

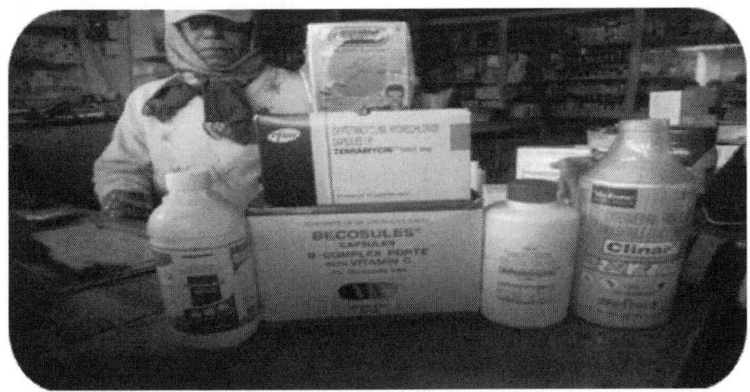

Medicines Generally Used

Human Resource Management

- ✰ It is felt by many that the curriculum development process in the universities should involve a more participatory approach.
- ✰ Farmer development is again dependent on trained manpower, which is produced by the universities. Appropriate technologies which can be transferred are the product of good researchers who are in turn the products of universities.
- ✰ Transfer of technology is another important activity in aquaculture development. This process depends on two factors:
- ✰ the inherent usefulness and adaptability of the technology itself, and
- ✰ efficiency of the extension officers/workers in preparing the recipients, *i.e.,* the prospective fish farmers, so that they can fully assimilate the technology and effectively utilize it.

Prohibition of Poaching Activities is also Necessary to Preserve our Natural Resources

- ☆ Humans acquire the FZT (Fish-Borne Zoonotic Trematode Infection) infection by eating raw or inadequately cooked fish. The production of FZT-free fish in aquaculture is a key component in establishing a sustainable program to prevent and control the FZT transmission to humans. Interventions in aquaculture should be based on knowledge of the main risk factors associated with FZT transmission.
- ☆ Aquatic environments contain dissolved compounds and various organisms that both promote and constrain aquatic animal health. Aquaculture management demands continuous monitoring of the aquatic environment to ensure the conditions are conducive to good health and growth.

Conclusion

Hazards and risks have the potential of affecting people and all human activities have this inherent capacity. Aquaculture is no exception. However, available information is largely on hazards and risks in developed countries. Unfortunately, in the developing countries where 87 per cent of the global aquaculture production takes place both workers, employers, government and consumers have tended to ignore policies (where present), which could ensure safe aquaculture practice. In some countries policies to ensure the enforcement of regulatory standards have not been instituted. This has resulted in unregulated aquaculture with the attendant adverse impact. It is our view that aquaculture stakeholders should work in unison to provide guideline and policies that would promote an environmentally friendly and sustainable industry. The volume and quality of the water supply available to the aquaculture facilities represents one of the key factors affecting success.

Chapter 10

Role of Ichthyophagus Birds on Culture Based Fisheries in Aqua Sector: Their Preservation and Control

Introduction

There are over 200 species of birds identified all over the world. Majority of all birds better suited to stay in marsh or estuary. Open water and concentration of culture aquatic animals attract predatory birds. Several bird species have been identified as "problem species" on aquaculture farms. Predatory birds are highly mobile, efficient and adaptable pray on aquatic cultivable fishes, also cause injury to the fishes (leading to secondary disease), spread diseases.Common types of predatory birds includes Double-crested Cormorant (*Phalacrocorax auritus*), American White Pelican (*Pelecanusery throrhynchos*), Hooded Merganser (*Lophodytes cuculatus*), Great Blue Heron (*Ardeahe rodias*), Great Egret (*Casmerodius albus*), Snowy Egret (*Egret tathula*), Tricolored Heron (*Egretta tricolor*), Green-backed Heron (*Butorides striatus*), Black-crowned Night Heron (*Nycticorax nycticorax*), Yellow-crowned Night Heron (*Nycticorax violaceus*), White Ibis (*Eudocimus albus*), Belted Kingfisher (*Cerylea lcyon*), Osprey (*Pandion haliaetus*), and various species of gulls and terns (Stickley, 1990). Not only the damage, primary infection, secondary infection of fishes and it can cause the damage equipment and nets, resulting in escape of fishes through damage. Due to predation on fishes by the predatory birds so much economic loss, which have led to the development of several strategies to reduce predation (Draulans, 1987; Littauer, 1990; Cezily, 1992; Littauer *et al.*, 1997). So it is significantly need the prevention and control of predatory birds in aqua field. Bird predation can have a significant economic impact on aquaculture operations. Birds may also negatively affect aquaculture production by transmitting or transporting diseases, weed seeds, and parasites from pond to pond or from one facility to another. A wide variety of birds are known to frequent aquaculture facilities. The presence of birds at a facility, however, does not necessarily mean a predation problem exists. Many bird species benefit from associating with aquaculture facilities and can exist there without interfering with fish production. Thus, proper identification of bird species is important to recognizing damage and taking responsible action, when necessary.

1. Loss of Fishes Due to Predation of Birds

Bird predation can have a significant economic impact on aquaculture operations. Birds may also negatively affect aquaculture production by transmitting or transporting diseases, weed seeds, and parasites from pond to pond or from one facility to another.

Due to predation on fishes by the predatory birds so much economic loss, graphical presentation of fish consumption by birds in Kg/month is shown in the following graph (Figure 10.1). *Peligans* are the highest predatory fishes; 15kg/month is consumed by this bird. *Cormorants, Terns* are the lowest predatory fishes, 1.5kg/ month consumed by this fishes.

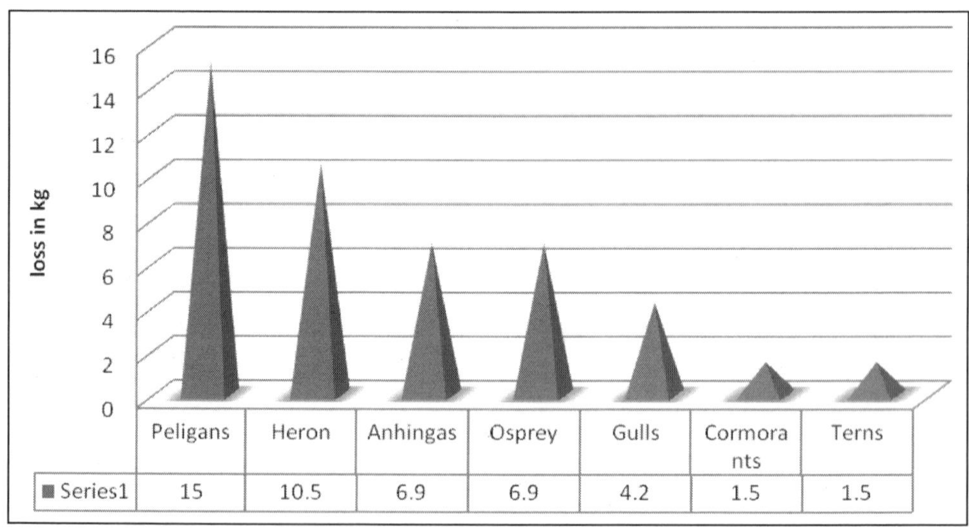

Figure 10.1: Graphical Presentation of Fish Consumption by Birds in kg/month.

2. Common Ichthyophagus Birds

Common classification of ichthyophagus birds:

1. Those that dive from air and catch fish with bill: Examples are kingfishers; terns and pelicans are rarely in small ponds,
2. Those that dive from air and catch fish with feet: Examples are eagles and ospreys are rarely in small ponds,

3. Those that dive from water surface: Examples are cormorants, mergansers, various fish-eating ducks; rarely in small ponds,
4. Those that wade in pond; or catch fish from edge of pond and floating plants, *etc.*: Examples are herons (includes egrets).

Birds that Dive from Water Surface

Cormorants

Description

Body length 32 inches (81 cm). Wingspan 52 inches (132cm). Short legs and webbed toes; short tail often used as prop when perched. Wings with body often nearly submerged, but with neck erect and bill pointed at upward angle.

Feeding and Other Habits

Dives from surface; pursues fish underwater. Daytime feeder; hunts singly or in flocks that may number 500 to 600 birds. Birds in feeding flock form line that advances on fish. Runs across water to take off. Nests in colonies; often perches on trees, poles, rocks, or buoys that overhang or project from water. Worldwide there are more than 30 species are present.In North America the double-crested cormorants is the most problematic predators.Cormorants have long, flexible neck and a straight hook tipped beak (Figure 10.2).Do not have water repellent feathers. Normally solitary feeders. Breeding colonies are primarily located in specific coastal areas.It consumes between 0.25kg and 0.5 kg fish per day.

Phalacrocorax carbo

Phalacrocorax fuscicollis

Figure 10.2

Anhingas

It is also called snake birds or water turkeys (Figure 10.3). Do not have water repellent feathers. Have straight bill not hooked one.Consume about 0.23 kg fish per day.

Anhinga anhinga

Anhinga melanogaster

Figure 10.3

Pelicans

Capture fish near the surface with their scoop like pouch (Figure 10.4).Out of seven species the white pelican is predominant. This bird capture their prey while swimming. They drive fish into the shallow where the birds encircle them and capture them in their pouches.White pelicans are among the largest birds in the world. They consume approximately 0.5 kg of fish per day. Brown pelicans are typically only ion marine habitats.White pelicans inhabit both marine and freshwater environment.

Pelecanus crispus

Pelecanus onocrotalus

Figure 10.4

Description

Body length 62 inches (158 cm). Wingspan 108 inches (274 cm). Very large white bird with black wing tips and enormous, orange bill with pouch. Young grayish colored. Swims with great buoyancy; floats high on water.

Feeding and Other Habits

Feeds from water surface; does not dive. Plunges head underwater and catches fish in pouch. Daytime feeder; hunts singly in deep water; in shallow water forms small flocks that drive fish toward shore. Takes off from water with feet kicking in unison. Nests in colonies usually at sites remote from people.

Birds that Wade in Pond

Herons, Egrets, Ibies and Stroks

These are prey on some extent of fish and shell fish.Members of this order are long necked and relatively long legged. They stalk their prey often by wadding through shallow water then striking with a rapid thrust of the bill.Large herons and egrets (great blue herons and great egret) consume between 0.14-0.35 kg of fish per day.Peak feeding at dusk or at night (April and August month).Ibises are common predators of crustaceans such as Cray fish.Strokes tend to prey on invertebrates and sick or weakened fish.

Herons

Description

Body length varies, 11 to 38 inches (28 to 97 cm). Long legs and toes, short tail, large wings. Wingspan varies with different species. Neck usually long, bent in S shape when flying; beak long, pointed. Plumage variable: all white, brown, gray, blue; or pattern of stripes.

Feeding and Other Habits

Various methods used depending on species. Feed mostly during day or at dusk; black-crowned night-herons feed mostly at dusk or night. Most nest in colonies; often perch or nest in trees near water. Depending on species, hunt singly but can occur in large, loose flocks that tend toward even distribution across foraging area.

Tigrisoma mexicanum

Botaurus stellaris

Figure 10.5

Kingfishers

They are small birds that are territorial and solitary, day time hunters.Plunge dive to catch their prey.Worldwide there are more than 80 species of kingfisher. They are problematic around ornamental and tropical farm.They consume from 0.02 -0.07 kg of fish per day.

Families: Alcedinidae, Halcyonidae, Cerylidae.

Ex: Oriental Dwarf Kingfisher (Figure 10.7), Green Kingfisher (Figure 10.8), Pied Kingfisher, White-throated Kingfisher.

Figure 10.7: Oriental Dwarf Kingfisher.

Figure 10.8: Green Kingfisher.

Gulls and Terns

Worldwide there are more than 80 species of gulls and terns. Gulls have hooked beaks while terns have strongly and pointed. Terns are solitary day time hunters, dive headfirst into the water to capture prey. Gulls also capture prey while swimming and can scoop prey from the water without plugging. Gulls usually feed in flocks, scavenge readily and consume aquaculture feed off the surface of ponds and raceways. Gulls consume between 0.07 and 0.14 kg per day while terns consume only about 0.05 kg per day.

Family: Laridae, Genera: Larus, Xema, Pagophila, Rissa.

Gulls

Description

Many species; in general, stout body, total length 12 to 32 inches (30 to 81 cm). Long, pointed wings; wingspan 36 to 65 inches (91 to 165 cm). Bill slightly hooked; webbed feet. Wings white or dark on upper surface with dark tips. Immatures darker, mostly brown or grayish.

Feeding and Other Habits

Variable methods: plunges from flight, perches on raceway walls; swoops down to scoop up prey; bobs while swimming. Daytime feeder. Highly social; breeds in colonies. Feeds in flocks.

Ex: *Larushe ermanni* Common Gull, *Larus livens* Glaucous Gull, *Larus glaucescens* Western Gull, *Larus marinus* Kelp Gull (Figure 10.9), *Larus canus* (Figure 10.10)

Figure 10.9: *Larus marinus.*

Figure 10.8: *Larus canus.*

Ducks

They prey o fish and shellfish.Loons, grebes, mergansers and other diving ducks capture prey by diving and swimming underwater.Many diving ducks feed extensively on molluscs and cause heavy losses in cultured sp of oyster, clams and mussels.Mergansers prey on small fish and crustaceans and can became problematic for farms if in large flocks

Family: Sternidae

Example: *Sterna aurantia*

Duck is the common name for a number of species in the *Anasplaty rhynchos.* (Figure 10.11)

Figure 10.11: *Anasplaty rhynchos*

Birds that Dive from Air and Catch Fishes with Feet

Hawk, Osprey and Eagle

- ✰ They are solitary hunters.
- ✰ They are not major concern for fish farms but have been know to cause problems by preying upon brood fish.

Order: Accipitriformes, Family: Accipitridae

The Accipitridae: The family which includes the "trure hawks" (Accipiters) as well as eagles, kites, harriers and buzzards.

Figure 10.12: Osprey

Figure 10.13: Hawk

Ospreys

Ospreys (and eagles) apparently need an unknown amount of space around a pond and an unknown amount of pond water surface in order to fly in and position themselves to grab fish near the surface with their feet. Also, the larger fish they hunt and catch are not normally found in small ponds. Local ospreys teach their young to fish on largest pond, which is about 150 feet long, and has trees at each end.

For ex: The "osprey" or fish hawk (Figure 10.12), red tail hawks (Figure 10.13)

Erget

Great Egret (*Ardea alba*), and the smaller Snowy Egret (*Egretta thula*) visit the larger ponds.

Black Birds

Description

Many species: in general, small-bodied, total length 6 to 16 inches (15 to 40 cm); sharp, pointed bill. Plumage iridescent black; some species have brightly colored areas of yellow, red or orange on head or wings. Females smaller-bodied; plumage brownish, often with streaked breast.

Feeding and Other Habits

Catches prey in shallow water at edges, or climbs down raceway screens to water surface. All species daytime feeders. Gregarious; flocks number from few birds to thousand.

Other Predatory Birds (Figure 10.14)

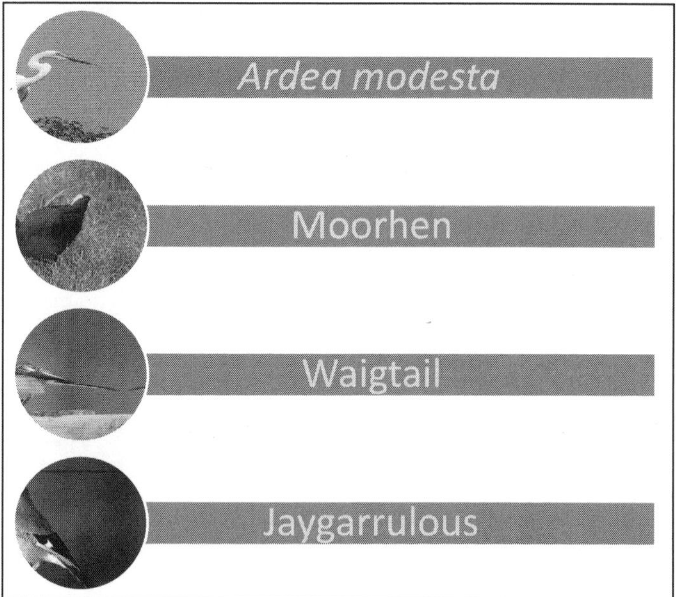

Figure 10.14

Some Miscellaneous Facts

The heron is an important intermediate host in the lifecycle of the cestode*Diphyllobothrium*sp, a parasite prevalent of cage trout farms.Otters are final hosts for the digeneanHaplorchissp a common parasite of Tilapia.Birds are also act as a vectors of pathogenic bacteria and virus. So birds are transfer the harmful diseases to fishes one to another.

Prevention and Control Methods

Several methods of deterrence and exclusion have been proposed. At its simplest level, the presence of dogs or scarecrows can deter predators and scavengers. In Israel, concerted efforts have been made to deter cormorants from visiting ponds, principally through use of scaring tactics such as scare cannons, sirens and shooting (Sly and Frankenberg, 1995). However, while causing an effective decline in numbers, the harassment techniques have impaired breeding of the protected pygmy cormorant *P. pygmaeus* (Be'er, 1995). According to Curtis *et al.* (1996), has given overview of techniques for reducing bird predation at aquaculture Facilities.

Method to discourage or prevent bird depredation have included completely enclosing the facilities, perimeter fencing, overhead wires and lines, automatic cannons or exploders, pyrotechniques, scarecrow, alarm or distress calls, water sprays, lights ultra light radio-controlled, model aircraft.Complete enclosure is the only method to entirely prevent birds depredation but is usually limited to raceways and ponds smaller than 5 ac (2ha).Birds are quickly or habituate to scary devices. So, these devices must be moved every few days and if possible, set so that they go off at random intervals and varying intensities.

Solutions or Techniques for Reducing Birds Damage

Barriers

All physical barriers should be visible to birds to maximize the effectiveness of the barrier and to minimize the potential for accidental injury or entrapment.Two types of physical barriers can be used for controlling bird predation at aquaculture facilities;

1. *Complete enclosures*: Which totally exclude predators from gaining access to cultured stocks, Complete enclosures are extremely effective against all birds but are more expensive than partial exclosure systems.
2. *Partially-covered systems*: Which interfere with predators' feeding behavior.

1. Total Exclusion

Total exclusion techniques is the only completely effective method for eliminating bird predation at aquaculture facilities. Total exclusion is the complete enclosureof fish tanks, raceways, and/or ponds with screen, cage or net.

Total exclusion is impractical for most large ponds due to the difficulties of spanning large distances. Other problems associated with exclosures are that they may negatively affect aesthetics of the site, hinder other management operations, and have a high initial cost. Due to the long-term benefits that total exclusion can provide, however, this method may be cost-effective for some facilities. So better go for partial exclusion.

2. Partial Exclusion

If total exclusion is not used or is not feasible at a facility, then the goal of damage control should be to reduce losses to an acceptable level while incurring the lowest possible cost. Facilities managers should recognize that some loss is unavoidable, even with the best predation management strategy.

Overhead Lines and Wires

Overhead line/wire systems are most effective against flying predators such as terns, gulls, cormorants, and ospreys, rather than against wading birds. Sides and ends should be protected as birds may attempt to enter the area by these routes. Raceways and ponds can be covered with heavy-gauge monofilament lines or high-tensile galvanized or stainless steel wires suspended horizontally in a grid pattern or in one direction over the water's surface. Overhead line/wire systems, like total exclusion systems, are impractical for large ponds due to the difficulties of spanning large distances. Another problem is that birds may learn to avoid the overhead lines/wires. Aside from replacing an occasional broken wire and maintaining adequate wire tension, overhead line/wire systems typically require little maintenance. Sides and ends should be protected as birds may attempt to enter the area by these routes (Figure 10.15).

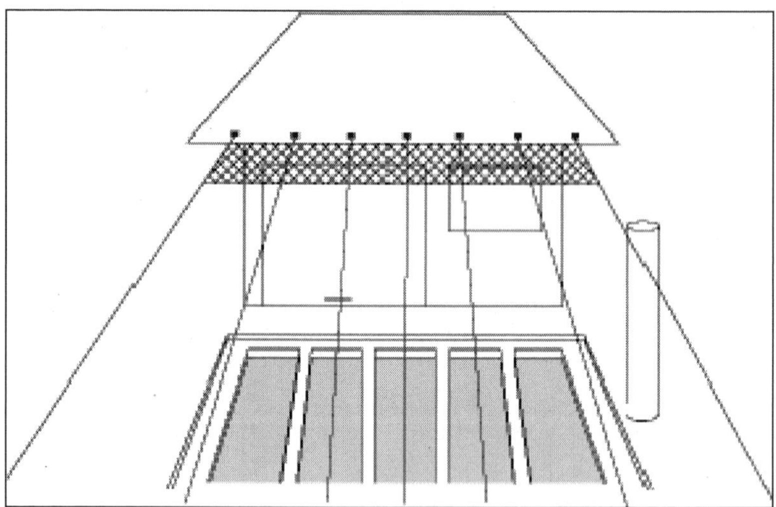

Figure 10.15: Netting where Wires Attach to Building Prevents Birds from Walking Down the Roof and Jumping through the Wires into the Facility.

Perimeter Fencing and Wires

Wires or perimeter fencing around ponds or raceways can provide some protection from wading birds, although birds may eventually learn to avoid these obstacles. These systems are largely ineffective against flying predators. For ponds, fencing that is at least 3 feet high should be constructed in water that is 2-3 feet deep. Small fish may be prevented from entering the shallow water by constructing the fence with a small mesh material, although this may require the removal of algae buildup from time to time. Construction of inward-angled or vertical barriers, typically made of plastic netting, chicken wire, or monofilament lines, around raceways or ponds may prevent predators from foraging from the edge of a holding structure or from entering raceways and ponds from the side. In general, wading birds prefer to land on solid ground before wading into the edge of ponds. Fences should be high enough to prohibit feeding from the wall.

Electric Wires and Fencing

Use of electric fences has had varying levels of success largely depending on the design of the culture facility and the type of bird species involved. In this method, electric wire/fencing of the type specifically designed for agricultural fencing applications is placed around the perimeter of ponds and/or raceways. This method is more effective than unelectrified wires or fencing as birds cannot push against the wires/fence. In another application, electric wires may be strung on supports that suspend them over the water's edge near the natural shelf that often forms in shallow areas of the pond margin. This system discourages wading birds from feeding on fish while walking along the shelf; however, if pond bottoms slope too gradually from the bank, wading birds may still be able to fish on the water side of the fence. Great care must be taken when installing and using electric fencing to ensure its safe operation. Operators should be knowledgable about electric fencing or should get qualified, professional help prior to installing this type of system. Electric fences should only be charged with commercial electric fence charges that send brief pulses of electricity through the fence. Charges must be nonlethal to humans and birds. Like perimeter fencing and wires, this technique is most effective against wading birds. Also like perimeter fencing/wires, birds may learn to avoid these obstacles. This system is ineffective against gulls, terns, cormorants, and pelicans that typically fish in the central part of the pond. Other problems include maintenance and preventing the system from becoming grounded, commonly caused by blowing debris and interference from vegetation. A combination of electrified or unelectrified perimeter fencing or wires combined with overhead lines/wires may be successful at deterring both wading and flying predators.

Frightening Methods

Frightening techniques rely on sight and sound stimuli to discourage birds from remaining at a site by making the birds believe the site is dangerous for them. A wide range of fear-provoking devices are commercially available for scaring bird predators. The success of a frightening program may vary depending on a range of factors, including the bird species involved, how long the birds have been at the site, the type(s) of technique(s) used, the duration and frequency of their use, the location of the site relative to roosting and loafing sites, and the proximity of alternative food sources.

Noice

1. Distress Calls

*Recorded bird alarm or distress calls may be of some use in frightening birds away from an area.*Reactions of birds to recordings of species-specific distress calls depend on the species, the time of year and day, size of the area, location, and distance of the birds from the broadcasting equipment.

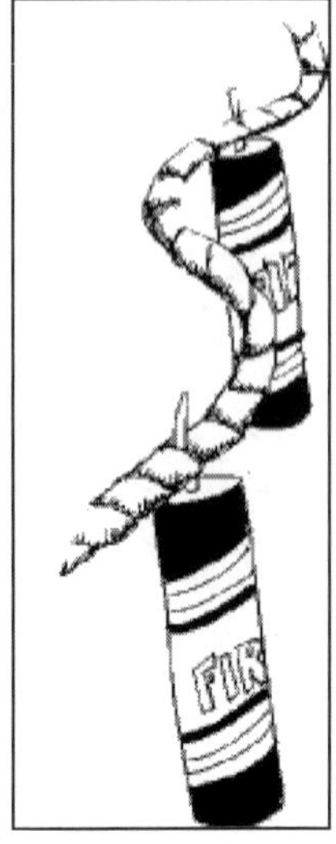

Figure 10.16

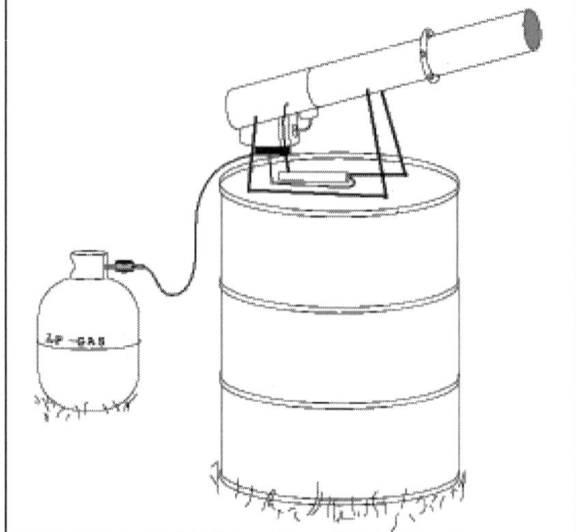

Figure 10.17

Pyrotechnique Devices

- **Cracker shells**
- **Whistle bombs, screamer rockets, Bangers**
- **Rope firecrackers**: Rope firecrackers are relatively inexpensive tools that are useful in frightening birds (Figure 10.17).
- **Live ammunition or blanks**

2. *Automatic Exploder*

Propane gas or acetylene gas is used to operate a small cannon that is equipped with an electronic timing mechanism (Figure 10.16)

3. *Electronic Noise Makers (Figure 10.18)*

Visual Scar Devices

(a) Lights

A variety of light-emitting devices can be used to confuse, frighten, temporarily blind, and interfere with the activities of night-feeding bird predators such as great blue herons and night herons. Typically, only short-term success is achieved with light devices as the majority of birds quickly become accustomed to them. A number of the more commonly-used devices are described below.

- **Construction of flashers:** Flashing, amber-colored construction lights are placed at intervals around raceways and ponds to deter approaching predators.

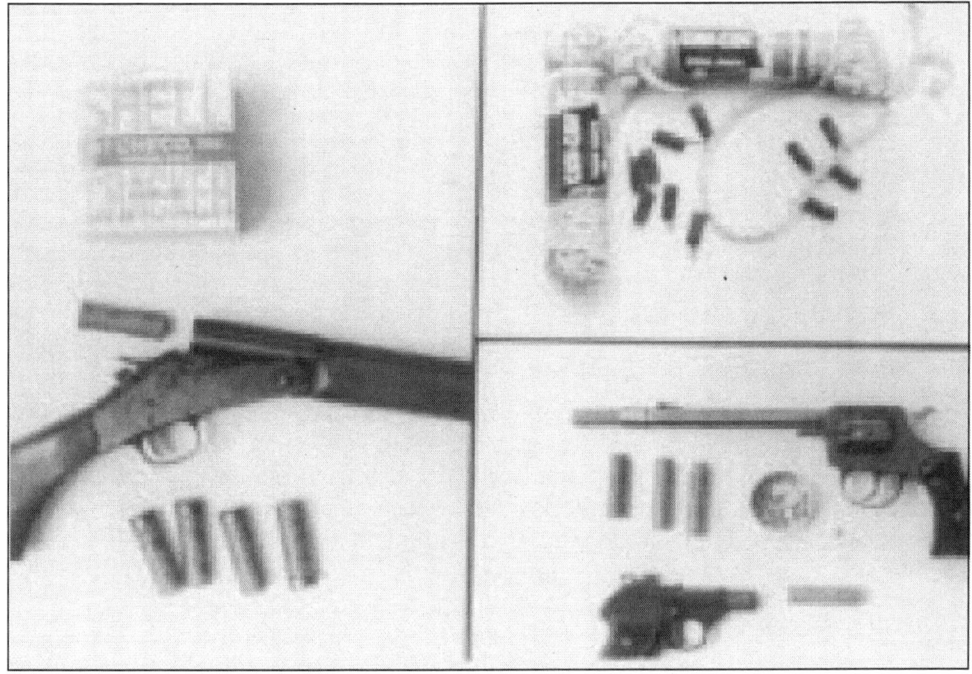

Figure 10.18

- **Area lights:** Bright lights, such as street lights or flood lights, are placed in specified areas. The use of motion-detecting mechanisms improves the effectiveness of these techniques.
- **Revolving beacons:** These devices project a very bright, revolving beam of light.
- **Store lights:** A high-i ntensity intermittent light is emitted from these devices.

(b) Bird Scaring Device

predators quickly habituate to these scary images and their effective ness is short lived unless used in conjugation with other deferent method.

(c) Scarecrows, Effigies, Predator Models

Models or silhouettes of humans and/or predators are placed in strategic locations at a facility. Dressing models in similar clothing to facility personnel or in hunter orange may improve performance. Pop-up versions and models with moving parts are available and are more effective than stationary units. The location of models should be changed frequently. The success of these methods may be increased with the addition of pyrotechnics fired within close proximity of the models.

(d) Mirrors, Reflectors, Streamers

Objects with shiny surfaces, such as balloons, pie tins, pinwheels, and reflective ribbons ortape, are placed around a facility. Success with these methods typically is minimal and short-term.

(e) Vehicles

A vehicle parked in a strategic location may be effective if birds are easily scared by a vehicle driven around the facility. The vehicle should be moved occasionally to reduce habituation. Occasional use of pyrotechnics and effigies near the vehicle may enhance effectiveness.

(f) Radio Controlled Airplanes or Boats

Radio-controlled scale models of airplanes and/or boats provide noise along with a visual stimulus. Planes appear to be most successful when used as birds attempt to land at a site. One plane operatorcan effectively cover a 200-300 acre area. The cost of using these methods tends to be high and their use is restricted by surrounding obstructions and weather conditions. Additionally, some birds may dive to avoid harassment and the devices run the risk of crashing.

(g) Water Spray Devices

Rotating or stationary water sprinkler devices can be placed in or around raceways or ponds. Water spray devices provide both sight and sound stimulation. The water spray limits the visibility of fish in the water and may repel certain birds, especially herons and gulls. Increased water pressure and intermittent water spraying instead of continuous spray increase the effectiveness of this technique. Still, birds often become accustomed to the water spray and feed among the sprinklers.

(h) Patrols or Visltation

Patrols/visitation, on foot or in a vehicle, may frighten and disrupt birds. The effectiveness of this method may be enhanced by increasing the frequency of patrols/visitation and by broadcasting noises or using pyrotechnics at the same time as the patrols/visitation.

(i) Dogs

The presence of a dog/dogs has been used to deter birds from landing at a site. This method has achieved varying levels of success.

Conclusion

The predatory birds are the most problematic predators around aquaculture facilities. First they cause physical damage; it can cause the secondary infection of the bacteria, virus and fungal diseases, most of the birds consume the cultivable fishes. The predatory birds can also cause damage to the nets and equipment's. So much economic loss due to consumption of cultivable fishes and as well as damage to the equipment's and nets. And the predatory birds act as a the vectors for transmission of diseases. Sometimes they can act as a transfer of foodborne diseases through raw fish. So significantly need require for the prevention and control of predatory birds in aqua fields for sustainable aquaculture. It may cause environmental problems and loss of birds species. They should be protected by either federal and state laws or international treaties. The lethal control of the birds are strictly avoided. Prevention is the better method to prevent the loss of cultivable fishes. Dogs are best for preventing the predatory birds in aqua field, so many farmers also following the same method.

Chapter 11
Current Issues in Aquaculture Welfare

Introduction

Aquaculture: A Growing Industry

It is estimated that global production from capture fisheries and aquaculture supplied about 132.2 million tonnes of fish in 2003. Of this total, 32.9 million tonnes were from aquaculture production (FAO, 2004). Aquaculture has seen a worldwide expansion over the past 20 years and it seems that growth is set to continue (Naylor *et al.*, 2000). World total demand for fish and fishery products is projected to expand by almost 50 million tonnes to 183 million tonnes by 2015, and it is expected that out of this increase, 73 per cent will come from aquaculture, accounting for 39 per cent of global fish production (FAO, 2004). Alongside, and perhaps partly due to this rapid expansion, the welfare of farmed fish has received increasing attention. Fish welfare is an important issue for the industry, not just for public perception, marketing and product acceptance, but also often in terms of production efficiency, quality and quantity (Broom, 1998). However, in a number of areas, conflicts between welfare and production exist, where procedures are seemingly associated with diminished welfare at the level of the individual fish. Such practices have been the subject of much commentary and discussion in recent years (Bernoth, 1991). However, although organisations such as the Farm Animal Welfare Council (FAWC), the Scottish Executive, and the Humane Slaughter Association (HSA) have published recommendations regarding the welfare of farmed fish (FAWC, 1996), this is a growing area where empirical research is lacking and these reports identify the need for an improvement in scientific knowledge on which to base future guidelines and potential legislation where necessary. It is, therefore, essential that there is good communication between the latest published scientific research, veterinary advice and the practical experience of the aquaculture industry (Scottish Executive, 2002).

With the continued growth of the aquaculture industry and increasing scientific discussion over the potential for negative events to give rise to suffering, research into the welfare of cultured fish is vital. How we define and measure welfare is cause for debate, particularly in fish. However, research into the effects of aquaculture procedures on welfare is crucial to produce data and recommendations for best practice and future legislation. Both behavioral and physiological measures of welfare are necessary for correct interpretation and while there is a wealth of knowledge on the physiological consequences of many aquaculture practices it is now equally important for us to understand the behavioral responses to these practices and to relate them to fish welfare.

Concepts of Welfare and Measurement

Animal welfare is by no means a straightforward concept. The two major issues here are the meaning or definition of animal welfare and how best to objectively measure it (Mendl and Paul, 2004). The production of accurate and objective guidelines for the levels of acceptability in all aspects of fish welfare will only be possible with the resolution of these issues. FAWC have, therefore, based their guidelines on the "Five Freedoms" framework, which defines ideal states rather than specific levels of acceptable welfare (FAWC, 1996). Freedom from hunger and thirst, discomfort, pain, injury, disease, fear and distress, as well as the freedom to express normal behaviour, provides us with a logical framework with which to assess welfare issues. Physical health is the most universally accepted measure of welfare and is undoubtedly a necessary requirement for good welfare. It is important to remember that poor health can be both a cause and a result of poor welfare. However, for many, good animal welfare goes beyond just physical health and also involves a lack of mental suffering. This aspect of welfare therefore seeks to understand subjective experiences of non-human animals and proposes the conscious experience of suffering in these animals (Broom, 1991b). This is a controversial issue, particularly when it comes to fish. Concepts of animal welfare have traditionally been applied to those which are considered to have the ability to experience pain, fear and suffering and as such have been associated with species with a higher level of cognition when compared to fish. However, there is scientific debate regarding the ability of fish to experience pain and fear. While some reviewers have argued that fish lack essential brain regions or any functional equivalent, making it untenable that they can experience pain and fear (Rose, 2002), others suggest that there is anatomical, physiological, and behavioural evidence that make it conceivable that nociception in fish is experienced and that they have the potential to experience suffering in the form of pain and fear (Dunlop and Laming, 2005). It is unlikely that animals with a different brain structure to humans would experience anything like the emotions that humans feel when experiencing pain and fear and it is impossible to know exactly what a fish experiences. However, if an animal experiences suffering or discomfort, the nature of the pain or fear they perceive is no less important (Ashley and Sneddon, in press). If we are to give fish the benefit of the doubt here, it follows that while health is essential for good fish welfare, good health does not necessarily mean good welfare (Broom, 1991a). However, our ability to scientifically and objectively measure welfare beyond the levels of disease, injury and ill health is a further complication (Mendl and Paul, 2004).

There is no single measure of welfare and although a wide range of physiological, biochemical and behavioral measures are used to assess welfare, none of these are considered reliable in isolation and multiple measures need to be taken (Broom, 1997). This is partly due to the need for careful interpretation of the functional context of many of the measures used (Dawkins, 1998). Physiological stress responses such as cortisol release, for example, are autonomic responses that indicate activity or arousal rather than being specific to poor welfare. These measures may, therefore, be affected by a number of different parameters (Dawkins, 1998). Interpretation of behavioral measures in isolation is also subject to doubt (Dantzer, 1986, 1991). Behavioral and physiological measures are intrinsically linked and are dependent on one another for correct interpretation with regard to welfare (Dawkins, 1998). Appropriate combinations of welfare measures will, therefore, be specific to a given situation, but the greater our understanding of both the behavior and physiology associated with a given situation the better equipped us are to answer questions of welfare.

To this end the aim of the current review is to summarize the science based knowledge associated with selected key issues relating to the welfare of fish in aquaculture. Collectively this work serves to improve our interpretation of the effect of aquaculture practices on fish welfare, and provides the foundation on which government and public bodies may base future guidelines and legislation. Perhaps due to the nature of the aquaculture setting and the importance of product quality, a large body of research has focused on the physiological aspects of welfare in fish. However, as the debate on the ability of fish to suffer develops, behavioral aspects will become increasingly important for balanced welfare assessment. Therefore, where possible, this review aims to highlight the value of behavioral research in improving welfare.

As fish welfare is a growing area of research this is by no means a comprehensive review covering all aquaculture fish species or indeed all situations where welfare may be a concern. One of the major themes of this subject is that different species have differing biological and environmental requirements and show differing responses to aquaculture conditions. Therefore, much work remains to be done, particularly considering the growth in the aquaculture of species such as cod (*Gadus morhua*), halibut (*Hippoglossus hippoglossus* L.) and tuna (*Thunnus* sp.) (FAA, 2004).

Stress: A Possible Indicator of Poor Welfare

There is an extensive literature on the biology of stress in fish and the physiological and behavioural responses of fish to a wide variety of physical, chemical, and biological stressors, including those seen in aquaculture (Wendelaar-Bonga, 1997 and Conte, 2004). Measures of physiological stress response naturally feature prominently in studies of welfare. However, stress response is an adaptive function in the face of a perceived threat to homeostasis and, as suggested above, stress physiology does not necessarily equate to suffering and diminished welfare. In the short term, stress responses serve a very important function to preserve the individual. Welfare measures in aquaculture are, therefore, largely associated with tertiary effects of stress response that are generally indicative of prolonged, repeated or unavoidable stress (Barton, 2002). These include both the direct and indirect maladaptive effects of chronic, long term stress, such as reductions in growth (Barton *et al.*, 1987), suppressed reproductive function (Contreras-Sanchez *et al.*, 1998), diminished immune function (Einarsdottir *et al.*, 2000) and disease resistance (Pickering, 1992; Balm, 1997). So while stress responses will not tell us all we need to know about the welfare of fish, concurrent deleterious effects in several of the above areas may provide a strong indication that welfare is poor (Broom, 1988; FSBl, 2002).

The primary stress response in fish involves the release of catecholamines and activation of the hypothalamic-pituitary-interenal (HPl) axis. Corticotropin releasing factor from the hypothalamus acts on the pituitary to synthesize and release corticotropic hormone, which in turn stimulates the synthesis and mobilization of glucocorticoid hormones (cortisol in teleosts) from the interrenal cells (Schreck, 1981). Both catecholamines and cortisol initiate secondary and tertiary stress responses and many of the adverse changes described above are associated with the long-term effects of high cortisol levels (Schreck *et al.*, 2001). HPI activation results in energy source mobilization, depletion of glycogen stores, and an increase in plasma levels of glucose, along with high muscle activity, anaerobic glycolysis and an increase in plasma lactate. Therefore, the levels of both glucose and lactate in the plasma are often used alongside cortisol to assess stress levels (*e.g.* Arends *et al.*, 1999). Measures of the expression of stress related genes might provide useful and sensitive biomarkers to diagnose stress and improve welfare in the future (Gomati *et al.*, 2004). Chronic stress exerts a strong effect on haematology (Montero *et al.*, 2001), metabolism (Mommsen *et al.*, 1999), neuroendocrine function (DiBattista *et al.*, 2005b), and hydromineral balance and osmoregulation (Wendelaar-Bonga, 1997). As will become apparent, these factors have subsequent relevance to welfare following a stressful event.

Avoidance of the maladaptive consequences of prolonged, repeated and long-term stress is clearly a central welfare goal in aquaculture. As such, the assessment of potential methods to reduce stress responses in aquaculture species is an active area of research. An example is the investigation into the feasibility of selectively breeding fish to minimize their responsiveness to stressors (Pottinger, 2003). Although the process of fish domestication is at a relatively early stage, some cultured species (*i.e.* salmonids) are inherently less easily disturbed in comparison to wild strains (Huntingford, 2004; Overli *et al.*, 2005). It appears that cortisol response to stressors varies between individuals and that this characteristic has a moderate to high heritability (Pottinger *et al.*, 1992, 1994).High responding (HR) and low responding (LR) lines of rainbow trout (*Oncorhynchus mykiss*) have been generated by selection for consistently high or low cortisol response to a standard confinement stressor (Pottinger and Carrick, 1999). While lines of HR and LR selected fish have shown divergent cortisol response to confinement stress (Pottinger *et al.*, 1992), they have also shown other clear behavioural and physiological differences. For example, LR fish show both a lower cortisol response to stressors and also a quicker resumption offeeding when placed in a novel environment (Overli *et al.*, 2004). However, while LR and HR fish displayed similar growth rates when maintained in monoculture, fish from the LR line grew larger than did HR fish when maintained in mixed groups (Pottinger and Carrick, 2001). LR fish were found to be more aggressive and competitive than HR fish; LR fish showing a strong tendency to become dominant over HR individuals in size matched staged fights for social dominance (Pottinger and Carrick, 2001). Non-aggressive locomotor response to a smaller con specific intruder was also higher in HR fish (Overli *et al.*, 2002b). The research on these fish points to these behavioral differences being closely linked or associated with not only cortisol response but also central neuroendocrine signaling systems (Overli *et al.*, 2005). Behavioral and physiological responses to stress have common control mechanisms in the brain and the monoamine neurotransmitters serotonin (5-hydroxtryptamine, 5-HT), dopamine (DA) and norepiniphrine (NE) play a vital role here, as well as being involved in aggressive behavior (Winberg and Nilsson, 1993). It would seem that selection for stress responsiveness in rainbow trout is also associated with changes in the function of these brain monoaminergic systems (Overli *et al.*, 2001) as HR trout reacted to stress by an increase in concentrations of 5-HT and DA in the brain stem and NE in the optic tectum and telencephalon, whereas LR did not (Overli *et al.*, 2005). Therefore, cortisol response to stress may be part of a complex trait that incorporates several behavioral and physiological responses. Such traits may represent different strategies or styles of coping with stress (Overli *et al.*, 2004). It will be important to understand whether it is possible to select for stress responsiveness alone or only as part of this complex trait. Further work is required to elucidate the mechanisms of integration between behavioural profile (stress response, aggression, *etc.*), cortisol response and monoamine influence (Overli *et al.*, 2002a), as well as the effect of the co-occurring traits on welfare in the aquaculture environment. While these investigations are likely to provide welfare benefits in their own right, the feasibility of manipulating stress responsiveness by selective breeding remains unclear.

Beyond inherited stress responsiveness, a positive conditioning procedure, pairing a stressor of lowering water levels with feeding, has been shown to moderate the severity of response of juvenile chinook salmon (*Oncorhynchus tshawytscha*) to subsequent transportation as well as enhancing survival compared to controls (Schreck *et al.*, 1995). Although repeated inescapable stressors would be detrimental to welfare, if this method was employed using short term, acute mild stressors to condition fish, an improvement in welfare may be possible.

Diet may also play an important role in stress sensitivity. African catfish (*Clarias gariepinus*) receiving a diet with high supplementation of ascorbic acid (Vitamin C) during early development showed lower stress sensitivity (Merchie *et al.*, 1997), although common carp (*Cyprinus carpio*) fed large doses of Vitamin C showed a more pronounced cortisol increase in response to stress when compared to fish fed recommended levels of the vitamin (Dabrowska *et al.*, 1991). Juvenile gilthead sea bream (*Sparus aurata* L.) fed a Vitamin E deficient diet showed faster elevation of plasma cortisol levels in response to stress and a lower survival rate than control fish (Montero *et al.*, 2001). Feeding rainbow trout (0. *mykiss*) glucan in low doses, several weeks prior to a stressor, such as transport, shows potential for reducing the immunosuppressive effects of stress (Jeney *et al.*, 1997).

Stress response is an adaptive function in the face of a perceived threat to homeostasis and stress physiology does not necessarily equate to suffering and diminished welfare. In the short term, stress responses serve a very important function to preserve the individual and long term stress, such as reductions in growth, suppressed reproductive function, diminished immune function and disease resistance.Although repeated inescapable stressors would be detrimental to welfare, if this method was employed using short term, acute mild stressors to condition fish, an improvement in welfare may be possible.

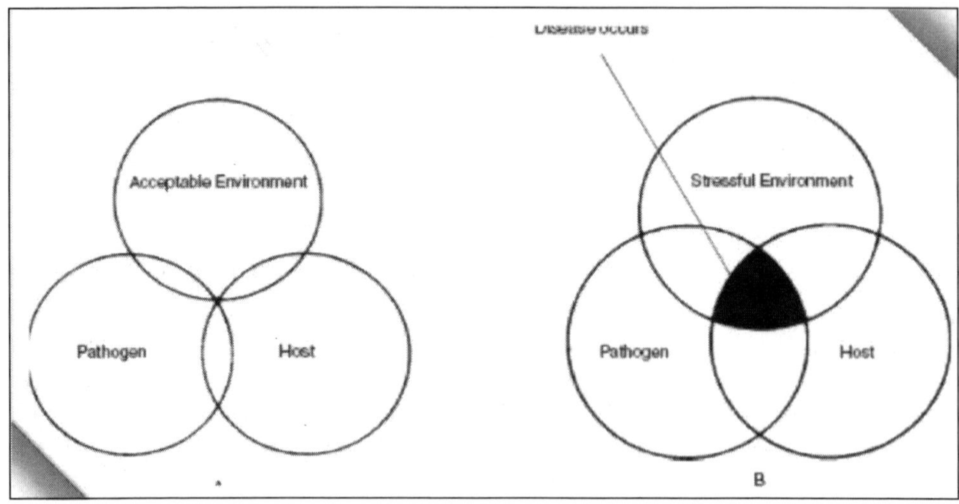

Disease, Host, Environment-Stress Diagram

Health: Fundamental to Welfare

Health is a fundamental measure of welfare. However, poor health may also represent an indicator of further welfare concerns as chronic stress can suppress immune function. For example, Iguchi *et al.* (2003) investigated the influence of rearing density on disease susceptibility in ayu (*Plecoglossus altivelis*). Fish in high experimental rearing densities exhibited more elevated serum cortisol concentrations and more suppressed immunoglobulin M concentrations when compared to fish in lower densities. A higher mortality, possibly caused by bacterial coldwater disease, was seen in the more stressed, high-density experimental group. Poor health itself can also lead to further diminished health and welfare through a number of different mechanisms, including impaired acute (adaptive) responses to further stressors, stress related reductions in immunocompetence, reduced feeding and negative social interactions (*e.g.* see Tort *et al.*, 1998a,b). When investigating the effects of infestation with an ectoparasite in rainbow trout (*O. mykiss*), Ruane *et al.* (1999) found that the parasitized fish showed an impaired response to a subsequent acute confinement stress when compared to non¬parasitized controls. The stress associated with ectoparasite infestation became apparent following confinement, with the parasitized fish showing immunosuppression.

Stress is, therefore, a major factor in the health of farmed fish (Wedemeyer, 1997). While illness is not always connected to poor environmental conditions (FSBl, 2002) the aim of effective aquaculture is to minimize exposure and, importantly, susceptibility to pathogens and parasites, and minimize sources of physical damage. Aquaculture practices present many situations where stress and physical injury can strongly increase susceptibility to naturally occurring pathogens. Prevention is the ideal and this can only be achieved through good management protocols, and optimal system design (Rottmann *et al.*, 1992).Infectious diseases in cultured fish, treatment, and the role of stress in fish disease are reviewed by Wedemeyer (1997), Post (1987), Toranzo *et al.* (2005), and Nougayrede (1995). A full description of each of the many different parasitic, bacterial, viral, or fungal diseases seen in aquaculture is beyond the scope if this review. However, there are several areas that provide a clear illustration of the reciprocal links between health and welfare and how scientific research can lead to improved welfare.

For example, diseases associated with low temperatures over winter periods have been described in a number of different species (Salte *et al.*, 1994) and although some of these winter diseases are clearly associated with specific bacterial pathogens (*e.g. Vibrio salmonicida*, the causative agent of coldwatervibriosis in salmonids), research on gilt head sea bream (*S. aurata*) indicates that severe immunosuppression in winter months may play an important role (Tort *et al.*, 1998a,b). Beyond immunization with vaccines (Toranzo *et al.*, 1997), addressing this immunosuppression may be the key to improved welfare and survival. Tort *et al.* (2004) have shown that an adapted diet providing a supplementary dosage of vitamins and trace minerals to assist the immune system, with a high palatability and nutrient density to maximize feeding and nutrient intake during the winter period when feeding rate is low, provides some degree of immune protection and may help to counteract and reduce some of the effects of winter syndrome. Different feeding regimes after an outbreak of coldwater vibriosis in Atlantic salmon (*Salmo salar* L.) have been shown to have a strong impact on the development of the disease (Damsgard *et al.*, 2004) and the level of dietary iron and omega-3 fatty acids also affect survival of this disease and furunculosis. This may be related to the levels of nutrients that are available to the pathogen (Blazer and Wolke, 1984). As iron is an essential growth factor for most bacteria, deprivation through low levels in the diet (Rorvik *et al.*, 2003) or through food deprivation (Damsgard *et al.*, 2004) may increase fish survival rates.

Fin rot describes a variety of lesions including erosion, splitting of fin rays, nodular thickening and extensive loss of tissue and is a common problem in farmed Atlantic salmon (*S.salar* L.) and rainbow trout (*O. mykiss*) (Turnbull *et al.*, 1998). It can be caused by abrasion with the environment *e.g.* nets and cages; (Ellis, 2002) but often occurs as a result of aggressive interactions, which may also increase susceptibility to infections through the effects of chronic stress (Turnbull *et al.*, 1996). Secondary infection with a range of bacteria can occur, such as *Aeromonas salmonicida*, the causative agent offurunculosis (Turnbull *et al.*, 1996). The use of injectable vaccines for diseases such as furunculosis has superseded less effective antibiotic treatment (FAWC, 1996). However, immunization is likely to be stressful to the fish as it involves handling, anesthesia, and injection. Some vaccines require an adjuvant for adequate efficacy, but these are associated with inflammation, granuloma and pigmentation at the site of injection, impaired growth and reduced carcass quality (Midtlyng, 1997).Adhesions in the peritoneal cavity and between organs have also been described (Hastein, 2004), although these may have more determinants than the type of vaccine and vaccination procedure (Vagsholm and Djupvik, 1999). While the vaccines do improve overall welfare, the development of vaccines with equal efficacy and reduced side effects or alternative oral vaccines may provide further improvements (Bogwald *et al.*, 1994).

Sea lice are a serious welfare concern. A number of different species of parasitic copepods, often collectively referred to as sea lice, erode the fish skin causing tissue damage and may also act as a vector of other diseases (Johnson *et al.*, 2004). Indirect and direct losses due to sea lice in salmonid aquaculture globally are estimated to be greater than US$lOO million annually (Johnson *et al.*, 2004). These are mostly from the costs of treatments and management strategies, reduced growth rates, and the costs of carcass downgrading at harvest. Medical intervention such as dichlorvos or hydrogen peroxide has traditionally been used. However, some sea lice have developed resistance to these chemicals (Jones *et al.*, 1992) and alternative treatments for sea lice are currently being sought (Mordue and Pike, 2002). A vaccine against fish lice may be plausible (Woo, 1997) and as susceptibility to lice has been shown to have a genetic component in salmon (*S. salar* L.) and brown trout (*Salmo trutta* L.), selective breeding towards louse resistance may provide welfare improvements (Jones *et al.*, 2002). However, selective breeding takes time and although theoretically possible, financial and practical constraints may limit this approach (Glover *et al.*, 2005).

Biological control of lice is an alternative and cleaner wrasse (Labridae sp.) have been stocked commercially with farmed salmon (*S. salar* L.) to reduce parasite levels (Tully *et al.*, 1996). However, this situation appears to provide poor welfare for the wrasse as they may be attacked and eaten by the larger salmon. Chronic unavoidable predator stress is, therefore, also a concern. Adequate refuges should be provided for the wrasse along with food supplements when natural sea lice levels are low (Treasurer, 2002). FAWC recommends removal of wrasse prior to grading and during times of food deprivation (FAWC, 1996).

Viral diseases are now one of the main threats to fish health in aquaculture. Viral diseases of particular importance in salmonids include infectious pancreatic necrosis (Park and Reno, 2005 and Roberts and Pearson, 2005), infectious haematopietic necrosis (Bergmann *et al.*, 2002), viral haemorrhagicsepticaemia (Fichtner *et al.*, 1998), infectious salmon anaemia (Hovland *et al.*, 1994), sleeping disease (Graham *et al.*, 2003), and heart and skeletal muscle inflammation (Kongtorp *et al.*, 2004). Traditional vaccines developed over the past 20 years have shown only moderate success and there are relatively few commercial vaccines and specific therapeutics with adequate efficacy (lfremer, 2002). Therefore, increased knowledge of these diseases along with the development of alternative anti-viral treatments (Lorenzen and Olesen, 1997) and the potential for selection for disease resistance (Midtlyng *et al.*, 2002) are crucial to future improvements of welfare.

Non-infectious production related problems are recurrent in aquaculture and are of welfare concern. These include deformities of the heart (Poppe and Taksdal, 2000), and spine (Vagsholm and Djupvik, 1998). These deformities undoubtedly affect welfare both directly, through poor health, and indirectly, through reduced ability to compete for food. Many of these deformations are thought to be caused by hereditary and/or environmental factors, as they are not associated with specific infectious agents. For example, high temperatures during incubation of eggs may be a causal factor associated with heart deformities in Atlantic salmon (*S. salar* L.) (Poppe and Taksdal, 2000). Affected fish tolerate stress very poorly and show a high mortality rate during stress, such as transportation, due to impaired cardiovascular function, cardiac failure, or heart rupture (Brocklebank and Raverty, 2002). Spinal deformities in farmed Atlantic salmon (*S. salar* L.) are also common (Vagsholm and Djupvik, 1998) and both genetic and environmental factors appear to make up a complex aetiology. Vitamin C deficiency (Halver *et al.*, 1969), an excess of dietary leucine (Choo *et al.*, 1991), the use of oxytetracycline (Toften and Jobling, 1996), and environmental pollutants (Bengtsson, 1975) have been associated with spinal deformity. Vagsholm and Djupvik (1998) found a number of risk factors throughout the salmon rearing process and recommended that risks of spinal deformities could be reduced by increasing the smolt weight to around 0.1 kg at seawater introduction, vaccinating the smolt just before sea water introduction and rising at sites where salinity and temperature vary as little as possible. Genetic factors also play a substantial role in the spinal deformities seen in salmon and it is recommended not to select breeders from families showing high incidences of deformed fish, to prevent the increase in the genetic susceptibility in the population (Gjerde *et al.*, 2005).

Aquaculture Procedures that Compromise Welfare

Overview

Stressors in aquaculture are unavoidable and reducing stress and its harmful effects is a fundamental goal for successful growth and production as well as welfare. The effects of a wide range of aquaculture practices on the stress physiology of fish are well documented (Pickering, 1992).Different species display a wide variation in physiological responses to stressors associated with aquaculture. Elevations in plasma cortisol levels can differ by as much as two orders of magnitude among different species of fish following identical stressors (Barton, 2000a,b, 2002). There is, therefore, economic and welfare related pressure to ensure that both the frequency and the effects of a number of key management practices are reduced to a minimum. In particular, research into optimum handling and transportation methods and species-specific responses can help to achieve this aim.

Handling

Handling and transport are inherently stressful events (Barton *et al.*, 1980). Removal from the water elicits a maximal emergency physiological response and should only be carried out when absolutely necessary. Air exposure for 3 min resulted in a 50-fold increase in plasma cortisol levels within 30 min in gilthead sea bream (S. *aurata*) (Arends *et al.*, 1999). Care must be taken at all stages to avoid abrasions and removal of scales and the fish's protective mucous coat, which serves as a physical and chemical barrier to infection as well as being important in osmoregulation and locomotion. Where applicable this should involve the use of wet hands and species appropriate nets, keeping the fish moist during handling. Excessive weight loading on fish at the bottom of nets and braille's should also be avoided (Conte, 2004). The use of a braille lining allows some water to be retained in the net and, therefore, provides some protection from abrasion. Moving water along with the fish should cause fewer injuries and appears to be the least stressful technique (FAWC, 1996). As such, the use of fish pumps and transfer pipes appears to be preferable for welfare. However, effective management should ensure that the design of the system provides appropriate speed and delivery rate, minimizes potential for abrasion, and minimizes the time spent in the pipe. Prolonged periods in pipes, particularly in warm conditions, should be avoided and pipes should be flushed through to ensure no fish are left inside after use (HSA, 2005).

Excessive crowding of fish prior to management procedures can be stressful, with potential decreases in oxygen levels and water quality, increased chances of abrasion, and rapid changes in light intensity (HSA, 2005). While some species adapt well to high loading densities (Barton *et al.*, 2000; Ruane and Komen, 2003) others show prolonged elevated cortisol levels following confinement procedures (Barnett and Pankhurst, 1998) have also shown that a fish's capacity to respond to acute stressors such as crowding may be altered by the effects of long term holding conditions. Gilthead sea bream (S. *aurata*) held at high density over 14 days showed a suppressed cortisol response to acute handling, likely the result of negative feedback from chronically elevated cortisol (Barton *et al.*, 2005). Stress during crowding may also affect response to further stressors such as net capture (Schreck *et al.,* 1989). Recovery periods are therefore often beneficial, particularly during transport (Iversen *et al.,* 1998). A cumulative effect on cortisol levels was seen when turbot (*Scophthalmus maximus* L.) were confined a second time, 4 h after an initial net capture (Waring *et al.,* 1997). A 24 h recovery period between multiple handling procedures avoided this cumulative effect. Stressors associated with crowding can be kept to a minimum by appropriate management techniques. HSA guidelines (2005) suggest good crowding management includes a slow and gentle technique, assessment of water quality, the addition of oxygen to the water if levels fall below a critical 6 mg/l, and, perhaps most importantly, close monitoring of the behaviour and activity of the fish. A simple scoring system is given to help train staff to recognize acceptable levels of activity and stress (HSA, 2005).

Where handling is prolonged such as during stripping and milking, sedation should be used to reduce stress response (FAWC, 1996). Although the grading of fish can be stressful through capture and handling, limited grading may be advantageous to welfare by reducing size differences, thus preventing aggression, reducing feeding competition, and removing maturing fish from a population. As the following sections will describe, stocking density and feeding methods have a strong influence on the levels of social interactions, dominance hierarchies and, subsequently, growth in captive fish. The use of species appropriate feeding techniques and stocking densities can, therefore, limit heterogeneous growth within a group of fish and thus the need for frequent grading (Volpato and Fernandes, 1994).

Transport

A wide variety of transportation techniques are used within aquaculture, but all should aim to minimize stress, optimize water quality and oxygen levels, and minimize the buildup of metabolic wastes and ammonia (see food deprivation below). Transportation involves capture, loading, transport, unloading and stocking and so can induce large stress responses that can affect fish over a prolonged recovery period (Specker and Schreck, 1980). In Atlantic salmon (S. *salar*L.) smolts, many of the disease outbreaks take place during the first months of transfer to the sea following transport in well boats (Iversen *et al.,* 2005). This study found that the loading process was a more severe stressor than the transport itself, with plasma cortisol returning to resting levels during the time in the well boats in four out of five transports. Only minor plasma cortisol increases were observed during unloading. lt may be that the well boats provide an important recovery function. Specker and Schreck (1980) also found that the greatest stress response occurs during loading and the first few hours of transport in Coho salmon (*Oncorhynchu skisutch*) smolts. A recent study, using electromyogram telemetry, indicates that rainbow trout (*O. mykiss*) show vigorous swimming activity and elevated oxygen consumption during transport (Chandroo *et al.,* 2005). While activity levels returned to baseline within 48 h, beyond this period, swimming performance, measured as critical speed and endurance, was still affected. The provision of a recovery period following transport is clearly important for welfare and subsequent survival (Erikson *et al.,* 1997). The net loss of sodium and chloride ions in freshwater fish, associated with the stress of transport means that certain species may benefit from dilute salt solutions or a current during, before or after transportation (Carmichael, 1984). Anaesthetic agents may also be used to sedate fish prior to transport and/or slaughter (HSA, 2005). At low doses these agents may reduce activity and stress. Iso-eugenol based anaesthetics show good potential in reducing stress (Tort *et al.*, 2002) and may be used to sedate fish in New Zealand, Australia, Chile and Korea. However, this method has yet to be licensed in Europe and North America.

Careful acclimation, to new environmental temperatures, water chemistry and light levels (Mork and Gulbrandsen, 1994), is also important for wellbeing. As with all aquaculture procedures, stockmanship and effective husbandry skills are of paramount importance during transport. Adequate training, knowledge of the signs of stress and appropriate working conditions, are vital to fish welfare.

Food Withdrawal

Fish are often deprived of food before certain management procedures are carried out and this is designed to reduce physiological stress during the procedure. Temporary starvation prior to transport, treatment of disease, and transfer of smolts from freshwater to seawater, serves to evacuate the fish's gut and to reduce metabolism, oxygen demand and waste production. As well as being beneficial to welfare, a reduction in metabolism prior to slaughter may also alter the qualities of the flesh (Johansson and Kiessling, 1991). Prior to slaughter, Atlantic salmon (S. *salar*L.) are often deprived of food for some days or weeks. However, the FA WC and HSA recommend that salmon should not be totally deprived of food except during a period of up to 72 h before slaughter or a handling procedure. For trout these periods are recommended to be up to 48 h only. FA WC guidelines also suggest that fish should not be deprived of food for any other reason such as conditioning and adjustment of body composition (FAWC, 1996).

As fish are ectothermic, periods of food deprivation may be of less detriment than in endotherms (FSBI, 2002). However, fish do have motivational mechanisms for feeding when nutritional reserves are low (*e.g.* Metcalfe and Thorpe, 1992). It is not uncommon for Atlantic salmon (S. *salar* L.) and rainbow trout (0. *mykiss*) in the wild to survive long periods of food deprivation (*e.g.* over winter). However, individual salmon have been shown to vary in their ability to reduce metabolic costs when food availability is low (O'Connor *et al.*, 2000). During the return migration in freshwater prior to breeding, adult salmon may cease feeding altogether for extended periods (Kadri *et al.*, 1995). Such long periods of anorexia are not seen during other life stages, including those seen in aquaculture. However, salmon do show seasonal variation in food intake in aquaculture (Smith *et al.*, 1993). As detailed above, diet, particularly levels of vitamins and trace minerals, has an important influence on immuno-competence, the development of diseases after outbreak and the response of fish to stressful aquaculture events. There is potential for these factors to interact with brief food deprivation to affect welfare in either a positive or negative way.

There have been relatively few studies on the effects of starvation on stress physiology or behavior (Barton *et al.*, 1988) and the majority of work in this area concerns the effect of prolonged starvation on growth, muscle protein and fat composition (Sumpter *et al.*, 1991). The impact upon welfare offood depriving a fish that has previously been fed regularly is not known. Therefore, food depriving farmed fish for short periods under appropriate conditions (*e.g.* temperature and season) may not cause welfare problems (FSBI, 2002). However, it is important to consider other effects of starvation and reduced nutrition, such as changes in metabolic activity (Martinez *et al.*, 2003) and changes in behavior related to competition, and the potential for increased aggression. For example, reduced food abundance can cause changes in territorial behavior strategies and activity patterns in brown trout (S. *trutta* L.) (Alanara *et al.*, 2001).

Slaughter Techniques

Humane methods for animal slaughter are based on the principle that the animal is killed quickly with minimum fear and pain or suffering (FAWC, 1996). However, many aquaculture slaughter methods have been developed not to minimize stress but to achieve product quality control, efficiency and processor safety (Conte, 2004). There is an increasing awareness in the aquaculture industry of the need to ensure that slaughter takes place under humane conditions and that fish should be stunned prior to slaughter by a method that causes immediate loss of consciousness that lasts until death (FAWC, 1996). The development of new commercial technologies to produce humane slaughter techniques for aquaculture is an active area of research.

From a welfare point of view, the most important factors in slaughter technique are the methods of handling fish during transfer to the killing facility up to the point of stun and the immediacy of loss of consciousness caused by the stunning method (Southgate and Wall, 2001). The technique used to kill the fish should do so quickly following successful stunning, to avoid regain of consciousness. Minimizing pre-slaughter stress and the use of humane slaughter methods also improves product quality in a number of areas (Robb *et al.*, 2000a). As with transport, sedation with anesthetics prior to slaughter may reduce the stress associated with the event. So called "rested harvesting" also improves flesh quality (HSA, 2005).

Aquaculture slaughter techniques are very diverse and fish species vary in their response to different methods (*e.g.* sensitivity to oxygen deprivation; Morzel *et al.*, 2003). However, the most prevalent methods have been investigated in recent years in order to assess both the associated welfare and product quality. When investigating suitable slaughter methods, consciousness can be assessed by measuring visually evoked responses (VERs) and somatosensory evoked responses SERs (*e.g.* responses to pain stimuli) in brain activity using electroencephalography (EEG) (Robb *et al.*, 2000b). In certain species, behavioralindicators have been shown to correlate with brain activity but the use of such indicators alone may not be sufficient (Van de Vis *et al.*, 2003). Adequate training of operators in the identification of behavioral indicators of consciousness is vital (Kestin *et al.*, 2002).

Many of the current commercial techniques are unacceptable in terms of welfare (FA WC, 1996). Removal from water followed by asphyxiation in ice slurry is often used to slaughter rainbow trout (*O. mykiss*). This is likely to cause a prolonged period of distress before death, as the time to death when fish are removed from water has been shown to be as long as 14 min at a temperature of 2 °C (Kestin *et al.*, 1991; Robb *et al.*, 2000b; Lines *et al.*, 2003). Immersion in ice slurry or asphyxia is used to slaughter gilt-head bream (*S. aurata* L.) and in both cases loss of brain function is not immediate (loss of VERs occur at 5-5.5 min. Salmon and trout have previously been killed by immersion in CO_2 saturated water, which causes narcosis and loss of brain function (*e.g.* Erikson *et al.*, 1997). However, this only occurs over several minutes, during which the fish appear severely distressed (Robb *et al.*, 2000a,b). Following a minimum of 4 min immersion, the fish are removed and their gills cut. The CO2 stun method causes a reduction in flesh quality (Roth *et al.*, 2002) and concern over welfare has led to a reduction in its use. While some Scandinavian countries still use this method, it is banned in the UK for all but emergency kills. Gill cutting without prior stunning is also used and is unacceptable as VERs are not lost immediately.

Current commercial slaughter methods for eels consist of a "de-sliming" phase where salt or liquid ammonia are added to eels in a dry tank. This denatures the mucus that protects the skin and the process may take as long as 20 min (Robb *et al.*, 2002b). The eels are then washed and subsequently eviscerated. Although the whole process takes about 1 h, the point of death may occur after the procedure has been completed (Robb *et al.*, 2002b). Eel slaughter using live chilling followed by freezing in cold brine at -18°C does not meet the criteria for humane slaughter. However, Robb *et al.* (2002b) have shown that the use of electric stunning tongues applied to either side of the eels head can instantly and effectively stun the eel and that increasing the duration of the application can cause death. Electrically stunning batches of eels in freshwater combined with nitrogen flushing can also achieve immediate unconsciousness (Van de Vis *et al.*, 2003), therefore, humane slaughter of eels is possible.

When applied correctly, percussive or electrical stunning methods appear to achieve humane slaughter in Atlantic salmon (*S. salar* L.), gilt-head sea bream (*S. aurata* L.), turbot (*S. maximus* L.), and rainbow trout (*O. mykiss*) (Robb *et al.*, 2002a). The UK trout industry appears to be moving towards electric stunning as its preferred slaughter method (Lines *et al.*, 2003).

When carried out in water, electrical stunning can be efficient and humane as it can cause very rapid transition to insensibility (Robb and Roth, 2003). Fish need not be singled out, restrained, handled, or removed from the water before they are insensible (Robb *et al.*, 2002a). Death may then occur as a result of bleeding, asphyxiation or as a direct result of the stun. However it is important to use the appropriate electric field strength; too weak and the stun may only cause paralysis rather that insensibility, too high and carcass damage such as hemorrhages, blood spotting, and broken bones may result. (Robb *et al.*, 2002a). However, carcass damage is influenced by the frequency of the alternating current and damage free humane stunning is possible (Robb *et al.*, 2002a).

It is important to point out that in order to maintain humane slaughter techniques; the efficiency of percussive stun or electric stun methods relies on adequate training and working conditions for those undertaking the operation (FAWC, 1996). For example, it is very difficult to maintain accuracy when percussive stunning manually over a prolonged period on a commercial level. The advent of automated percussive stunning equipment, therefore, represents a welfare improvement. More recent models avoid the need for the operator to handle the fish as the design encourages the fish to swim into the entry channels. These machines, therefore, have potential to further improve welfare as they reduce handling to a minimum and keep fish in water until seconds before being stunned (HSA, 2005). Automated systems provide more accurate and consistent stunning, provided they are used and maintained correctly with an appropriate delivery system. Ineffective stunning may occur where fish are of incorrect size or deformed. Therefore, it is important that all fish are monitored and a manual stun applied if the initial stun is not effective (HSA, 2005).

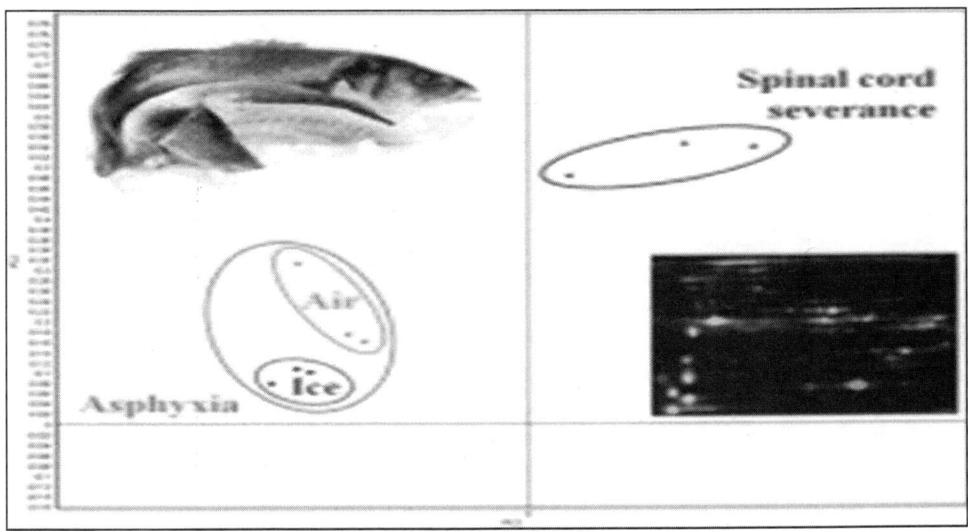

Stocking Density: A Complex Issue

Stocking density is a pivotal factor affecting fish welfare in the aquaculture industry, especially where high densities in confined environments are aimed at high productivity. Although rarely defined, stocking density is the term normally used to refer to the weight of fish per unit volume or per unit volume in unit time of water flow through the holding environment (Ellis, 2001). The stocking density at any point in time will increase as fish grow or decrease following grading. Stocking density is, therefore, hard to measure in the field. The concept of minimum space for a fish is more complex than for terrestrial species as fish utilize a three-dimensional medium (Ellis, 2001). Because fish are dependent on this medium for both physiological and behavioral needs the welfare concerns associated with stocking density should address both the carrying capacity of the holding environment and the spatial and behavioral needs of the species. Carrying capacity refers to the maximum number of fish that an environment can support through oxygen supply and removal of metabolic waste and will be determined by, amongst other things, the oxygen consumption rate of the fish and their response to metabolic waste products such as CO_2 and ammonia (Ellis, 2001). Beyond providing for the physiological needs, the FAWC recommends that fish "need sufficient space to show most normal behavior with minimal pain, stress and fear" (FAWC, 1996). Stocking density is, therefore, an area that illustrates both the significance of species differences and the existence of a complex web of interacting factors that affect fish welfare.

For species that live naturally in large shoals, low rather than high densities may be harmful, whereas for territorial species the opposite may be the case. Reductions in stocking density may reduce the physical spread of disease, but stocking density will also have an effect on many other aspects of welfare such as the water quality, particulate matter, fish to fish interaction, and fish to housing interaction (Ellis *et al.*, 2002). In different systems and with different species the way these factors combine to affect welfare will differ, making it difficult to provide blanket guidelines, much less legislation on maximum densities (Ellis *et al.*, 2002). Within species there may even be age dependent differences in stocking density effects (Jorgensen *et al.*, 1993).

As the above might imply, studies in both laboratory and aquaculture settings show that the effect of stocking density on measures of welfare varies between species. For example, sea bass (*Dicentrarchus labrax* L.) showed higher stress levels at high densities, as indicated by cortisol, innate immune response, and expression of stress related genes (Vazzana *et al.*, 2002; Gomati *et al.*, 2004). High stocking densities in juvenile gilthead sea bream (*S. aurata* L.) also produce a chronic stress situation, reflected by high cortisol levels, immunosupression, altered metabolism (Montero *et al.*, 1999). In contrast, Arctic charr (*Salvelinus alpinus*) feed and grow well when stocked at high densities but show depressed food intake and growth rates at low densities (Jorgensen *et al.*, 1993). These differences did not seem to be due to the formation of dominance hierarchies in the low-density group. However, beyond certain high densities, stress related behavioral responses do re-occur (Ellis *et al.*, 2002). In halibut (*H. hippoglossus* L.), a non¬social flatfish that spends most of its time resting on the sea floor, tolerance for high stocking density appears to be stage dependent. Greaves and Tuene (2001) showed that small juvenile halibut had better growth and fewer injuries from aggressive encounters at the highest densities whereas Kristiansen *et al.* (2004) observed abnormal activity and surface swimming, reduced food consumption, and reduced growth rates with increased stocking density in adult halibut.

There is considerable evidence for decreasing welfare associated with high stocking densities in rainbow trout (*O. mykiss*). The effects of high density include decreased growth, reductions in food intake, food conversion efficiency (Refstie, 1977) and nutritional status, fin erosion, gill damage, a reduced immune capacity (Ellis, 2002), and alterations in swimming behavior (Anras and Lagardere, 2004). When reviewing some 43 studies examining the effects of density on such parameters Ellis *et al.* (2002) noted that the magnitude of the effects tended to be dependent on study specific conditions and increased disease incidence did not seem to be a generic effect of density. High densities are associated both with increased competition, aggression, and physical injury (due to increased contact between fish and between fish and the housing/net), and with degradation in water quality. However, it seems that excessive aggressive behavior and a poor feeding response may also occur when trout are held at very low densities (Ellis, 2002). Although the relative contributions of aggression and degradation of water quality vary with conditions, experimental evidence indicates that water quality is the key factor in relation to density affecting welfare in rainbow trout (Ellis *et al.*, 2002).

Recently, Turnbull *et al.* (2005) carried out a study on the effect of stocking density on Atlantic salmon (*S. salar* L.) in marine cages under commercial production conditions. A multivariate analysis was used to combine four commonly used measures of fish welfare, including condition of the body and fins and plasma concentrations of glucose and cortisol, into a single welfare score. This score was consistent with the evaluation of welfare by experienced farmers, and, therefore, appears to be a good measure of welfare in real environments. High stocking densities were associated with a reduction in welfare but only above a threshold of 22 kg m^{-3}. This suggests that below this threshold increasing density did not reduce welfare in these situations. When the effects of stocking density and sampling period were taken into consideration, the reduction in welfare at high stocking density did not seem to be dependent on reduced water quality or associated with increased rates of social interaction. Also the time since the last stressful husbandry event was not significantly associated with welfare. However, the sampling period and the location of the cage was a significant predictor of welfare even when adjusted for stocking density (Turnbull *et al.*, 2005). These results show us that a range of other factors combine with stocking density to affect welfare and these factors will vary from site to site and perhaps with time. A density threshold for one set of conditions may, therefore, not to be relevant to another. This study also serves to illustrate the value of qualitative assessment of welfare by experienced animal keepers. Integration of observations of behavior, feeding response and gross appearance can, at the very least, aid interpretation of other quantitative measures of welfare. The simple scoring system to monitor crowding behavior in salmonids given by HSA is a good example of how such observations can improve welfare (HSA, 2005). It has been proposed that such qualitative assessment of behavior in pigs has validity and important potential for future welfare measurement (Wemelsfelder *et al.*, 2001). As concepts of welfare in fish develop, the value of a similar approach in aquaculture should not be underestimated.

It is clearly very hard to propose within species maximum stocking densities for all situations. However expressed, stocking density alone will not directly predict welfare. A complex matrix of factors appears to influence the effect of stocking density on a range of welfare measures. It has been suggested that a more productive approach to ensure welfare is to study the specific and dynamic components of the overall effect such as behavior, water quality, health, stress physiology, and gene expression (Ellis, 2002).

Behavioral Approaches to Improve Stocking Density Effects

Species relevant behavioral observations in particular may provide further understanding of density effects and perhaps produce practical steps to improve overall welfare as well as health. For example, behavioral observations of Atlantic halibut (*H. hippoglossus* L.) in experimental tanks suggest causal mechanisms behind the reduced growth and feeding with increasing densities in adults. Although the natural activity patterns of this largely solitary fish species are not well documented it is thought that they spend extended passive periods on the sea floor. Kristiansen *et al.* (2004) observed that with increasing density the tendency of the fish to leave the bottom increased and that there was a negative correlation between the subsequent vertical surface swimming and growth. Negative growth rates were seen in many individuals with high activity levels, indicative of poor husbandry and poor welfare. The origins of the surface swimming behavior are unclear, though it may represent an effort to move away from an area or unfavorable situation. If this is the case and the behavior repeatedly fails to achieve this aim, as in a high density restricted aquaculture situation, then this may represent a further welfare concern. The increased tendency to leave the bottom when fish are crowded also means that fish landing after swimming are more likely to come into contact with resting fish and initiate further activity. Considering these observations the authors suggest measures that may reduce the effects of density such as the use of multiple overlapping floors, alternative substrates, and an altered social environment with regards to fish size and species.

Acoustic telemetry has been used to monitor the swimming activity and positional data for rainbow trout (*O. mykiss*) within three different rearing densities on an experimental fish farm using circular tanks (Anras and Lagardere, 2004). At the lowest density the fish showed a preference for the outer circles of the tank (where the feeder and water inlet were positioned) and spent long periods in a stationary position. However in the highest stocking density there was a greater residence in the inner circle of the tank (where the water drained) and swimming patterns were chaotic, probably due to crowding and contact between fish, as suggested by observed fin damage in this group. Under natural conditions and low density rearing, rainbow trout show low intensity swimming activity during the day and a drop in nighttime swimming activity. However, in the high density experimental group high levels of activity were recorded and the day-night rhythm was not seen. Remote measurement of behavior in this way will enable better interpretation of parallel changes in health and stress physiology.

Atlantic salmon (*S. salar* L.) in sea cages do not occupy all the available space and seldom distribute themselves randomly (Juell *et al.*, 2003). Swimming depth and behavior is strongly influenced by environmental factors such as temperature and light (Oppedal *et al.*, 2001). Artificial photoperiods are used in cage rearing of Atlantic salmon as this may increase growth rates and postpone sexual maturity (Oppedal *et al.*, 1997). The swimming depth of Atlantic salmon (*S. salar* L.) is strongly influenced by light levels to enable the maintenance of schooling behavior, which relies on visual contact (Ferno *et al.*, 1995). Recent studies, using echo-integration to observe swimming depth and fish density, have shown that both natural light and the photoperiod and positioning of artificial lighting exert a powerful influence on the vertical positions of the fish throughout the day and, as a result, affect the "real" or observed fish density (*e.g.* density within the shoal rather than the cage as a whole) (Juell *et al.*, 2003). These studies have shown that the swimming depth and density of caged salmon can be manipulated by the use of underwater lamps and that appropriate positioning of these lamps can be used to avoid crowding, particularly in the upper water layers, which may be considered sub optimal with regards to welfare (Juell and Fosseidengen, 2004).

Aggressive Behavior

Stocking density naturally has a large effect on social interactions between fish. In species where social hierarchies are formed, such as salmonids, these may lead to chronic social stress (Ejike and Schreck, 1980), reduced feeding, changes in metabolism, down regulation of digestive function and suppression of growth in low ranking fish and higher growth rates in dominant individuals (Abbott and Dill, 1989). Also, the high levels of activity associated with agonistic behavior and chronic social stress (*e.g.* Abbott and Dill, 1985) may result in greater energy expenditure, elevated metabolic rates (Sloman *et al.*, 2000), decreased food utilization efficiency (Li and Brocksen, 1977) and impaired immune function (Pottinger and Pickering, 1992). Beyond social stress, conspecific attacks are often the primary cause of fin damage, particularly in those fish that are already immune suppressed and so secondary infection may follow (Turnbull *et al.*, 1996, 1998). Food availability is therefore an important factor when it comes to stocking density (Holm *et al.*, 1990) and how captive fish are fed has a strong influence on the levels of social interactions (Alanara and Brannas, 1996). For example, aggressive interactions are a major cause of injuries to the eyes, tails and pectoral fins, causing secondary infections and mortality in farmed Atlantic halibut (*H. hippoglossus* L.). The majority of damaging contact between fish appears to occur during feeding. Greaves and Tuene (2001) observed that in experimental tanks some interactions were misdirected feeding attempts. However, the majority of attacks were directed at fish that had won food pellets from aggressors. The majority of aggression occurred early in hand feeding sessions and so, to avoid aggression, the authors recommend hand feeding where food is spatially dispersed but concentrated in time. The use of automatic feeders throughout the day to prevent the buildup of appetite and allow all fish the opportunity to feed to satiation may achieve this aim (Greaves and Tuene, 2001).

Andrew *et al.* (2002) have shown that varying feeding method can alter aggressive interactions in Atlantic salmon (*S. salar* L.), gilthead sea bream (*S. aurata*

L.) and European sea bass (*D. labrax* L.) in sea cages. The behavior of control fish that were fed to apparent satiation at set times was compared to those that were fed on demand using interactive feedback systems. Similar effects on swimming speeds and feeding intensity were seen in all three species and suggested stronger and significantly higher levels of aggressive scramble competition in the control fish.

It is clear that stocking density, the nature and level of food availability, and social behavior act together to affect welfare. The effect of hierarchy and stocking density on the use of demand feeders has been investigated in rainbow trout (*O. mykiss*) and Arctic charr (*S. alpinus*, Alanara and Brannas, 1996). At low densities, in both species, top ranked individuals were able to dominate the feeders, performing the majority of trigger activations. However, this ability was reduced at higher densities, where a small group of fish dominated the activation of the feeder trigger (Alanara and Brannas, 1996). As discussed, such domination of a food source is likely to lead to heterogeneous growth rates with low ranking individuals gaining little access to food (Alanara *et al.*, 1998). However, it seems that in both species, some low ranking individuals may apply an alternative strategy to attain adequate growth by feeding at night when dominant individuals are less aggressive (Alanara and Brannas, 1997). Thus, changes seen in activity and behavior with increasing densities, described in the previous section, are likely to be intrinsically linked to the availability of food experienced by the individual and the levels of aggression and social hierarchy within a group.

The chronic social stress and high cortisol levels observed in subordinate fish are associated with a suppression of aggressive behavior, locomotory activity and feeding levels (Overli *et al.*, 2002a). These changes in behavior are likely to be related to cortisol-induced changes in brain monoamine levels, particularly 5-HT and DA (DiBattista *et al.*, 2005b). Increased brain 5-HT activity is associated with inhibition of aggression and subordinate behavior in fish, while increased DA activity is associated with aggressive, competitive behavior and dominance (Winberg and Nilsson, 1993). Increased dietary levels of the 5-HT amino acid precursor, L-tryptothan, have been shown to suppress aggressive activity in juvenile rainbow trout (*O. mykiss*) without affecting food intake or plasma cortisol levels (Winberg *et al.*, 2001). Similar dietary supplements have been found to decrease aggressive behavior in cod (*G. morhua*) (Hoglund *et al.*, 2005) and size heterogeneity and cannibalism in juvenile grouper (*Epinephelus coioides*, Hseu *et al.*, 2003). It has been suggested that this may be a promising management strategy, since the effects of the dietary L-tryptothan will be most pronounced in the most aggressive, dominant individuals that eat the most food and thus strong dominance hierarchies and heterogeneous growth rates may be avoided (Winberg *et al.*, 2001). Such a strategy maybe particularly useful during periods of reduced feeding where competition is high.

An alternative approach to reducing aggression may be possible through the presence of larger conspecifics. The presence of a large dominant fish may suppress aggression among its smaller companions. This was illustrated by a small-scale experiment on Atlantic salmon (*S. salar* L.) parr where 0+ fish that had been housed with a 1 + fish showed significantly reduced aggression amongst themselves and significantly increased growth rates when compared to controls (Adams *et al.*,

2000). Although the larger fish attacked the smaller ones, the attack rate was an order of magnitude lower than that seen between smaller fish in the control group. The observations suggest a more stable social hierarchy with less overt aggression. Whether this situation represents improved welfare and a less stressful social environment for the small fish requires further investigation. Here again, caution should be taken in interpreting stress physiology or arousal in relation to welfare and suffering. While stress measures may tell us little about potential suffering here, behavioral observation may provide further insight (Dawkins, 2003).

Background or substrate color may be another factor that affects levels of aggression (*e.g.* in salmonids). Dark body coloration has been suggested to signal social subordination in Atlantic salmon (*S. salar*) and Arctic charr (*S. alpinus*, O'Connor *et al.*, 1999). A study investigating the levels of aggression seen in the latter species showed that in pairs of fish interacting on a white background, both fish showed initial pale coloration and high levels of aggressive behavior (Hoglund *et al.*, 2002). Fish interacting on a black background showed initial dark coloration and a lower frequency of aggressive interactions. While one of the fish on the white background became subordinate and took on darker body coloration, the subordinate fish on the black background did not show any additional darkening. This would suggest that the use of dark substrate has potential for reducing aggressive behavior and social stress in Arctic charr (*S. alpinus*). Background color also affects agonistic behavior in Nile tilapia (*Oreochromis niloticus*) (Merighe *et al.*, 2004). This work serves to illustrate that knowledge of socio-biology of different aquaculture species may help to improve welfare.

Abnormal Behavior: A Welfare Concern?

Changes in behavior can be useful welfare measures where they represent clinical symptoms of poor health or provide early warnings when other symptoms are sub clinical, such as altered swimming behavior associated with parasitic infections (Post, 1987). Beyond this, a wide number of stressors have been shown to cause changes in fish behaviors such as activity, avoidance, predator-prey interactions, feeding, shelter seeking, aggression, and learning (Schreck *et al.*, 1997). These changes may represent an adaptive response mechanism. However, chronic or unavoidable stressors may deny this function and so behavioral changes may become maladaptive. The expression of abnormal behavior in fish kept in high-density aquaculture is, therefore, of potential welfare concern. However, caution should be taken in interpreting the expression of abnormal behavior alone in terms of welfare and suffering as, other than those where physical injury occurs, abnormal behaviors do not necessarily indicate poor welfare and might even be compatible with good welfare (Dawkins, 1998).

It has been suggested that the vertical swimming behavior in Atlantic halibut (*H. hippoglossus* L.) described above (Kristiansen *et al.*, 2004) and the circular shoaling behaviour seen in Atlantic salmon (*S. salar* L.) in sea cages (*e.g.* Oppedal *et al.*, 2001) may represent abnormal behaviour or a kind of stereotypy comparable to the pacing seen in some zoo animals (Lymbery, 2002). Stereotypies are fixed sequences of behavior performed repetitively in the same way with no obvious function and have been used as indicators of reduced welfare (Mason, 1991a,b). While some stereotypies cause injuries and are clearly of welfare concern, the consequences of others are not so clear and there is no one to one correspondence between stereotypy and welfare (Barnett and Hemsworth, 1990). The causes of stereotypies and what they tell us about welfare are a matter of debate (Dawkins, 1998). There are suggestions that some stereotypies are mechanisms that help animals to cope with environmental change, as reduced behavioral and physiological signs of stress have been observed with such behaviors (Dantzer, 1991). The origin of a stereotypy may be previously adverse conditions where a behavior pattern was repeatedly elicited, and so they do not necessarily show current suffering. Levels of stereotypy cannot be used to indicate levels of wellbeing (Mason, 1991 b). However the repeated expression of a behavior that represents failed attempts by the animal to alter its environment is thought to indicate poor welfare at some stage. Irrespective of whether they represent a stereotypy, the loops of vertical swimming behavior observed in Atlantic halibut (*H. hippoglossus* L.) at high experimental densities are, possibly, a welfare concern. Understanding the origin of these apparently functionless behaviors is the key to interpreting how they relate to welfare in fish. For example, it has been suggested that the surface breaking behavior of Rays (Raja sp.) kept in public aquaria represents stereotypic behavior (Casamitjana, 2004). However, there appears to be a temporal link between this behavior and scheduled feeding and it has been suggested that the behavior is a method of foraging adapted for the captive situation (Scott *et al.*, 1999a). Changing the method of food delivery away from the surface has been found to significantly reduce the level of the behavior. The proposals to reduce vertical swimming patterns in Atlantic halibut (*H. hippoglossus* L.) described above may have analogous results.

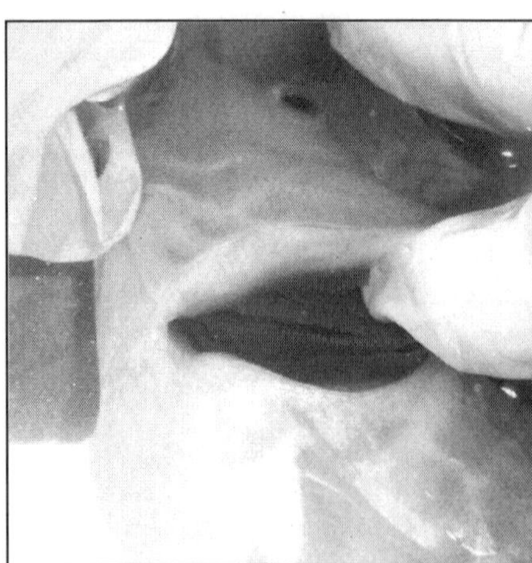

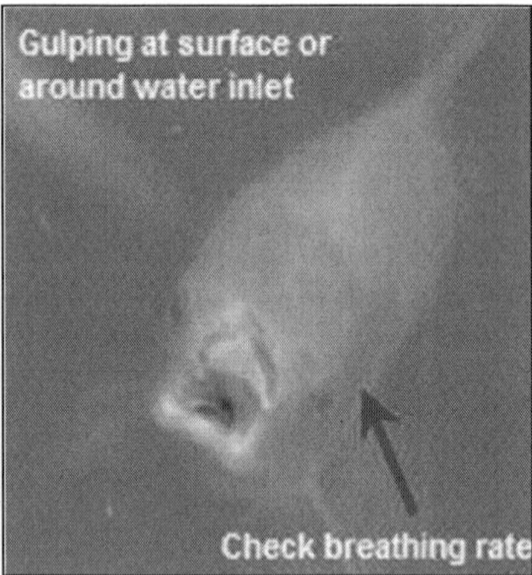

Although wild salmon do mass in shoals when preparing to migrate back to their home river to spawn, the circular shoaling seen in salmon in sea cages has been described as abnormal when compared to fish in the wild at a similar life stage (Lymbery, 2002). However, without further investigation, it is unknown whether both the activity level and shoaling behavior are indeed functionless in this situation or whether they form an adaptive functional response to the captive environment with no detriment to welfare. The above examples illustrate the importance of investigation into the origins and reinforce of such behavior as well the behavioral and physiological consequences of denying this behavior. It is interesting to note that swimming depth of Atlantic salmon (*S. salar* L.) in sea cages may represent a compromise between predator avoidance, hunger, bright light avoidance, and the ability to maintain this shoaling behavior (Fraser *et al.*, 1994).Investigations into the consequences of the breakdown of a shoal in low light levels may be useful.

The Freedom to Express Normal Behavior

Freedom to express normal behavior by providing sufficient space, proper facilities and company of an animal's own kind is one of the five freedoms defined by the FAWC (FAWC, 1996). In this sense the stocking densities and confinement seen in the aquaculture industry may be a concern and the inability to express natural behaviors, such as migration, may cause suffering. However, the "naturalness" of an animal's behavior does not necessarily have a direct positive or negative correlation with welfare. This becomes clear when we consider that much natural behavior are shown in response to adverse conditions such as predator avoidance (Dawkins, 1998). There appears to be no way of predicting the importance of a particular behavioral pattern to a given species and the effect its deprivation has on the welfare of an animal. The only way to assess whether the behavior is necessary or not for welfare is through empirical tests to discover the mechanism of control and the behavioral and physiological consequences of denial of expression (Dawkins, 1998). The example of Atlantic salmon (*S. salar* L.) migration has been used to good effect here (FSBl, 2002). If the control mechanism for migratory behavior is based simply on continuous swimming and the search for improved feeding grounds, then supplying these things in the captive environment may reduce the motivation to migrate and avoid related suffering. However, if based on an intrinsic drive to move to new areas, confinement might well cause suffering (FSBl, 2002).

Even if we are to accept the subjective experiences of fear, pain, and suffering in fish, further research is required to enable us to distinguish between stimuli that are aversive, as opposed to those that are stimulating, as well as those that are important reinforces. To this end choice tests that allow fish to express their natural preferences (Olsen, 1986), particularly in the context of the aquaculture environment (*e.g.* depth choice seen in Juell *et al.*, 2003), should be used in combination with the physiological measures described above (Broom, 1991a,b). With appropriate interpretation, such experiments may tell us how certain practices cause detriment to welfare and which behaviors are important for welfare.

Aimed at reducing stereotypies and deleterious behavior, and increasing natural behaviors, environmental and feeding enrichment strategies have been used in both zoos (Carlstead *et al.*, 1991). However, it should be clear that any enrichment does provide an improvement in the biological functioning and welfare of the animal. Therefore, given the argument above, the assessment of potential enrichment programs for fish in aquaculture is very difficult. For example, the use of a dark substrate has the potential to reduce levels of aggression in Arctic char (*S.*

alpinus, Hoglund *et al.*, 2002). Also, when red porgy (*Pagruspagrus*) were exposed to a crowding stressor in different color background tanks, fish on a light background color showed a reduced stress response when compared to those on a black tanks and background effects may be enhanced by color illumination (Rotllant *et al.*, 2003; Van der Salm *et al.*, 2004). Thus, species relevant changes to rearing environments may have clear welfare benefits. Feeding and environmental enrichment during rearing appears to affect behavior and may improve survival rates of hatchery reared Atlantic salmon (*S. salar* L., Brown *et al.*, 2003) and Chinook salmon (0. *tshawytscha*, Maynard *et al.*, 1996) when released to augment wild populations. Following release into the wild, steelhead trout (*O. mykiss*) reared in enriched environments containing submerged structures, overhead cover, gravel substratum, and underwater food delivery, showed altered competitive ability and aggression when compared to conventionally reared fish (Berejikian *et al.*, 2001). In terms of social dominance and aggressive behavior, fry raised in enriched environments were more similar to naturally raised fish than were fish raised in a baron hatchery environment (Berejikian *et al.*, 2000). Conventionally reared fish showed unnaturally high levels of intra-group aggression that did not translate to successful acquisition and defence of territories when released, whereas fry reared in a natural stream environment appeared to tolerate the presence of conspecifics. Fish reared in an enriched environment showed an intermediate behavioural profile in this respect. Whether such environmental manipulation can be used to alter the behavioural profile to improve welfare (*e.g.* reduce aggression) within fish farms remains to be seen and will surely be dependent on the social behaviours of individual species.

Conclusions

It is clear that the areas of welfare concern discussed do not exist in isolation but interact to affect the welfare of the individual. Stocking density, diet, feeding technique, and management procedures all have strong effects on stress levels, subsequent stress tolerance, health, and the presence of aggressive behavior. In turn, these effects feedback to one another to further influence welfare. Research into the interaction between stress physiology and behavior has also shown strong links between aggression and stress sensitivity (Overli *et al.*, 2005). Future investigations into the neuroendocrine control mechanisms involved and the existence of "coping styles" may provide further insight into how multiple factors combine to effect welfare within a given situation. The very nature of aquaculture demands high standards of management and stockmanship to avoid the adverse economic effects of stress and disease. However, knowledge of the effects of handling, transport, and food deprivation on stress physiology should be used to make species specific recommendations for best practice. Continued research into reducing stress response through diet, conditioning and perhaps selective breeding will help to improve welfare standards. Further research into the most effective treatment and/ or management of sea lice, fin rot, viral diseases, and non-infectious production related problems, is crucial to future improvements of welfare. While it has been shown that humane slaughter is possible, the key goal now is to implement these techniques commercially.

A complex matrix of factors influences the effect of stocking density and the relative importance of these factors is case specific. Therefore, the use of research into how environmental parameters, physiology, and particularly behavioral factors should be productive in improving welfare and studies on the farm must continue to be carried out to put laboratory findings into context.Whilebehavioral observations are useful as early indicators of poor welfare and health, caution should be taken in the interpretation of abnormal behavior and the lack of freedom to express natural behavior. Investigation into the origin of apparent abnormal behavior may improve welfare, and the use of choice test experiments within the aquaculture environment may help us to interpret physiological response to aquaculture stressors as well as the importance of expression of natural behaviors. Regardless of judgments on the ability of fish to suffer, these studies will help improve welfare. However, they may, in addition, provide further insight into the subjective experiences of fish, their capacity to suffer and subsequently the importance of fish welfare consideration.

Chapter 12
Problems of Undrainable Ponds

Introduction

In India ponds are relatively small and shallow bodies of impounded water with limited wind action. They may be called perennial if they retain water the year round or temporary/seasonal, if they do so seasonally. They may be further classified as drainable ponds and undrainable ponds, depending upon the drainage facility by gravity. Drainability imparts a very desirable feature to a pond and some authors prefer to call only the drainable type of ponds as fish ponds.

General Morphometry of Undrainable Pond

The water area is usually less than 2 ha and in case of small ponds with high embankments, wind action is limited. Oxygen and nutrients in the water column and at the sediment-water interface are transported by the limited convection currents. The water depth in most ponds remained less than 2 m. The photosynthetic processes are light-limited in these ponds and oxygen levels may not be sufficient for fish.

Chemical Environment and Nutrient Status

The slightly alkaline waters (pH 7.0–8.8) show total alkalinity values of 52–244 ppm.

The ranges of NH_4-N, NO_3-N and PO_4-P concentrations remain low. The level of most basic nutrients in the sediment interstitial water are thousand times higher than that of water column.

Various Problems and its' Solutions

Deficiency of Dissolved Oxygen

The most common and serious hazard in the undrainable pond is the depletion of dissolved oxygen level in the water. Gulping for air, especially during the early morning hours, is the most common behavioural symptom

Solution

Add freshwater in the pond by pumping it from nearby sources.Agitate the pond water by splashing, beating with bamboo poles, repumping with the help of a pump or using a mechanical churner/aerator. Apply potassium permanganate at the rate of 2–3 ppm.Lime at the rate of 200 kg/ha and rake up the bottom.

Pre-Stocking Management

Pre-stocking management aims at proper preparation of ponds to remove the causes of poor survival, unsatisfactory growth, *etc.* and also to ensure ready availability of natural food in sufficient quantity and quality for the spawn/ fry/fingerlings to be stocked. Pre-stocking part of the management involves the following sequential measures.

Eradication and Control of Aquatic Weeds and Algae

Aquatic weeds are unwanted plants that grow within the water body and along the margins. Unlike in temperate climate, the pond fish culture in tropics face serious problems due to weed infestation and frequent appearance of algal blooms. They remove a large quantity of nutrients from the water, which otherwise would go into the production of planktonic growth. Even the poor fish crop that is produced in weed chocked water is difficult to harvest. The fishes are subjected to stress due to dissolved oxygen depletion and wide fluctuation between the dissolved oxygen values of the day and night. Decomposition of the dead aquatic weeds further creates the oxygen problem. Dense growth of the submerged weeds restrict fish movement and interfere with fishing operations. Filamentous algae often get entangled in the gills of the fish and suffocate them to death. Floating weeds such as water hyacinth, pistia, *etc.*, very often cover the entire water surface cutting off light drastically, thus resulting in critical reduction in primary productivity of the pond. Common aquatic weeds creating problems in fish culture ponds are broadly classified according to their nature of occurrence, into four major groups. They are floating, emergent, submerged and marginal. In addition, algal blooms and mats also create serious problems in terms of dissolved oxygen and production of certain toxic materials in some cases. Aquatic weeds of common occurrence in undrainable ponds are grouped in the following table.

Common Aquatic Weeds in Undrainable Ponds

Groups of commonly occurring aquatic weeds, algal bloom and algal mats in undrainable pondsAquatic weeds, algal bloom and algal mats		
Groups	Scientific Name	Common Name
Floating	*Eichhornia crassipes*	Water hyacinth
	Pistia stratiotes	Water lettuce
	Salvinia cucullata	Water fern
	Spirodela polyrrhiza	Duck weed
	Lemna minor	Duck weed
Emergent	*Nymphea mexicana*	Banana water lily
	Nymphea tuberosa	Fragrant water lily
	Nelumbo spp.	Lotus
	Nymphoides spp.	Floating heart
Submerged	*Hydrilla verticillata*	Hydrilla
	Najas marina/minor	Najas
	Potamogeton crispus	Curly leaf pondweed
	Vallisneria spiralis	Eel grass
	Ottelia spp.	
Marginal	*Ipomea aquatica*	Ipomea
	Jussiaea spp.	Water primrose
	Typha anqustata	Cat-tails
	Cyperus spp.	Cyperus
Algal blooms	*Microcystis aeruqinosa*	Microcystis
	Anabaena	Blue green algae
Algal mats	*Pithophora*	Horse hair clump
	Spiroqyra	Filamentous algae

Control measures for all the above mentioned classes of weeds and blooms fall into four major categories, *viz.* preventive, manual and mechanical, chemical and biological. Any of these methods or at times a combination of methods may be taken up depending on the nature of infestation, pond condition, cost involvement and availability of required inputs.

Preventive Control

Taking into consideration the high cost of controlling aquatic weeds, certain preventive measures are to be followed to reduce the chances of their infestation. The preventive measures have to be taken well in advance. The measures include trimming of pond margins, dewatering and desilting of old ponds, uprooting or burning of dried marginal weeds during the summer and providing barriers to prevent the entry of floating weeds.

Manual and Mechanical Control

Manual removal of aquatic weeds is an age-old practice and holds good even today in rural areas. The free floating groups of weeds are either hand picked or dragged by wire or strong coir rope nets. In bigger ponds they should be removed part by part from the marginal areas and finally the centrally located weed mass is dragged towards the banks and lifted out. Certain small and light floating weeds such as spirodela, lemna, azolla, wolffia, *etc.*, are easily skimmed out by twisted straw ropes or fine meshed nets. The manual removal of submerged weeds from a heavily infested water body is relatively much more difficult. They are either pulled by hand or hand-drawn bottom rakes or uprooted with bamboo poles having a cross piece tied strongly at the terminal end. Repeated cutting of the aerial shoots and leaves of rooted emergent plants are also useful. Implements used for manual control are mostly hand scythes for cutting, and hand forks, strong nets and bamboo poles with terminal cross piece for twisting and uprooting (Fig. 35).Mechanical devices used for clearance of rooted submerged weeds are steel cables, cutting chains and diesel operated winches (Mitra, 1956).

Chemical Control

The manual removal of weeds from heavily infested large water bodies is difficult and time consuming. Under such conditions certain commercially available chemicals (herbicides) can provide an efficient means of eradication of undesirable aquatic plants. Total kill and disintegration of weeds can be achieved by this method ensuring full return of the nutrients back to pond soil and water for production of natural fish food. As a matter of fact there is not a single chemical known so far which can eradicate all types of weed infestation. Therefore, one must know the weeds and its species, appropriate herbicide and its rate and time of treatment. In larger ponds where dense infestation covers a substantial portion of the water, the herbicide should be applied part by part if the pond is already stocked with fish. As discussed earlier most herbicides are selective in nature and when applied to a mixed population of weeds, growth of some tolerant weeds may be encouraged at the cost of susceptible ones; likewise, when surface or floating weeds are destroyed, the submerged weeds develop. Under such conditions subsequent application of appropriate herbicide should be taken up.

Floating Weeds

Water hyacinth is one of the most important weeds of this group. Depending on its degree of infestation, they are categorized in three groups, *viz.* small, medium and big, based on their wet weight per unit area. The recommended doses of the herbicide 2–4-D are 2,7 and 12 kg/ha for small (13 kg/m^2), medium (23 kg/m^2) and big (35 kg/m^2) (Ramchandran, 1969; Patnaik and Das, 1983). Addition of a detergent (0.2 per cent concentration) to the aqueous solution gives better results. The dilution for better coverage has been estimated at 400 l/ha. The foliar spray (spraying over the leaves) is undertaken with the help of a foot pump/hand pump sprayer with a three-action nozzle. Field application of herbicide, especially towards the interior of thick water hyacinth infestation, is a difficult task. In such cases a pair of stout bamboo poles should be laid on the top of the infestations so that the operators can walk over them. Normally, the complete kill of plants takes around 25 days. This chemical is available in two suitable forms as sodium and amine salt.

Hand Tools Used for Manual Control of Aquatic Weeds

Water lettuce which often causes a serious problem in fish ponds can be controlled with 0.1–0.2 kg of paraquat/ha. This infestation could also be controlled by foliar spray of aquous ammonia (1 per cent) at the rate of 50–75 kg/ha along with 0.2 per cent of any commercially available detergent as a wetting agent.

The aquous ammonia is broadcast as foliar spray over the infestation with a foot pump sprayer and a small funnel–shaped sprinkler 3–4 cm in diameter, provided with 10 pin-sized holes pierced on the diaphragm covering the mouth of the funnel. The stem of the sprinkler is connected to the sprayer through a 30 m long polyethene tube, so that the sprayer is kept on the shore and only the sprinkler is taken inside the infested area in a boat. The area to be treated inthe field is divided into small plots (20–30 m^2size) and solution is sprayed at the rate of 5000 1/ha

Salvinia forms a thick surface mat in ponds and can be conveniently controlled by the application of foliar spray of paraquat at the rate of 1 kg/ha. Usually it takes 30–40 days for the weeds to be killed and settled in the pond.Smaller floating weeds, *e.g.* Spirodela, Lemna and Azolla can also be cleared with 0.1 kg/ha of paraquat.

Emergent Weeds

Water lily, lotus, and floating heart can be cleared by spraying the herbicide 2–4-D at the rate of 8–10 kg/ha with detergent (0.25 per cent). The chemical is diluted at the rate of 300 l/ha and sprayed through a foot pump sprayer.

Submerged Weeds

Ottelia, Vallisneria, Hydrilla, Najas, Potamogeton and *Ceratophyllum* can be controlled by paraquat at the rate of 3–4 ppm within two weeks. It can also be controlled by application of anhydrous ammonia at the rate of 15–20 ppm.

Marginal Weeds

Ipomea, Jussiaea, etc., could be controlled by spraying the herbicide 2–4-D at the rate of 8 kg/ha.

Algal Blooms and Mats

Due to overdose of fertilizers or enrichment of the water through treated sewage or agricultural fertilizer, the minute algal cells multiply fast turning the pond water bright green or sometimes brickred. Some of the more harmful blooming algae are microcystis, anabaena and euglena. A number of chemicals have been employed to control these algal blooms. Copper sulphate is perhaps the oldest and a very widely used algicide. The recommended doses are 0.2 to 1.0 ppm, but it is not very effective in ponds having high pH (pH above 8.6), Microcystis bloom is cleared with 0.3 to 0.5 ppm of Diuron. Simazine also clears the bloom in 16–20 days and the rate of application is 0.3–0.5 ppm. Both the chemicals do not have harmful effect on fish. It has been observed that the sudden kill of blooms is likely to cause oxygen depletion which might cause mortality of fish. In order to avoid this a prophylactic dose of diuron (0.1 ppm) should be applied in the very early stage of bloom development. Usually the chemical is sprayed over the affected portions of the ponds. The common mat forming algae which occur in fish ponds are Spirogyra, Pithophora, Oedogonium and Cladophora. Although repeated netting can reduce the infestation to a considerable extent in nursery and rearing ponds, application of Diuron at the rate of 0.3–0.5 ppm is recommended. Various chemicals and the dose of application is summerised in the ready reckoner given below.

Biological Control of Aquatic Weeds

Another important controlling method is by introduction of weed-eating fishes.Common carp, gourami, tilapia, pearl spot, the grass carp and a species of \ puntius are the fishes of known weed-eating habits. Grass carp is the most effective biological control agent against most of the submerged and floating weeds except the water ferns. Grass carp normally consumes choiced aquatic weeds, at least 50 per cent of their body weight in a day. About 300–400 fish, each of about 0.5 kg weight, are enough to clear 1 ha of Hydrilla infested water body in about a month. Normally Hydrilla infestation density ranges from 5–25 kg/m^2 (Alikunhi and Sukumaran, 1964).

Eradication of Unwanted Fish

Predatory fish prey upon the spawn, fry and fingerlings of carps and the weed fish compete with carp for food, space and oxygen. Therefore predatory and weed fish should be completely eradicated from nursery, rearing and stocking ponds before these ponds are stocked. The commonly encountered predatory and weed fish in undrainable ponds are listed below.

Absolute removal of these unwanted fish by thorough and repeated netting is not possible and hence dewatering and poisoning the pond are the only alternative methods. If situation permits, dewatering should be the preference as it ensures complete eradication of unwanted fishes and disinfects the pond bottom. Dewatering also offers the opportunity to desilt the pond bottom. However, where it is not possible, which is true in most situations, the pond should be treated with fish poison. From an economic point of view the poisoning should be done during pre-monsoon season when the water level is usually low, requiring the minimum quantity of poison material. The date of poisoning, however, should be fixed about three weeks before the anticipated date of stocking. Seasonal ponds which dry up during summer months need not be treated with fish toxicants.

Ready Reckoner for Chemical Control of Aquatic Weeds

	Weeds	*Herbicide*	*Brand Name*	*Dose*	*Additives*
1.	Water hyacinth pistia and other floating weed	2-4-D (sodium salt/amine salt)	Taficide Hexamar Fernoxone	2–12 kg/ha	0.1–0.2 per cent detergents
2.	Lotus, waterlily trapa, *etc.*	-do-	-do-	8–10 kg/ha	0.25 per cent detergent
3.	Marginal weeds	-do-	-do-	8 kg/ha	0.25 per cent detergent

	Weeds	Herbicide	Brand Name	Dose	Additives
4.	Salvinia	Paraquat	Gramoxone	1.0 kg/ha	-
5.	Pistia, spirodelalemna, azolla, *etc.*	-do-	-do-	0.1–0.2 kg/ha	0.1 per cent detergent
6.	Submerged weeds (Ottelia, vallisneria, hydrilla, najas, potamogeton, ceretophyllum, *etc.*)	-do-	-do-	4 ppm	-
7.	Pistia	Aquous ammonia	Dry ammonia gas	50–70 kg/ha	0.2 per cent detergent
8.	Submerged weeds	Anhydrous ammonia	Dry amomia gas	15–20ppm	-
9.	Rooted submerged weeds	Copper sulphate	-	35kg/ha	-
10.	Algal blooms/mats	Copper sulphate	–	0.2–1.0 ppm (not very affective at high pH	-
		Simazine	-	0.3–0.5 ppm	-
		Diuron	Karmex	0.3–0.5 ppm	–

Common Weed Eating Fish and the Weeds of their Preference

Fishes	Names	Feed upon
Common carp	*Cyprinus carpio*	Tender shoots
Gaurami	*Osphronemus goramy*	Tender shoots of submerged weeds and filamentous algae
Pearl spot	*Etroplus suratensis*	Filamentous algae
Grass carp	*Ctenopharyngodon idella*	Submerged weeds *e.g Hydrilla Najas, Ceratophyllum, Potamogeton, Ottelia* and duck weeds
Silver carp	*Hypophthalmichthys molitrix*	Algal bloom

Common Predatory and Weed Fish of Undrainable Ponds

Predatory Fish	Weed Fish
Channa spp.	*Puntius* spp.
Clarias batrachus	*Oxygaster* spp.
Heteropneustes fossilis	*Gudusia chapra*
Pangasius pangasius	*Amblypharyngodon mola*
Mystus spp.	*Laubuca* spp.
Ompok spp.	*Esomus danricus*
Wallago attu	*Osteobrama cotio*
Glossogobius giuris	
Mastocembelus spp.	
Amphipnous cuchia	

Recommended Doses of Fish Poison

Poison	*Dose (kg/ha/m)*
Bleaching powder	350 – 500
Mohua oil cake	2 500
Anhydrous ammonia	20 – 30
Powdered seed of Croton tiqlium	30 – 50
Root powder of Milletia pachycarpa	40 – 50
Seed powder of Milletia piecidia	40 – 50
Seed powder of Barrinqtonia acutanqula	150
Tea seed cake (*Camellia sinensis*)*	750

Fish Toxicants

Although a number of chemicals and plant derivatives are available in the market which are poisonous for fish, only a limited number of such toxicants are safe and suitable for fish culture purposes. Based upon the following criteria a suitable fish poison is selected.

- Poisoned fish should be safe for human consumption
- Least adverse effect on the pond biota
- Toxicity period should be of short duration
- Should not have residual effect
- Easy commercial availability
- Simplicity of application
- Cost considerations.
- Mohua oil cake, bleachng powder and ammonia are considered suitable.

Application of Toxicants in Ponds

Mohua Oilcake

Of all the fish poisons of plant origin, the most extensively used fish toxicant in undrainable ponds is oil cake of Mohua (Basia latifolia). It kills all the fish species within a few hours when applied at the rate of 250 ppm (CIFRI, 1968). It contains about 4–6 per cent of active ingredient, the saponia, which on dissolving in water haemolyses the red blood cells and thus kills the fish (Bhatia, 1970). The required quantity of mohua oilcake should be soaked in water and uniformly broadcast over the entire pond surface. Following this operation, repeated netting should be done to ensure proper mixing of the poison and removing the affected fishes which are suitable for human consumption. The toxicity of doses up to 250 ppm lasts for about 96 hours (Jhingran and Pullin, 1985) and subsequently it serves as organic manure in the pond. It should be applied at least two weeks before stocking the ponds.

Bleaching Powder

Bleaching powder or Calcium hypochlorite ($CaOCl^2$) is another practical and safe fish toxicant. It kills all the predatory and weed fish of the pond when applied at the rate of 25–30 ppm (Tripathy *et al.*, 1980). However, during storage, significant chlorine content is lost and hence it is always safer to use the commercially available bleaching powder at the rate of 35–50 ppm or 350–500 kg/ha/m of water. Fish kill occurs within 1–3 hours and the toxicity lasts for 3–5 days. Plankton and benthic fauna start developing from the 7th or 8th day after treatment. Chlorine content of the bleaching powder thoroughly disinfects the pond which is essential in undrainable ponds where disinfection by sun drying is not at all possible. Disinfection of the pond is one of the essential measures for maintaining proper health condition of the fish. Besides, it also satisfies the lime requirement of the pond soil.

The method of application is also relatively simple. The powder is mixed with water and uniformly spread over the entire water surface. Distressed and dead fish are removed by netting. Chlorine killed fish are safe for human consumption.

Ammonia

Anhydrous ammonia when applied at the rate of 20–25 ppm kills the predatory and weed fishes. Besides, it also controls the aquatic weeds and later acts as nitrogenous fertilizer. Toxicity of ammonia lasts for 4–6 weeks.

The nursery ponds require subsequent poisoning for selective killing of the larger planktonic copepods. These copepods are predatory in nature and instead of serving as food for the delicate spawn and early fry, they attack and prey upon them resulting in poor survival. For this reason 4–5 days prior to stocking of spawn, the pond should be treated with malathion at the rate of 0.25 ppm (active ingredient) for selective killing of the planktonic copepods. This treatment significantly increases the survival in nursery ponds (Kumar *et al.*, 1986). Such treatment is not required in rearing and stocking ponds.

Calculation of Dose

The required quantity of poison can be calculated using the following formulae.

For rectangular ponds

Required amount of poison in kg.

For circular ponds

= Required amount of poison in kg.

Eradication of Predatory Insects

Many aquatic insects in their larval and/or adult stages, prey upon fish hatchlings and fry and also compete with them for food. The common insect predators are beetles, bugs and dragonfly nymphs. Among beetles, diving beetle (Cybister), water scavenger beetle (Sternolophus) and whirling beetle (Gyrinus) are more dangerous forms. Back swimmers (Anisops) appear in swarms in manured ponds during rainy season and cause heavy damage. Other predatory members of this group are water scorpion (Laccotrephes), giant water bug (Belostoma) and water stick insect (Ranatra). Dragonfly nymphs are highly predatory on carp spawn. Proper prepration of nursery ponds for stocking with spawn thus also aims at total eradication of such predatory insects. The basic method is to apply a thin oily film over the pond surface which chokes the respiratory tubes of aquatic insects. The spawn and fish food organisms remain unaffected. Some of the common treatment methods are presented in the following table.

Pond treatment methods for eradication of predatory aquatic insects

Treatment Method	*Dose/ha*
Soap oil emulsion	56 kg vegetable oil + 18 kg soap
Diesel oil	50 – 60 1
Kerosene oil	80 – 100 1
Turpentine oil	75 1
Diesel emulsifier	Diesel 50 1 * emulsifier 37.5 ml + water 2 1.

Fertilization of Ponds

Fertilization schedule involving both organic and inorganic fertilizers starts 10–15 days prior to stocking and is prepared on the basis of nutrient status and chemical environment of the pond soil and water.

Basis of Fertilization

In undrainable ponds where the frequent change of water is a remote possibility, the physico-chemical properties of pond water governing the biological production cycle are more or less a reflection of the bottom soil. The organic and mineral constituents of the soil play their part in releasing the required nutrients into water for pond productivity through chemical/biological processes. Pond bottom soil also provides suitable substrates and necessary environment for the microbial decomposers - the living fertilizer factory of the pond. Thus it is the soil condition and its nutrient status that forms the basis of pond fertilization by using either organic manure or inorganic fertilizer or a combination of both. Important characteristics of pond soil which influence fertilizer use is briefly described here.

Texture of the Soil

The texture of pond soil, *i.e.* mechanical composition of the soil comprising sand, silt and clay and organic matter content, basically influences the economy of both inherent and added nutrients. Sandy and very clayey soil are not desirable as in the former the nutrients are lost due to heavy leaching; while in the latter, due to high adsorption capacity, the nutrients from the water are trapped. Clay minerals and organic matter of the bottom mud are both colloidal in nature and thus exhibit colloidal properties like adsorption and cation exchange phenomenon. Sandy soils, on the other hand are low in colloidal substances and also deficient in organic humus. These are important considerations for deciding the application of fertilizers and manures.

Soil pH

As in water, pH of soil is also one of the critical factors affecting pond productivity. Under anaerobic condition the decomposition of organic matter is slow and the products of decompositions are mainly reduced compounds and short chain fatty acids thus making the soil strongly acidic. Soil pH also influences transformation of phosphorus into available forms and controls the adsorption and release of essential nutrients at the soil-water interface. Both for soil and water a slightly alkaline pH is considered favourable for fish ponds.

Availability of essential mineral nutrients such as phosphate, nitrogen, potassium, carbon and calcium is a consideration which determines the quality and quantity of fertilizers to be applied. Nitrogen is required in large quantities as it is the basic and primary constituent of protein and chlorophyll. Although, phosphorus is required in a small quantity compared to nitrogen, it is considered as the single critical element for maintaining aquatic productivity. Banerjee (1967) classified the undrainable ponds into low, medium and highly productive groups, on the basis of their nutrient status considering mainly nitrogen, phosphate and organic carbon.

Fertilization Schedule

Proper analysis of soil and water is essential before deciding on the fertilization schedule. Detailed recommendations have been made in the chapter on pond environmental monitoring.

Nutrient status of high, medium and low productive ponds

Available Nutrients

Productivity Level	*pH*	*N (mg/1000 g soil)*	*P_2O_5 (mg/1000g soil*	*Organiccarbon (Per cent)*
High	6.6– 7.5	50 or more	6 – 12	1.5 or more
Medium	5.5– 6.5	25 – 49	3 – 5	0.5 – 1.4
Low	Below 5.5	Less than 25	Less than 3	Less than 0.5

Liming

Diurnal changes in pH values ranging from pH 5 during the night and pH 11 during the day due to community respiration and photo-synthesis is a common experience but such wide variations impose stressful conditions for the fish. An adequate level of calcium in the pond provides a buffering system in the pond.

Liming helps to raise the total alkalinity level and consequently the reserve CO_2 will increase the availability of carbon for photosynthesis by raising the bicarbonate concentration in water. This raised level of reserve CO_2 will also prevent biological decalcification.

Mechanism of Buffering Action of Lime

Depending on the pH of the soil, the dose of the liming should be Adjusted as per the following table. Alkalinity can also be used as an indicator of the need for lime in fish ponds. The total dose of lime calculated as per the table, need not be applied at one time. It may be divided into 3–4 doses and the first dose may be applied about a week prior to the manuring of the pond. It helps in faster mineralisation of organic matter in the pond sediment and acts as a prophylactic agent as well. The same dose is applicable for nursery, rearing and stocking ponds. However, as and when needed during the culture period, additional doses of lime can also be applied.

Requirement of Lime for different Types of Pond Soils

Soil pH	*Soil type*	*Requirement of Lime (kg/ha)*
4.0 – 4.9	Highly acidic	2000
5.0 – 6.4	Moderately acidic	1000
6.5 – 7.4	Near neutral	500
7.5 – 8.4	Mildly alkaline	200
8.5 – 9.5	Highly alkaline	Nil

Manuring

Organic manuring besides being important as means of adding the nutrients, is also equally important for improving the soil texture. A combination of organic manures and inorganic fertilizers is considered more effective than using either of these alone. However, in nursery ponds, use of mineral fertilizers is not recommended as the application may cause blooms of algae which may persist and may harm the young fry. Cow dung at an initial dose of 10 000 kg/ha may be applied in the nursery ponds about two weeks prior to anticipated stocking. If the pond is poisoned by mahua oil cake, then the dose should be restricted to 5 000 kg/ ha. If two or more crops of fry are to be produced during the season from the same nursery ponds, then the pond should be fertilized with 2 000 kg/ha of cattle dung about a week before each subsequent stocking. In case of poultry manure the dose should be only 33 per cent of the cattle dung. Rearing ponds are initially manured with the raw cattle dung about two weeks prior to stocking. The rate of application is between 5 000 – 7 000 kg/ha in 5 instalments. If the pond is treated with mohua oil cake then the dose of organic manuring is reduced to half. Dose of inorganic fertilizers may be regulated as per pond soil productivity determined by detailed analyses. In the absence of soil testing facilities a general recommendation should be followed. In such cases inorganic fertilizers are applied at the rate of urea 140 kg/ha and triple superphosphate 60 kg/ha in 4–5 instalments.

In stocking ponds a combination of organic and inorganic fertilizers is considered more effective. Initial manuring with organic manure at the rate of 20 per cent of the total requirement is done 15 days prior to stocking and the remaining 80 per cent of the organic manure is applied in 11 equál monthly instalments during the rearing period. However, if mohua oil cake is applied earlier, the initial manuring is not essential. The total quantity of inorganic fertilizers to be applied is decided according to soil type tand applied in equal monthly instalments. The monthly instalments of organic and inorganic fertilizers are applied alternately allowing a gap of a fortnight between the two applications. Nitrogenous fertilizers are selected on the basis of soil pH.

Amount of Fertilizers Required for Ponds having High, Medium and Low Levels of Productivity

	Pond Productivity Levels		
	High	*Medium*	*Low*
Rate of application of fertilizer (kg/hg/y)			
Cattle dung	5000–8000	8000–10000	10000–25000
Urea (43–45 per cent)	112–155	156–225	226–260
Ammonium sulphate (20.5 per cent)	225–330	-	-
Calcium mmonium nitrate (20.5 per cent)	-	350–500	501–650
Single super phosphate (16–20 per cent)	156–219	220–315	316–405
Triple super phosphate (40–45 per cent)	54–75	76–110	111–145

In the absence of proper soil testing facilities fertilization schedule in stocking ponds may be followed as per the following table.

Generalized Fertilization Schedule for Stocking Ponds (CIFRI, 1985)

Item		*Quantity(kg/ha)*	*Periodicity of Application*
A.	Cattle dung	2 000	Initial dose
	Cattle dung	1 000	Monthly
B.	Urea (pH 6.5–7.5) or	25	Monthly
	Ammonium sulphate (pH above 7.5) or	30	Monthly
	Calcium ammonium nitrate (pH 5.5–6.5)	30	Monthly
C.	Single super phosphate or	20	Monthly
	Triple super phosphate	8	Monthly

Stocking of Grow-out/Stocking Ponds

After proper preparation, the pond should be stocked with 100–150 mm long fingerlings of desired carp species. In case the fingerlings are not available, the pond can also be stocked with advanced fry or early fingerlings in absolutely predator-free ponds. The stocking rate depends primarily upon the volume of water and on the oxygen balance of the pond. Quality of available natural fish food in the pond and the capacity of the farmer to provide supplementary feed, are also matters for consideration. Usually a pond having average water depth of 1.5–2.5 m should be stocked at the rate of 5 000 fingerlings/ha. The volume of water available for fish in an undrainable pond should not be less than 2 m^3/fish if there is no provision of artificial aeration. In composite fish culture, rearing of six species of carps, *viz.* catla (Catla catla), rohu (*Labeo rohita*), mrigal (*Cirrhinus mrigala*), silver carp (*Hypophthalmichthys molitrix*), grass carp (*Ctenopharyngodon idella*) and common carp (*Cyprinus carpio*) is considered to be the ideal combination. However, depending on the availability of quality fingerlings of these carp species, three or four species combinations can also be taken up. Ratio of different species in the combination is also equally important. However, there are certain general guidelines for selecting species combinations.

Different Species Combinations and their Stocking Ratios for Composite Fish Culture

Species Combination	*Surface Feeder*		*Column Feeder*	*Bottom Feeder*		*Macrophyte Feeder*
	Catla	*Silver Carp*	*Rohu*	*Mrigal*	*Common Carp*	*Grass Carp*
3	40	-	30	30	-	-
4	30–40	-	20–30*	15–20	20–25	-
6	10–15	20–30	15–30*	15–20	20–25	5–15

*Lower units in shallow ponds.

Availability of weed in the pond or in the vicinity decides the stocking density of grass carp. In older ponds where the soft sediment layer of the pond bottom is usually very thick and anaerobic in nature, the ratio of bottom feeder and especially the common carp should be kept at a higher level. Likewise, the relative density of column feeder-rohu should be kept on the high side in deeper ponds than in shallower ponds, whereas ponds showing consistently higher zooplankton population should have a higher ratio of surface feeders. Based on the performance of individual species in the combination and availability of seed, combinations can be modified in subsequent years. Silver carp, however, should be stocked 1 or 2 months later. Interspecies competition for food between catla and silver carp to some extent is the key point for such differential stocking. The stocking pond also should have a desired level of plankton population of about 30–50 ml/m^3.

Method of Stocking

Stocking of spawn, fry and fingerlings should be done very carefully to avoid any post-stocking mortality due to shock or infections. To minimize post-stocking mortality the fry/fingerlings should be slowly and gradually acclimatized to the temperature and quality of the water in the stocking pond. To do so, open the mouth of the seed transport bag/container and gradually add the pond water in phases and after 15–20 minutes slowly dip and tilt the bag/container in the pond so that the spawn/fry/fingerlings are free to swim out. Stocking should preferably be done in the cool evening hours. Apply prophylactic treatment to seed prior to their release so as to avoid any post-stocking infections.

Post-stocking Management

Post-stocking management involves harnessing the pond productivity in the form of natural fish food, maintenance of pond environment congenial to the cultivated fish and fish husandry, mainly feeding and health care.

Feeding

Soon after stocking, the fish start grazing natural food available in the pond irrespective of their stage of life cycle. Spawn feeds voraciously on plankton. Therefore, immediate steps must be taken for providing supplementary feed. In the case of nursery ponds where spawn are reared for about a fortnight up to fry stage, supplementary feed is broadcoast on the pond surface in the form of fine powder daily in the morning hours at prescribed rates.

Rates of Daily Supplementary Feeding at Various Stages of Culture

Stage	*Daily Feeding Rate*
Spawn to fry	4–8 times of the initial body weight
Fry to fingerlings	50–100 per cent of the initial body weight
Growers	1 – 2 per cent
Brood fish	1 – 3 per cent

The following schedule of feeding should be followed for nursery ponds

Feeding Schedule for Nursery Ponds

Period (Day from the date of stocking)	*Rate of Feeding*	*Amount of Feed for 0.1 million of Spawn*
1 – 5	4 times the total initial weight	560 g/day
6 – 12	8 times the total initial weight	1 120 g/day
13	No feed	-
14	Harvesting	

At the time of stocking, the spawn of 0.65–0.75 cm average length weigh about 0.0014 g each, and a mixed collection of 0.1 million weigh about 140 g.Grass carp is fed its preferred aquatic vegetation or green animal fodder as per the following table.

Feed for Grass Carp during Various Stages of Life Cycle

Stage	*Feed*
Fry (1.7 – 3.9 cm)	Soft macrophytes suchas *Azolla, Wolffia, Lemna* and *Spirodella, etc.*
Fingerlings (4.0 – 15.0 cm)	Hydrilla, Ceratophyllum, Vallisneria, Najas, Chara, *etc.*, in addition to those mentioned above.
Juveniles/Adults (above 15.0 cm)	In addition to above, green animal fodder such as barseem, napier, hybrid napier, elephant grass, tender leaves of vegetables and trees such as soobabul, drumstick, *etc.*

Feeding Enclosure for Grass Carp

The form in which the supplementary feed is given is also important. In the nursery ponds the feed should be provided in finely powdered form and may be broadcast over the pond surface. In the case of rearing, stocking and brood stock ponds, the supplementary feed mixture should be mixed with enough water to make a dough and applied into feeding trays fixed in the ponds. Better results can be obtained if the feed mixture is pelletized and fed to fish.The pellets may be of the sinking or floating type, but both types should be water stable. The sinking type of pellets are put in feeding trays fixed in the pond.

The standing crop of fish is estimated every month on the basis of sample netting for growth and health check and feeding schedule is adjusted accordingly. Periodical netting should be done strictly on a monthly basis and with the help of hand nets and spring balance. the average weight of each species should be recorded. The average weight of individual species, monthly increment in weight, total standing crop and amount of feed to be given should be estimated on the basis of data thus available.

The feeding tray should be cleaned daily before the application of fresh feed. Fish normally stop feeding if they are sick or the temperature is far below normal. In such situations a proper health check is required and the feeding rate is adjusted. Grass carp should be fed until they stop eating. Usually they consume aquatic vegetation, about 50 per cent of their body weight on a daily basis.

Periodic Fertilization

The next step in post-stocking management is the periodic fertilization which ensures replenishment of nutrients and consolidation of the energy base for microbial decomposition activities. The desired total quantity of fertilizers are best applied in small equal doses at periodical instalments throughout the rearing period so as to ensure maximum utilization of these fertilizers. The mode, sequence and timing of application of fertilizers are important or achieving best results. Lime should be applied first followed by the organic manure and finally the inorganic fertilizers an the same order is followed subsequently. These fertilizers should be applied only when the physical conditions of the water are most suitable such as plenty of sunlight, adequate oxygen, optimum temperature, adequate water level and low wind velocity. Turbid water with a high content of suspended solids are not preferred. Fertilizer should be sprayed or distributed properly over the water surface during the day time when the top layer of water is warmer and lighter. Inorganic fertilizer application must be stopped temporarily when the nitrate and phosphate content of water show a level of 0.5 ppm or above at any stage during the periodic pond environment monitoring. Similarly, organic manuring may also be stopped if the soil organic carbon level goes beyond 2 per cent. However, normal application may be resumed after the specific nutrient level goes down. Care should be taken to see that the phosphatic fertilizers dissolve properly in the water since powdered orgranular fertilizer may often solidify after coming in contact with water. It is more effective if doses are divided further so that application is more frequent. The results are encouraging when organic manures are applied in daily doses in pons. The desired amount of cattle dung is mixed with water and uniformly spread over the entire pond surface. In nursery ponds the first manuring is done two weeks prior to stocking and if more than one crop is nursed, fresh manuring should be done a week prior to every subsequent stocking.

Hand Net and Spring Balance

A periodical fertilization schedule is summarized in Table. The rate of fertilization by organic and inorganic manures has already been discussed.

Periodicity of Fertilization in Nursery, Rearing and Stocking Ponds

Ponds	*Manure*	*Periodicity*
Nursery ponds	Organic manure	3 weeks
Rearing ponds	Organic manure and Inorganic fertilizer	3 weeks - daily
Stocking ponds	Organic manure and Inorganic fertilizer	Monthly

Pond Environmental Monitoring

General Considerations

Proper pond management involves a regular and steady supply of nutrient for sustained production of fish food organisms. The supply of nutrients could be from within the pond itself or from outside. It is also required to regulate the physico-chemical parameters of the pond ecosystem within the safe tolerance limits of the cultured fish species. This necessitates periodical monitoring of pond environment and taking corrective measures in time. Olah and Sinha (1984) have developed a practical monitoring system of perennial undrainable ponds which offer the monitoring of basic architecture and production processes of such pond ecosystems in tropical monsoon lands. The system needs simple instrumentation, little working time and labour and reveals sufficient information about the actual nutrient level of pond sediment and water. Most of these parameters can be easily measured at the pond site while some require laboratory facilities. The monitoring system gives reliable guidelines for fish farmers to optimize fish production.

Parameters to be Monitored

It is essential for extension workers to name and code-number the ponds in their area. Such coding may be based either on postal district/unit/village farmer's name, *etc.* The fish farmer should record the following information on his fish farm:

Nature of pond: Perennial or seasonal; nursery pond, rearing pond or stocking pond.

Water area: Measurement of the water area is essential in order to know the size of the pond for proper fish stocking and quantifying the production processes. This can be done easily with the help of a bamboo pole of known length.

Age: Age is one of the most important parameters, since it has direct relevance with the productivity of the pond which usually varies from one year to several hundred years.

Management: Management status should record the existing management techniques and its level (intensive or extensive). The species of fish present, details of culture activities, stocking structure and density, fertilization, feeding, harvesting, marketing, *etc.*, need to be recorded. To obtain qualified data on the organic carbon and biogenic nutrient load it is necessary to know the number of livestock and human population associated with the particular pond.The fish farmer should also monitor the following parameters on a routine basis.

Water colour: The visual colour of the pond water is a simple but important reflection of the basic production processes.

Water transparency: Water transparency measured with a Secchi disc is intended to quantify the result of those processes which determine and modify the visual colour. However, a low transparency may result either from high turbidity alone or from dense algal population and thus cannot reflect the correct trophic or production level of the water. However, the Secchi transparency readings together with the visual colour provide valuable information on the productivity of the water.

Water depth: The primary water source is usually the rainfall during the monsoon. After the rainy season the water level gradually decreases which results in a very shallow water column by the end of the dry season. The water depth can be measured with a 4–5 m long bamboo pole fitted at its base with a wooden disc of 25 cm dia.

Soft sediment depth: A soft sediment layer is usually present in the pond bottom. The depth of this layer can be measured with a 6–8 m long bamboo pole having a wooden disc of 10 cm dia at its base.

Solid sediment depth: In older ponds, in addition to the soft sediment layer, a solid sediment layer with a low water content is also present. The thickness of the layer can be measured with a 6–8 m long bamboo pole with a sharp end. The total thickness of the soft plus solid sediment layers has a direct relation to the age of the fish pond, at times the sediment layer measures more than 2 m. Such thick sediment, having a rich nutrient content, is anaerobic in nature with slow bacterial decomposition and mineral cycling rates. This should be properly utilized for fish culture.

Chemical environment in the water column: The water is chemically characterized by pH, alkalinity, NH_4-N, NO_3N and PO_4-P measurements following standard methods. Normally the pH and alkalinity do not change from pond to pond on the same types of maternal soil. The measurements of NH_4-N, NCO_3-N and PO_4-P indicate the basic inorganic nutrient status of the pond.' Simple chemical parameters such as dissolved oxygen and pH may be measured using field kits. Slightly alkaline water (pH 7.0–8.5) and oxygen levels of 6–9 ppm indicate optimum condition.

Dawn oxygen: Fish ponds usually exhibit wide fluctuations in the dissolved oxygen content from day to night. This diurnal oxygen fluctuation is normally measured to calculate the community metabolism of the whole pond while quantifying the production and respiration processes in the ecosystem. A single measurement just before sunrise would be an important indicator of the risk of fish kill due to oxygen depletion. Desirable ranges of various pond environment parameters are presented in Table.

Desirable Ranges of Pond Water Quality Parameters

Parameters	Desirable Range
Water colour	Greenish brown
Transparency	25 – 50 cm
pH	7.0 – 8.5
Dissolved oxygen	5.0 ppm
Free carbon dioxide	15.0 ppm
Inorganic nitrogen	0.2 ppm
Inorganic phosphorus	0.2 ppm

A simple schedule for monitoring the important parameters is presented in the above table.

Fish Health Monitoring

In most of the situations, cultured fish are healthy even in the continuous presence of pathogens. However, when environmental stresses occur and the balance shifts in favour of the disease, the characteristic pathogens flourish. Under such circumstances if the fish fail to adjust adequately or if corrective measures are not taken timely, outbreak of diseases may occur. By resolving environmental problems and applying effective therapeutics, the original balance between the host and the pathogen may be restored. Thus a disease outbreak may often be a symptom of environmental imbalance and it gives a distress signal so that the adverse environmental conditions may immediately be corrected to prevent fish losses. The approach to health care in composite fish culture in undrainable ponds is essentially one of management of ecosystem and fish husbandry.

Host-Pathogen-Environment Linkage

Susceptible fish, the virulent pathogen and the aquatic environment in which they encounter each other are the three contributing factors in fish disease outbreaks (Snieszko, 1974). The fish itself possess a varied and complex defense system, the immune system, the potency of which determines the susceptibility or resistance to the particular pathogen under a particular circumstance. Several environmental components effectively influence the normal immune mechanism of the fish when their value exceeds the normal tolerance limits. A virulent pathogen, when present in the surrounding, is usually capable of causing an infectious disease to fish under stress. The causative agents of the disease and their fish hosts carry on their struggle in the aquatic environment and the environmental parameters which influence this encounter may shift the balance from one side to the other and often determine whether the host will overcome the infection or the pathogen will flourish.

Some of the infectious and parasitic agents can survive only in live fish, and in such cases the disease transmission is from fish to fish. Such disease-producing agents are true pathogens. Others are extremely adaptable organisms which can survive outside the fish and cause infections whenever fish are weakened or otherwise predisposed to disease due to environmental stress. Most of the fish disease agents belong to this category.

Health Monitoring Programme

Health protection of cultured fish is considered to be one of the most important aspects of modern aquaculture systems including the composite fish culture which requires a programme basically to check the health status of the fish quite frequently and employment of fish health management measures. This enables timely detection of any disease outbreak and taking up proper treatment measures at the initial stage. Otherwise, in advanced stages of the disease, control and treatment measures do not provide economical and effective.

A fish health monitoring programme should consist of the following components:

1. Daily observation of fish in each pond.
2. Sampling and examination of fish at regular intervals for health check and diagnosis of the disease if any.
3. Monitoring of pond quality and sanitation.
4. Sampling and examination of fish at the onset of distress, disease outbreak or mortality.

Effects of Environmental Changes on Fish-Pathogen Relationship

The sampling for health check of fry and fingerlings should be done at weekly and fortnightly intervals respectively, while in composite fish culture ponds it should be at least once a month. A thorough health check of fry/fingerlings is required 1 or 2 weeks before netting out for stocking in grow-out ponds or before transfer to another pond. Such an examination will provide sufficient info rmation for planning.

Diseased fish may exhibit either or both physical and behavioural signs, the most common of those are listed below:

Behavioural Signs

1. Slowing down or a complete stoppage of feeding;
2. Loss of equilibrium, swimming erratically or in spirals;
3. Surfacing for gulping air and scraping against the floor and sides of the pond.

Clinical Symptoms

- ✰ Excess mucous secretion;
- ✰ Change in normal colouration;
- ✰ Erosion of scales, part of fins, skin, *etc.*
- ✰ Decolouration or paling of gills;
- ✰ Abdominal swelling;
- ✰ Bulging of eyes;

- ☆ Presence of cysts, spots or patches over the body and gills, *etc.*;
- ☆ Appearance of lesions, haemorrhagic spots and greyish or brownish areas over the body.

Laboratory Examinations

Thorough visual examination for external signs of the disease should be followed by detialed but quick laboratory examination by pathomorphologica, pathoanatomical and microscopical studies of squash and smear preparation from different organs/tissues. Diagnostic procedures in brief are presented below table.

In situations where a disease problem is suspected, only those specimens exhibiting symptoms of distress or disease should be selected. Live moribund speciments are preferred, but if necessary, freshly dead specimens may also be collected for laboratory examination.

Methods for Diagnosis of Commonly Occurring Diseases of Asiatic Carps in Undrainable Ponds

Sl. No.	*Disease Agent*	*Method of Examination*	*Positive Indications*
A.	Parasites		
1.	Protozoa		
	Ichthyophthirius	Microscopy	Pin-head size white spots on the skin, fins and gills. Presence of ciliated trophozoites with relatively large horseshoe shaped nucleus.
	Trichodina	Microscopy	Presence of saucer-shaped actively moving ciliate parasites on body surface and gills.
	Myxozoans	Microscopy	Presence of cysts, spores on gills, body surface and/or in the squash preparations of kidney. spleen, air-bladder, *etc.*
2.	Crustaceans		
	Arqulus	Visual examinations/ microscopy	Haemorrhagic spots, lesions over the body and presence of parasites attached to fish body by means of suckers and hooks.
3.	Flukes		
	Gyrodactylus/ Dactyloqyrus	Microscopy	Presence of parasites in gills and skin.
	Diplostomum	Visual examination/ microscopy	Small pigmented black nodules over the body surface
B.	Fungi		
1.	*Saprolegnia*	Microscopy/visual examination	Body lesions associated with small white tufts of hyphae on fins and skin. Infected fish eggs fail to hatch and show presence of fungus mycelium protruding from the egg surface.
2.	*Branchiomyces*	Microscopy	Decolouration of gills, erosion of lamellae and presence of fungal hyphae in blood vessels.

Sl. No.	Disease Agent	Method of Examination	Positive Indications
3.	*Achlya*	Microscopy	Cottony outgrowths of fungal mycelium over the infected area.
C.	Bacteria		
1.	*Aeromonas hydrophila*	Culture/ microscopy	Dropsy condition and haemorrhages over the body.
2.	*Pseudomonas fluodrescens*	Culture/ microscopy	Clinical condition is usually indistinguishable from that of aeromonas. Haemorrhages over the body.
3.	*Flexibacter columnaris*	Culture/ microscopy	Appearance of external lesions on the body, head region and gill. Lesions initially begin as whitish or brownish patches with reddish zone around the periphery.
D.	Virus		
1.	Rhabdovirus of common carp	Cell culture/serum neutralization test	Common carp is prone to this disease showing dropsy condition.
2.	Rhabdovirus of grass carp	Cell culture/serum neutralization test	Only grass carp is prone to this disease exhibiting similar dropsy symptoms.

Smear preparation of selected tissues and organs may be made on the spot by smearing the material on a slide. Slides can then be dired, stained and examined immediately. Bacteriological media can be inoculated with materials from various organs, especially kidney, heart, *etc.*, employing aseptic techniques. On-site disease diagnosis permits the immediate application of chemotherapy or remedial measures to control or eradicate the disease. However, accurate diagnosis of disease is of utmost importance if proper treatment is to be applied and this is possible only through experience and training. At times, may disease conditions occur which cannot be properly diagnosed without specialized laboratory facilities and in such conditions samples should be sent to such laboratories under proper preservation, packing and shipment (Dey, *et al.*, 1982). As far as possible the specimen for examination to reference laboratories should be always sent live but when circumstances prohibit live delivery, specimens may be forwarded packed in ice. Specimens for parasitology examinations may be preserved in 5–10 per cent formalin solution. In case of larger specimens incision may be made to facilitate effective penetration of the fixative. The volume of fixative should be at least five times the volume of materials to be preserved.

Health Management Measures

Understanding and managing the undrainable pond environment is the key to successful fish health management and profitable fish culture, and to ensure this the knowledge of the role of various environmental components in the occurrence of disease outbreak is essential. The main thrust of such measures is directed toward:

- ✰ Minimizing the stress on cultured fish;
- ✰ Prevention of the introduction of serious disease agents;

- ☆ Confinement of disease outbreaks to affected areas;
- ☆ Minimizing losses from disease outbreaks.

The following important measures are the key components of successful fish health managements.

Surveillance and Maintenance of Water Quality

Abrupt and wider fluctuations in some of the environmental parameters such as dissolved oxygen content, pH, turbidity, temperature, additions of some chemicals, detergents, pesticides and naturlaly produced toxic substances such as hydrogen sulfide, ammonia, dinoflagellate toxins, *etc.*, often cause stress in fish and predispose them to infectious diseases. Anything that alters the environment of the fish is a potential stressor and efforts should be made to identify and avoid them. Undrainable ponds offer great protection against spreading of disease outbreaks by confining the outbreaks only to the affected ponds. However, the recent trends of intensification in aquaculture involve high stocking rates, increased feeding and fertilization programmes resulting in nutrient accumulation leading to appearance of algal blooms that lead to dissolved oxygen and other water quality problems. In older ponds, cases of excessive accumulation of organic matter have been observed, resulting in the appearance of bacterial bloom and related oxygen depletion (Radheyshyam *et al.*). For health and optimum growth, the dissolved oxygen level should not drop below 5 mg/1. Carbon dioxide concentration up to 20–30 mg/l may be tolerated by fish provided oxygen is near saturation. At lower levels of dissolved oxygen, toxicity of carbon dioxide increases. When pH values remain above 9.5 or below 6.0 for extended periods, fish will be under stress and may not grow well. Liming agents may be used for low pH corrections. Ammonia concentration above 1.0 mg/1 indicates organic pollution. Hydrogen sulfide toxicity increases with decreasing pH and it is harmful even at 1.0 mg/l concentration level. Making the pond environment more congenial and hygienic, eliminates the risk of stress and provides safety to fish. Proper and timely management of soil and water qualities by manipulating feeding, fertilization, liming, addition of clean water, bottom raking, aeration of water by recirculation or splahsing, *etc.*, reduces most of the environmental problems and provides congenial conditions for the health growth of fish. An interval of about 15 days between the pond poisoning and the stocking eliminates most of the pathogens from the environment.

A Model for Integrated Fish Health Management System

It is always advisable to stock the pond only with healthy and genetically vigorous fry and fingerlings so that they may have better growth rate and resistance towards diseases. Prior to stocking, samples of the stocking material should be examined to check their health status. This avoids any risk of introducing infected stock in the pond. However, the stocking materials should also be prophylactically treated before releasing into the pond (detailed under Chemoprophylaxis).

Overstocking may lead to biological crowding resulting in waste build up, decreased availability of natural food, depletion of dissolved oxygen, deterioration of water quality, *etc.*, and hence it is advisable to follow the recommended stocking density for nursery, rearing and stocking ponds.

Minimizing Handling Stress

The rougher the handling, the greater is the stress and the risk of disease (Kumar *et al.*, 1986). Care should be taken not to break the protective mucous coating of the skin. During summer months netting should always be done early in the morning and it is better to have minimum possible handling during hauling. High temperature during hot water causes increased metabolic activity and induces more stress upon them.

Measures in Pond Management

Poisoning of Pond

Wild fish population is one of the most potential sources of disease-producing organisms. Use of chlorinated lime (bleaching powder) is the most suitable material for this purpose, since it kills all the wild fish species, molluscs, tadpoles, frogs, crabs, *etc.*, and also disinfects the pond water and soil. It is applied at the rate of 40–50 ppm (Tripathy *et al.*, 1978). Mahua oilcake is also a widely used piscicide, but it fails to disinfect the pond. In nursery and rearing ponds it is desirable to have second poisoning with malathion at the rate of 0.25 ppm 4 or 5 days prior to stocking. It eliminates the larger copepods which do appear in large numbers after organic fertilization. These copepods prey upon young fish larvae and also serve as vectors or carriers of many infectious pathogenic organisms. Some of the common crustacean fish parasites also get killed. Malathion application has significantly increased the survival level in nursery ponds (Kumar *et al.*, 1986).

Disinfection of Appliances

All required appliances such as fry carriers, hapas, utensils, buckets, nets and gears, *etc.*, require thorough cleaning and disinfection before being put to use. Some of the pathogenic organisms are found adhering to them and may cause disease if they are allowed to come in contact with the host fish species. Disinfection can be done by washing or immersing in a concentrated solution of disinfectant. Some of the most effective and easily available disinfectants for such use are chlorine, sodium hydroxide, sodium chloride potassium permanganate, *etc.* Chlorine is probably the most widely used disinfectant in fishery management and is easily available as a solution of sodium hypochlorite and powder of calcium hypochlorite (bleaching powder). Solution of 1–2 per cent chlorine is active against bacteria, viruses and fungi but is extremely toxic to fish and hence their residues must be thoroughly rinsed from the disinfected items before being brought into contact with fish. Sun drying of nets, hapas, *etc.*, is also a practical method of disinfection.

Proper Feeding

In addition to the natural fish food which is made available by fertilization, an adequate amount of good quality supplementary feed is essential for maintaining healthy growth of fish. Any deficiency in quantity and quality of feed may cause various diseases by increasing susceptibility to many infections.

Prevention of Entry of Unwanted Fish

Most undrainable ponds lack proper embankments. Most of these ponds have channels in the embankments connecting them with outside waters during the rainy season. Most of the ponds lack even proper embankments. These channels are the vulnerable sites through which some of the wild unwanted fish species or other animals get entry to the pond. Fixing fine meshed screen into these channels may eliminate the risk of entry of unwanted fish species into the pond. Pond embankments may also be raised to prevent risk of inundation and entry of undesirable animals and fish species. Some fish eating birds, molluscs, *etc.*, serve as intermediate hosts for many parasites that infect fish. Tadpoles and frogs may also act as carriers of certain parasites and bacteria which ultimately may infect carp species and hence such animals should not be allowed in the pond.

Separation of Young and Brood Fish

Brood fish may serve as carriers of disease causing organisms without exhibiting any clinical symptoms. They sometimes become survivors of previous epizootics due to built up immunity but retain some of the pathogens. To avoid such risk, the best course is to separate the young ones from the adults.

Removal of Dead Fish from the Pond

Dead and apparently sick fish should be removed. A daily log of losses must be kept. Such records will provide valuable insight into the problems and may lead to their solution.

Holding the fish in a hand net and dipping it into a concentrated solution of the drug for one minute or less is used as prophylactic treatment in case of mild diseases. A short bath is useful when facilities for a rapid flow of water are available. Water flow is stopped and relatively high concentration of the drug is added. Exposure time should not be longer than one hour. A long bath is a very effective method for prophylactic treatment of pond fish for external parasites. The oral route is used in prophylactic treatment to prevent certain infections. It is generally conceded that feeding medicated feed to fish is a prophylactic rather than a curative measure.

Prophylactic use of streptomycin and penicillin at the rate of 25 mg of streptomycin sulphate and 20 000 I.U. of penicillin has been found to be very effective in preventing outbreak of columnaris disease in rohu in a field-oriented experiment (Kumar *et al.*, 1986). Feeding antibiotics with feed has successfully prevented the occurrece of CE (Carp Erythrodermatics) in European carp culture. Prophylactic treatment of pond with locally available organophosphorous insecticide (malathion) at the rate of 0.25 ppm of active ingredient successfully prevents occurrence of trematode and copepod infections.

Occasional application of potassium permanganate at the rate of 2 or 3 ppm is recommended for increasing dissolved oxygen concentration and hauling prophylaxis. Dip treatment in 500–1 000 ppm solution of potassium permanganate for a few seconds before releasing adult fishin ponds is also a very effective and practical prophylactic measure. Short bath for a few minutes in 2 or 3 per cent common salt solution is also a safe and inexpensive prophylactic measure against a wide range of parasitic an microbial pathogens.

Immunoprophylaxis

Immunization is becoming one of the most important ways of preventing communicable diseases in animals, including fish. Several commercial vaccines are now available and being used in many developed countries. Vaccines for some of the bacterial diseases of carps which do occur in undrainable pond culture systems are also available. These vaccines are against *Aeromonas hydrophila* and *Flexibacter columnaris.* Viral vaccine against Spring Viremia of Carp (SVC) is also being used on a commercial scale very successfully.

Appearance of Algal Blooms

- ✰ Algal blooms of *Microcystis* sp., *Euglena* sp., *etc.*, frequently occur causing serious problems again in terms of dissolved oxygen.
- ✰ Sometimes lead to a mass fish kill.

Remedy

Apply chemical algaecide at the rate of 3–5 ppm

Problem of no rain and plenty of rain

During drought the water level drops down to critical levels in some ponds, while during seasons of heavy rains the incidence of flooding is not uncommon

Solution

- ✰ Harvest the fish before such a situation is encountered.
- ✰ Repair and strengthen the dykes before the onset of rains
- ✰ Regular check on water level.

Leakages in Embankment

Sometimes leakages do occur in embankments and if not checked immediately they could cause extensive damage to the pond embankment.

Leakages can be checked by pushing sawdust, bran, *etc.*, into the upstream site. These are carried by water into the leak where it swells and stops the leak.

Pond Poisoning

- ✰ Due to the undrainable condition, some harmful gases form in the pond bottom which leads to fish mortality in extreme cases.
- ✰ Major Gases responsible for pond poisoning:
- ✰ Ammonia, Nitrite, Nitrate, Hydrogen Sulphide, Carbon Monoxide, Chlorine and Free Carbon-di-oxide.

Ammonia

Sources of Ammonia

- Decomposition of organic matter.
- Excretion by aquatic organisms.
- Denitrification.
- Death of phytoplankton blooms
- unexplained losses in fish hatcheries

Effect of Toxic Ammonia

Fish may experience convulsions, coma, and death. Hyperplasia-the gill filaments are swollen and clumped together, reducing the fish's ability to 'breath'. lethargy, loss of appetite, laying on the pond bottom with clamped fins, or gasping at the water surface if the gills have been affected.

Controlling Measures

- **Flush the pond with well water:** Ammonia can be flushed from ponds, although pumping the huge volume of water.
- **Add acid:** In theory, adding acid (such as hydrochloric acid) to water will reduce pH. This can shift the ammonia equilibrium to favour the non-toxic form.
- **By application of Zeolite.**

Nitrite

- Nitrite is an intermediate product in the biological oxidation of ammonia to nitrate (Nitrification).

Sources

- Nitrite is produced by autotrophic bacteria combining oxygen and ammonia.

Effects of Nitrites

Sub lethal effects depending on species, may Include gill hyperplasia, reduced activity and growth, liver may be affected.

Control

- Increase aeration
- Stop feeding
- Add salt, preferably chlorine salt

Hydrogen Sulphide

Hydrogen sulphide is produced by bacteria in anoxic waters (waters deficient in dissolved oxygen).

Toxic Effects of H_2S

- ✰ Sub lethal effects, including gill damage, Gasping, darkened gills, increased respiration rate, shimmy, anorexia,etc.

Control

- ✰ Keeping the system well oxygenated is the best way of stopping hydrogen sulphide being formed.
- ✰ Vigorous aeration.

Outbreak of Diseases

General Considerations

- ✰ In an undrainable pond system no addition of water or aeration is normally done and the accumulated wastes are not usually removed unless provision is made to desilt the pond after a couple of years.
- ✰ All such factors can cause deterioration of fish health and magnify the risk of outbreaks of diseases.

Common Diseases

All the Asiatic carp species cultured under composite fish culture undrainable ponds are prone to many communicable and non-communicable diseases, the most significant among them are described here under three groups:

1. **Microbial diseases**
2. **Parasitic diseases**
3. **Environmental and nutritional diseases**

1. Microbial Diseases

Virus, bacteria and fungal pathogens come under this category.

Viral Diseases

As far as carp species are concerned, mainly two viral diseases are of importance.

Spring Viremia of Carp (SVC)

Spring viremia of carp is an acute systemic viral infection caused by *Rhabdovirus carpio* (RVC).

Symptoms: The main clinical signs are gathering of fish at water outflows, dark colouration, lower respiratory rate, haemorrhages especially over the skin and gills, loss of balance, exophthalmia, inflamed vent, *etc.*

Rhabdovirus Diseases of Grass Carp

A new serotype of rhabdovirus similar to spring viremia of carp virus (SVCV) and Rhabdovirus carpio (RVC) has been isolated from grass carp

Control Methods

Control methods for viral diseases are restricted as there is no chemotherapeutic measures available at present. Application of antibiotics helps only in prevention of secondary infections.

Bacterial Diseases

Among infectious diseases the role of bacteria has been strongly emphasized as they present many practical problems in nursery, rearing and stocking ponds.

Columnaris Disease

- ✰ Columnaris disease is commonly occurring throughout the world and affects virtually all species of freshwater fishes.
- ✰ The first external sign of colunaris disease of the appearance of greyish white spot on the body of fish.

Control Measures

- ✰ Environmental improvements, especially increased oxygenation, control of organic addition, *etc.*, are the most valuable supportive therapy.
- ✰ Practical control of outbreaks of columnaris is possible with a number of drugs, including copper sulphate (0.5–1.0 ppm) and potassium permanganate (2–3 ppm) in pond treatment.

Fungal Disease

Fungal fish diseases also sometimes cause extensive losses. Species of the genera Saproleqnia sp., Branchiomyces sp. and Achlya sp. are usually implicated in fungal infections.

Control Methods

Malachite green (zinc free grade), formalin, potassium permanganate, copper sulphate, salt, *etc.*, are the common therapeutics for effective use.

2. Parasitic Diseases

- ✰ Parasitic diseases are usually encountered more frequently than microbial diseases.
- ✰ The presence of high level of organic matters in undrainable ponds, encourages multiplication of parasite organisms and resulting in extensive parasitic infection.

  Protozoan diseases are among the most significant of all parasitic diseases in carps. The following are the most important protoizoans parasitizing carp species in undrainable pond culture system.
- ✰ *Ichthyophthirius multifilis*: "Ich" or white spot disease is probably one of the most detrimental diseases caused by this parasite which affects all the species of Indian major carps

- *Ichthyobodo sp.*: A small flagellate external protozoan parasite of skin and gills causes considerable damage in fry and small fingerlings.
- Trichodina is another small saucer-shaped protozoan parasite that harbours gills and body surface of the host fish.

Protozoan Diseases

Treatment

- Pond treatment with 0.1 ppm malachite green and 25 ppm formalin gives a better result against *Trichodina sp., Costia sp.* and *Ichthyophthirius sp.*
- Pond treatments with 15–25 ppm formalin have yielded excellent results.

3. Environmental and Nutritional Diseases

- Diseases known to be occurring due to nutritional deficiencies and environmental disorders are of little importance.
- Proper monitoring and management of pond ecosystem and provision of adequate quality of supplementary feed will avoid occurrence of such diseases which sometimes appear in ill-managed ponds

Therapy to Fish Diseases

Selection of the therapeutic compound must be based on the following considerations:

- Effectiveness against the causative agent;
- Adverse effects, if any, on the host;
- Possibility of penetrating the site of infection;
- Effectiveness under the existing water chemistry;
- Local availability;
- Cost considerations;
- Convenience of application.

Conclusion

- Undrained pond has both advantages along with disadvantages.

  Reduction of diseases transmission, Reduction entrance of intermediate host like birds and snails.
- Nowadays shrimp culture is more acceptable than fish culture in undrainable pond (zero water exchange) due to the short period of culture time.

  Undrained pond required some research to solve the problems. It is also need the awareness among the farmers by conducting awareness programmes and training programmes.

Chapter 13
Poaching Problems and its Solution in Aquaculture

Introduction

Poaching is the illegal taking of wild plants or animals; the law concerned may be *e.g.* the law of property or local or international conservation and wildlife management laws. Fish poaching considered in this study are primarily illegal fishing (poaching) in EEZ waters, unregulated fishing in areas of Regional Fisheries Management Organizations (RFMOs) either by parties to those RFMOs or by non-parties to them, and all fishing in high seas areas not subject to RFMOs.

"China is the largest fisher in the world, and the illegal fishers would come second,"

What is Poaching?

Poaching is the illegal collection or killing of fish, wildlife or native plants. Some common examples include:

- ☆ Taking more fish and game than the law allows;
- ☆ Hunting or fishing in a closed area or during a closed season;
- ☆ Using prohibited gear or equipment;
- ☆ The unlawful collection of fish and wildlife,

Who Are Poachers?

Given British Columbia's diverse fish and wildlife resources, poaching is carried out by different individuals for different reasons. Generally, poachers are most likely to kill for profit. There are eager national and international markets of collectors and consumers who pay outrageous sums for rare trophies and wildlife parts. Whether they operate alone, or as part of organized crime, trophy and commercial poachers think that game laws only matter if they get caught. They kill with no regard for fish and wildlife, habitat or the rights of other resource users. While poaching for commercial gain is a significant problem, the taking of one or a few "extra" fish or animals, for reasons other than profit, is also a serious issue. While these individuals believe their bending or breaking of the rules has little impact on the overall health of the resource, they choose to ignore the total impact of all such acts

Fishing with Explosives

Techniques that Poachers Use

Poachers frequently rely on trapping devices; since traps can do the work untended, the traps can be abandoned if discovered and the poacher can make his illegal harvest at discreet times.

1. Dynamites are used to poach fish.

2. Poisoned water placed in broken earthen pots is another popular method for large carnivores that threaten livestock.
3. High voltage live electric wires connected to fences are used to kill large fish.
4. Poachers also set gill nets specifically to catch kinds of fish not usually targeted by commercial fishermen.

Electrofishing

Impact of Poaching

1. Several species could be stamped out in the fish farm.
2. The hunting of herbivores will directly affect the carnivores that are dependent on them. Thus the prey depletion is one of the most serious threats to the predators.
3. The affects of extinction of certain fish species can be seen on the marine ecosystem and how when the top predators are removed, the oceanic ecosystem degrades due to the imbalance.
4. The most obvious **economic impact** of illegal, unreported and unregulated (IUU) fishing on developing countries is the direct loss of the value of the catches that could be taken by local fishermen if the IUU fishing was not taking place. Available estimates place the economic loss of illegal fishing to be between $10 billion to $23 billion annually.
5. Fishing generally has the capacity to damage fragile marine ecosystems and vulnerable species such as coral reefs, turtles and seabirds. In fact, all eight sea turtle species are now endangered, and illegal fishing and hunting are two major reasons for their destruction.
6. It has a significant impact on the sustainability of both the targeted species and the ecosystem.
7. Fish poaching can also lead to increased pressure on endangered fish species.

Control Over Poaching

1. The first responsibility of all fish farm is to follow the rules for regulating fishing that are found in recent international documents. Many of those rules are in the Code of Conduct.
2. Those fishers who break the rules should face tough penalties. Governments should improve their ability to inspect fishing vessels, investigate when they believe that IUU fishing is taking place and prosecute violators successfully.
3. **Fishing Vessel Registration**: Before a country allows a fishing vessel to be registered and before it allows a vessel to fly the country's flag, the country should make sure that it has the ability to control the fishing activities of the vessel.
4. **Record of Fishing Vessels**: Each country should maintain a record of its fishing vessels. The information in the record should relate to the physical characteristics of vessels and their past fishing history.

5. **Authorization to Fish**: Flag countries should prohibit their vessels from fishing in any ocean area unless the vessels receive express authorization from the government to engage in fishing.

Vessels should receive authorizations to fish regarding some conditions, for example:

- ☆ what species can be caught;
- ☆ what gear can be used, and
- ☆ where and when the vessels may operate.

6. Strict implementation of law. It is a very efficient way of controlling inconspicuous poaching activities.
7. Coast guards and senior officials have to be trained on hunting patterns, method and ecology of hunting species.
8. Establishment of anti- poaching camp near important protected area.
9. Awareness programme should be done against poaching of endangered species.

Predator Control and other Effects on Fish and Shrimp Farms

The concentration of potential food resources in aquaculture facilities provides attractive foraging opportunities for birds and other predators of cultured animals, which are viewed as a nuisance by aquaculture producers. The construction of aquaculture ponds has created new aquatic habitat while suitable habitat for wildlife with an aquatic orientation, especially migratory waterfowl, has decreased elsewhere. Ponds are attractive to waterfowl and migratory birds as sources of food and shelter. Aquaculture structures such as net pens and shellfish rafts create a complex, three-dimensional environment in the water column that attracts and aggregates wild fish, their predators, birds, turtles, and marine mammals.

Aquaculture producers use a range of nonlethal and lethal techniques to control predation of cultured animals. Most methods are intended to harass wildlife and induce them.

Tips for Fish and Shrimp Farms

- To avoid grazing animals from entering and harming the fish, high banks must be built to protect the ponds.
- Always keep a check on potential predators, like snakes, which can harm the fish.
- The health condition of the fish must be regularly monitored to prevent the spread of fish diseases.
- The diseased fish must be quarantined and removed from the fish tank.
- Construct the cages according to the size of the water body. For deep water bodies, go for floating cages, while fixed cages best suit low depth water bodies.
- Install water bamboos in the depth of the water body to avoid poaching

Cows Avoid for Poaching

Guard Adjustment

Use of Bamboo Stripes

Net Used

Use Bamboo Stripes

Netting all over the Pond

Eichornia all over the Pond

Use Ducks

Solve the Poaching Problems

- ☆ Using Dogs
- ☆ Netting over small pond
- ☆ Use of bamboo stripes at regular distance
- ☆ Eichornia-to put eichornia for adjust side of the embankment
- ☆ Echo-sounder to avoid birds problem
- ☆ Guard adjustment
- ☆ Ducks to avoid poaching

Social Unrest, Distruption and Conflicts

The capital intensive nature of high density shrimp culture has favoured the entry of multinational corporates investors and national and local elites. With their survival at stake, villagers have organised themselves and started to fight back. Small farmers in Andhra Pradesh,India and Kerpan were arrested and jailed for defying plans to convert paddy to prawn farms. Shrimp owners in Bangladesh have hired guards to prevent poaching of shrimp, force landowners to sell their land and stop protests by villagers.

Conclusion

"Fish poaching" and "illegal fishing" may sound like misdemeanors, on the cosmic scale of crime, but they provide an astonishing mass of the fish people eat around the world. And they amount to a uniquely self-destructive problem that might one day solve itself by collapsing fish populations.

Chapter 14
Water Losses in Ponds and its Remedy

Introduction

Water is one of the most important natural resources and all forms of life are dependent on it. It is, therefore, very essential to properly conserve and manage this resource and regulate its use to obtain maximum benefits. Precipitation is the most important source of freshwater and there is need to manage the rain water so as to reduce the impact of moisture stress and obtain sustainability in agricultural and aquaculture production. Out of 400 million ha-m precipitation in the country, 70 million ha-m is consumed as evapotranspiration, 215 million ha-m infiltrates into soil, whereas 115 million ha-m is lost as run-off, resulting in drought in the catchment and flood downstream.

Ponds in India

India is a traditional user of ponds, these are small water bodies either man-made or natural which have been use since ages. They are traditional water harvesting structures and the water stored in these structures is commonly used for drinking water supply, washing/bathing for human and animals, irrigating crops and raising the fish. These small water bodies also play an important role in improving the ecological system and help to maintain biological diversity. In India, approximately 25 lakh ponds are used for a variety of applications. The traditional areas of ponds are in Andhra Pradesh, Karnataka, Tamil Nadu, Orissa, West Bengal and Madhya Pradesh.

Ponds are located on a variety of soil type, which exhibit a wide range of seepage characteristics. Seepage losses not only mean loss of useful water but it also leads to other problems such as trench in the embankments, water logging or increased salinity in the adjacent area. Seepage takes place at the bottom as well as sides of the farm pond.

Classification of Ponds

Ponds can be classified into 4 types:

1. Embankment ponds
2. Excavated ponds
3. Levee ponds and
4. Combination of watershed-levee ponds

Embankment Ponds

Also called a watershed pond or hill pond. A watershed is the drainage area around the pond within which rainfall runoff drains toward the pond. A dam or embankment is constructed in a depression between two hills and serves to impound water in a basin area on the upstream side of the dam. This type of pond is best suited in areas with slightly to moderately rolling topography. Embankment ponds usually depend on rainfall runoff or irrigation water to fill and then maintain water levels. This type of pond works well in the high country where there is more seasonal moisture and snow pack runoff, as it all goes back into the water supply. Pond size, shape, and depth are limited by the topography of the site and the size of the watershed draining to the pond. Generally, the steeper the slope of the pond site, the smaller the pond that can be constructed. Well-sited embankment ponds generally require the least amount of earthmoving per acre of water impounded compared to other types of ponds. Because construction costs are based largely on the amount of earthmoving, an embankment pond is generally the least expensive type of pond per surface acre of water to construct.

Excavated Ponds

Excavated or "dug," ponds are constructed almost entirely below original ground level. This construction method is usually used only for construction of small ponds (generally less than 1/2 acre) because of the large amount of earthmoving required in relation to the size of the pond. Excavated ponds require an external water source to fill and maintain the pond where springs, groundwater, or runoff are not sufficient. An excavated pond is usually the most expensive type of pond to construct on a per-acre basis.

Levee Ponds

Suitable for flat or nearly flat land, levee ponds are only partially excavated. Earth from what is to be the basin area of the pond is removed and used to construct the sides, or levees, of the pond that impound the water. The water level in a levee pond is higher than the original ground level. Water depth is usually similar throughout the pond and is regulated by the height of the outlet pipes and constructed levees. An externally pumped water source, such as a creek, will be necessary to fill and maintain this type of pond due to the absence of a watershed. Per-acre construction costs of levee ponds generally fall between those of watershed and excavated ponds.

Combination Watershed-Levee Ponds

An example of a combination watershed-levee pond would be a two- or three-sided levee pond that connects to an existing hill to form its other side. Depending on the site, the hill side of the pond can provide a significant amount of watershed runoff to the pond, thus reducing and, in some cases, eliminating the need for pumping water to fill and maintain the pond.

Source of Water

The three sources of water for filling ponds are: rainfall runoff, groundwater, and surface water. Each has advantages and disadvantages.

1. Rainfall Runoff: Rainfall runoff is the primary source of water for embankment ponds. It comes from the drainage or watershed area surrounding the pond. The best runoff water source for ponds is a watershed containing undisturbed, well vegetated cover such as timberland or grassland.
2. Groundwater: Groundwater can serve as a source of water to excavated ponds, or as a supplementary source to ponds with inadequate watersheds or excessive seepage.
3. Surface Water: Surface water from nearby irrigation ditches, springs, streams, rivers, or reservoirs with good water quality can be used as a pond water source.

Seepage and Evaporation Losses

Seepage takes place at the bottom as well as sides of the farm pond. It is normally high in sandy soils while it is low in loamy to clay soils. Further, seepage losses are higher in initial years of construction while it reduces slowly in one or two years due to normal siltation of clay particles coming along with the run-off water. Various sealant materials like cement, bentonite, polythene lining, brick lining, stone slab lining and few chemicals based on sodium have been tried and found effective. Normal water loss rate of 0.25 cm per day is tolerable for farm ponds. The factors like wind velocity, intensity of sunshine and humidity affects evaporation losses.

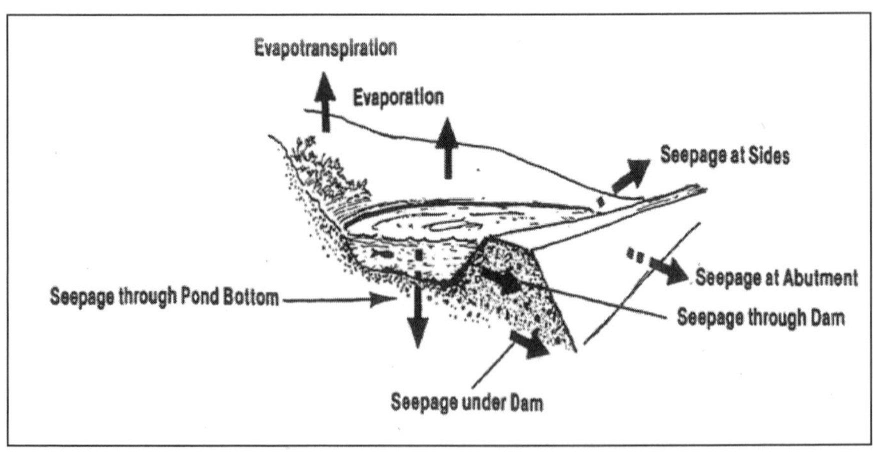

Aveneus of Water Losses from a Pond

Seepage

Seepage is the major cause of water loss from most fish culture ponds. It usually exceeds the evaporation rate. The principle causes of seepage are as follows:

1. Ponds built in areas where soil contains high amounts of sand or gravel.
2. Shallow soil with underlying fractured bedrock.
3. Soils with a high gypsum content in which voids develop as gypsum dissolves.
4. Improper construction methods including failure to compact clays in the bottom of the pond., failure to place a clay core in the levees, no anti-seep collars around drain pipes and improperly designed drains.

Water Losses Due to Seepage

If ponds are constructed on permeable soils, seepage rates will be very high. Reports of more than 1 inch of seepage loss per day are common. Even in the best situation some water will be lost through seepage. The typical rate is from 0.01 to 0.2 inch per day. As the viscosity of water decreases with increasing temperature, seepage rates for ponds may be up to 25 percent higher in summer than in winter.

A seepage rate of 0.2 inch per day from a 1-acre pond does not seem significant but this rate will result in a water loss of 6 inches per month or 6acre-feet in a year. If better site selection and proper construction techniques would reduce this to even 0.1 inch per day, more than 3 acre-feet of water would be saved in a year. Obviously seepage control should be of the highest priority during pond construction.

How to Protect Seepage Losses?

To protect against seepage, select a site for the ponds where soils have a high clay content and follow good construction techniques. Here are some seepage reduction techniques which may be used during initial construction or when renovating old ponds.

- ☆ Select the pond site based on soils engineering data for the subsoils. Subsoils should be classified as clay loam to clay texture with a liquid limit of 30 per cent or more, plasticity index of 15 per cent or greater, and more than 30 per cent passing a No. 200 mesh sieve. Sandy clays, clay loams, sandy clay loams and clays all fit these specifications.
- ☆ Compact the embankments in layers 6 to 8 inches thick.
- ☆ Compact soils on the pond bottom and sides at optimum moisture content. This can be done with heavy machinery during initial construction or it can be accomplished using sheep or goats to cut up and trample the bottom and levees when renovating.
- ☆ Install clay blankets over the pond bottom and sides. This clay should be carefully chosen, as some clays contain heavy metals which can be toxic to fish.

- Incorporate bentonite into the pond bottom when a suitable clay is unavailable. Bentonite is a clay product which swells when wet and fills the minute holes in the soils.
- Incorporate organic matter such as manure into the pond bottom to reduce seepage. This is basically the same technique used for centuries in making adobe bricks in the southwest.
- Place a pond liner over the pond bottom when clay soil is unavailable. These flexible membrane liners can be of rubber, polyethylene, vinyl plastic or woven fabric. They are very expensive, but may pay themselves in reduced pumping costs.
- The last 4 techniques are expensive; however, the need for them usually can be avoided with proper site selection and good construction techniques.

Placing plastic sheet over the pond bottom as clay soil is unavailable

Evaporation

- Obviously pond evaporation varies widely across the state, with higher rates occurring in the more arid and western region. Actually net evaporation (pond evaporation minus precipitation) is more important in determining water requirements.

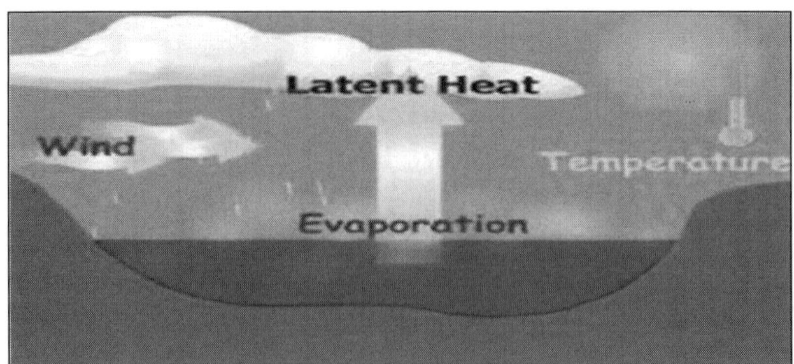

- ☆ Additional evaporation loss can occur when ponds are aerated. To guard against evaporation, management of aeration is important. Aeration is usually used in fish ponds to prevent low concentrations of dissolved oxygen. Aeration of ponds using 1-horse power per day increased the evaporation rate by 8 per cent per day. Nighttime aeration did not increase evaporation from pond surfaces.

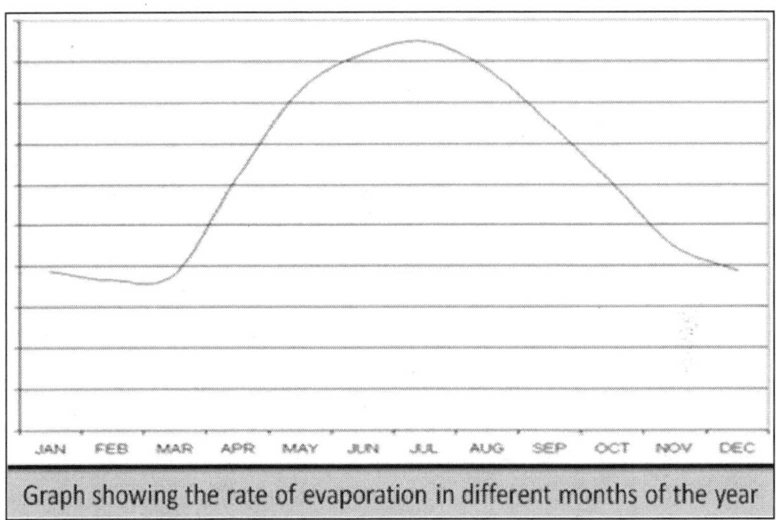

Graph showing the rate of evaporation in different months of the year

How to Reduce Evaporation Losses?

In warmer regions of the country, pond evaporation is a year-round battle. The summers are warm enough to speed up water evaporation in ponds. Because it's a natural occurrence, you can't prevent evaporation completely. However, you can drastically reduce how much water you lose in the warmer months by keeping your water shaded and at cooler temperatures. Use a combination of techniques to protect your pond from evaporation.

1. Add aquatic plants that will partially cover the water surface to your pond. Aquatic plants will not only shade the water below, they will also absorb some of the sun's energy, keeping the water cooler.
2. Plant tall bushes or trees around the outside of your pond to shade the water. Choose evergreen bushes and trees that drop few leaves or seed pods to keep the water clear. If you choose a tree or bush that can get messy, plant it farther away from the pond where it will still provide shade but will not drop leaves or seeds into the water. The leaves or seeds can change the chemistry of the pond and may even be toxic to pond fish.
3. Reduce the amount of agitation caused by any water features in the pond. Waterfalls and filters that splash and agitate the water cause the water to evaporate more quickly.

Evapotranspiration

The water loss to the air from a natural pond in which hydrophytes are growing consists of two parts, evaporation and transpiration. As both parts vary directly with the same meteorologic factors, the total loss, evapotranspiration, can be computed by means of a simplified mass transfer equation in which the only coefficient has been evaluated for that pond by means of a water budget. In addition to the variation with meteorologie factors, transpiration occurs at different rates throughout the growing season, depending on the stage of growth. Evaporation is affected by the amount of vegetation present. The mass transfer coefficient varies in accordance with these conditions. By assuming that the coefficient during the dormant season represents losses by evaporation alone, a basis is obtained for estimating the coefficient representing evaporation during the growing season. The computed evapotranspiration minus the computed evaporation gives the water loss attributed to transpiration alone.

Pond Leakage

Persistent leakage of a farm pond probably indicates that it was built in an area containing shale bedrock or other rock formations. If possible, drain the pond and apply a 12 to 15 inch layer of clay soil. During compaction, the pond bottom should be worked well with a sheeps foot roller. The clay should be compacted to a thickness of no less than 8 to 10 inches. Prior to this application, all tree stumps or similar structures should be removed and cavities completed filled. Investigations should be made to locate any cracks or crevices in the pond bottom. Again, compaction is important in producing a good seal. The deeper the pond, the thicker the seal should be. In the deeper area of the pond, where the depth exceeds 10 feet, the clay layer should be doubled.

Sheepsfoot Roller

The most popular method of sealing leaks in a pond involv es the use of bentonite. This is a type of clay material that can be applied in the water above the area of the leak. Bentonite will expand when in contact with water, thereby helping to impede flow through the sediment. According to sources, the most effective method is by applying it to the dry pond bottom followed by discing and compacting. Sources for bentonite can be located by contacting farm supply stores, construction companies, or well drillers. Application rates range from 1 to 2 pounds per square foot of pond bottom.

Pond Sealing Material

Another method would be to line the pond with a type of man-made sheeting. Liners are available from commercial dealers and are quite effective if installed correctly. If a liner is used, livestock should not be allowed access to the pond as the liner may be punctured.

Other Reasons for Water Loss

- ☆ Water also can leak through muskrat or crawfish burrows or through channels left by rotten tree roots. Older ponds can develop leaks around drain or overflow pipes. If the water drops rapidly to a certain level, carefully check the perimeter of the pond at the waterline for holes or discontinuities in the soil.
- ☆ Ponds built in shrinks well clays will develop deep cracks when dry. As they are refilled water runs through the cracks until the soil swells again. A considerable volume of water may also be lost during the initial filling of a pond as it soaks into the surrounding soil, giving the impression of a leak.
- ☆ It is not uncommon for new ponds to have high initial seepage rates that decrease as microbial activity and the accumulation of fine sediment and

organic matter help seal the pond bottom. If ponds that have sealed in this manner are dried or re-worked, they may again leak excessively until the seal re-establishes.

Management of Ponds to Reduce Water Loss

- Ponds require periodic maintenance.
- Cleaning: If the inflow water contains trash, such as weeds and other organic materials, the inflow will need to be screened. Most irrigation ditches that supply water will carry a considerable amount of silt at specific times of the irrigation season. Consequently, you may need to clean your pond every year or so to maintain the desired capacity. Great care must be taken during the cleaning process to avoid damage to the pond liner.
- Don't let trees grow on a new pond dam, and remove any smaller trees (less than six inches in diameter) found on older pond dams. Tree roots can eventually penetrate the core of the dam and cause excessive pond leakage. Cutting down larger, existing trees (over six inches in diameter) will result in deteriorating roots leaving large voids in the dam.

Conclusion

Water loss from a pond is a serious problem. For the farmer, it can mean little or no water for fish, irrigation or cattle. For the homeowner, a leaking or empty pond is unattractive and detracts from the landscape. Leaky ponds also increase water use, reduce fertilizer effectiveness, and can pollute groundwater.

Chapter 15
Algal Problems in Aquaculture and its Remedy

Introduction

Many will be familiar with algae in the form of green or brown encrustations on the rocks and glass in aquaria, as green water or as thin, tough green filaments which can prove extremely difficult to eradicate. It often comes as a surprise, therefore, to learn that some algae can actually cause diseases in fish. The three main nutrients that algae need are nitrogen, phosphorus and potassium. In an aquarium, these nutrients come from fish waste, uneaten food, the metabolism of beneficial bacteria, and the decay of other organics (*e.g.* plant debris and even dead algae). Algae are utilized diversely in aquaculture, but their main applications are related to nutrition. They are used as a food additive to supply basic nutrients, color the flesh or for other biological activities.

What are the Characteristics of Algae?

- ☆ Algae are photosynthetic creatures.
- ☆ They are neither plant, animal nor fungi.
- ☆ Many algae are single celled, however some species are multicellular.
- ☆ Many, but not all of red and brown algae are multicellular. Red and brown algae are common in marine environments. Some green algae are multicellular.
- ☆ The scum found on ponds are algae. They are the green hairy growth on things found underwater objects. Algae releases oxygen into the water as it manufactures its food.

- Algae forms the broad base on which the food pyramids in ponds and lakes is built. In manufacturing food, algae release oxygen, increasing the amount dissolved in the water.
- However, when algae becomes overabundant the decaying algae depletes oxygen levels. So during the summer, when conditions for growing algae are ripe, oxygen levels may decrease, causing "**summerkill**" for aquatic plants and animals.

Types of Algae

Green Algae/*Chlorophyceae*

Green Water Algae

- Caused almost invariably by too much light. Green water is a great source of nutrients (food) for species such as brine shrimp, *Daphnia* spp. and minute fry.
- Often accompanied by poor plant growth.It is an indicator of insufficient illumination.

Green Encrustations

- Two types: 'fuzzy" and beard.
- The fuzzy algae are not of great concern. It grows in short, single strands of about 2-3mm.
- Beard algae is more serious. Its individual strands have a very fine texture. It grows in thick, furry patches, up to 4cm.

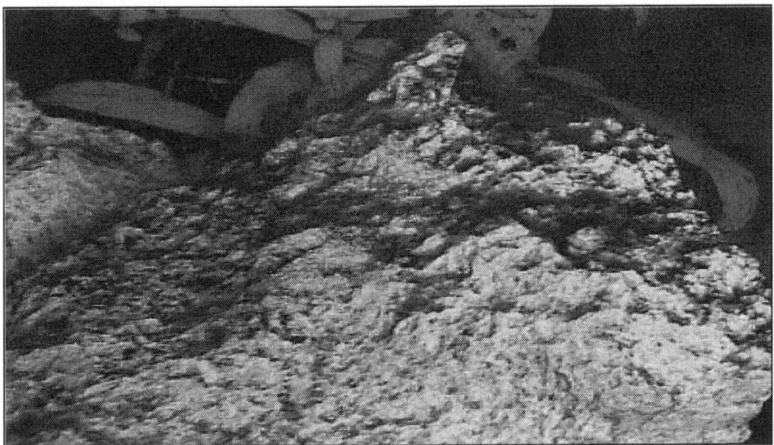

Brown Encrustations/*Phaeophyceae*

- Slimy appearance and feel on plants, décor and glass.
- Often accompanied by poor plant growth. It is an indicator of insufficient illumination.

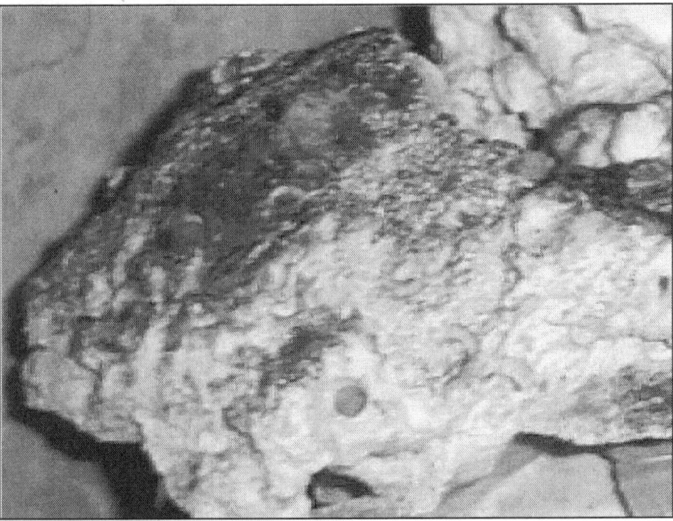

Red Algae/*Rhodophyceae*

- Red algae are unusual among the algae because they can include in their cell walls calcium carbonate which makes them hard and resistant to wear.
- Red algae that grow this way are referred to as "coralline" algae, because they are hard like corals.
- In addition to the common upright forms of coralline algae some species can grow in a thin mat over rocks and other hard structures.

Blue-green Algae/*Cyanophyceae*

- ✰ Blue-green algae appears as a slimy coating in a number of different colours.
- ✰ It may release toxins harmful to fish.
- ✰ It can fix nitrogen and may therefore occur in tanks with zero or very low nitrates (but possibly high levels of other nutrients, particularly phosphate).

Besides the main 4 types of algae there are different types of algae those are:

Spot Algae

- ✰ The small round green spots that appear on the sides of the aquarium are not algae.

  These are actually populations of diatoms, microscopic animals that secrete a hard silica shell on which green algae grows.

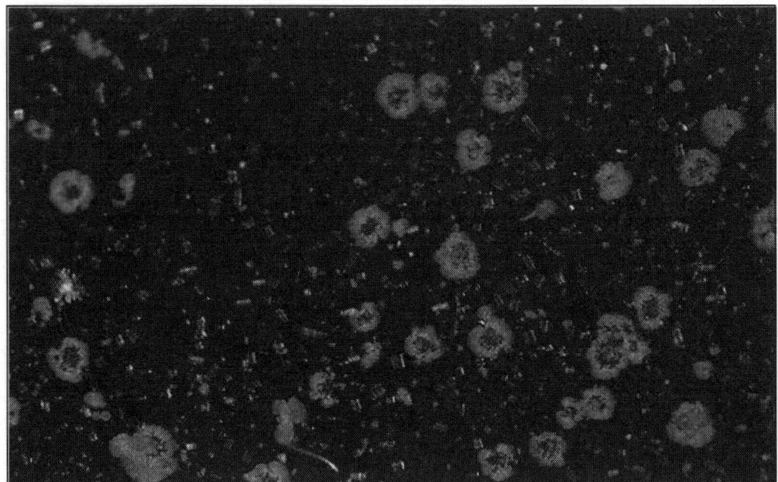

Filamentous Algae

- Two main types: hair and thread. Hair algae have shorter strands than thread algae (around 5cm in length) and tends to form matted clumps.
- Thread algae have much longer strands (as long as 30cm) and is often an indicator of excess iron.

Staghorn Algae

- As the name suggests, this branching algae looks like miniature greyish-green stag antlers.
- It is mostly seen nearer the tank surface, establishing itself on filters and décor.

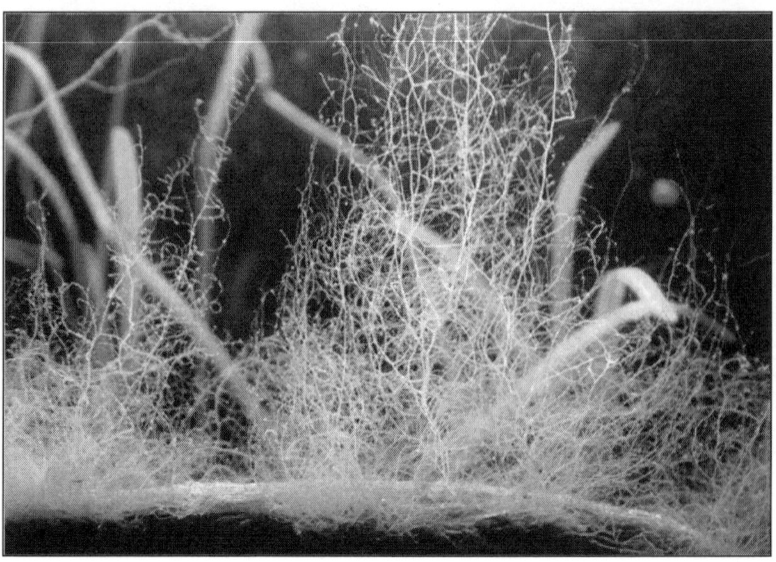

Brush Algae

- ☆ Brush algae is blackish (actually classed as a red algae) in appearance and forms feathery tufts about 2-3mm long.
- ☆ It is a nuisance on décor, filters and broad leaf plants especially.

Occurs as fine threads clumped together. Possibly one of the hardest to get rid off. Tends to favor water with a high carbonate hardness and high P^H.

Hair Algae

- ☆ Can be many different colors but is usually green or greyish.
- ☆ They are long thin strands sometimes growing 3-4 inches long.

- ☆ Normally due to overdosing of liquid fertilizers, especially ones high in iron and phosphates.

Branched Algae

- ☆ They are branching, tree-like plants which are attached to the substratum by calcified, root-like holdfasts.
- ☆ The plants are made flexible by having noncalcified sections (genicula) separating longer calcified sections (intergenicula).

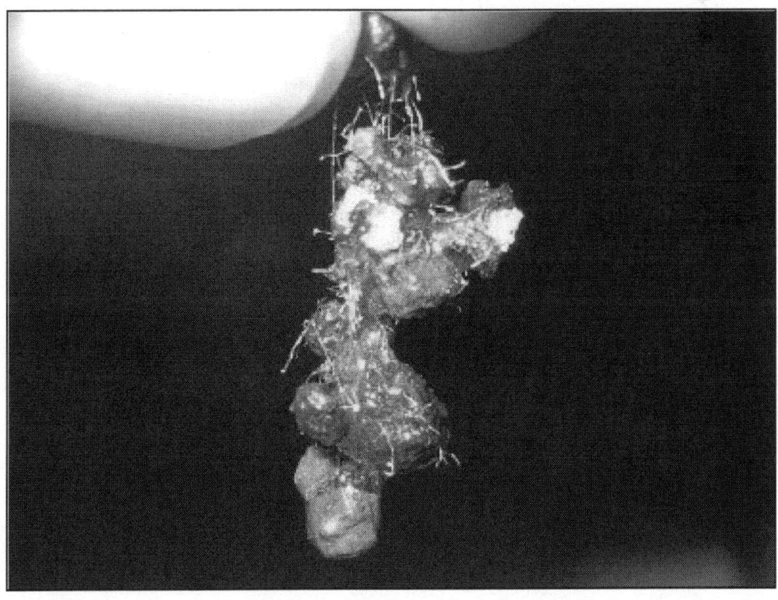

Harmful Effects of Algae

- When environmental conditions are favorable for their development, these cells may multiply rapidly and form high numbers of cells; this is called an algal bloom.
- Algal blooms can be any color, but the most common ones are red or brown. These blooms are referred to as red or brown tides.
- Most algal blooms are not harmful but some do affect fish and humans, as well as other animals like birds and marine mammals. These are known as Harmful Algal Blooms (HABs).
- Algae are vitally important to marine and fresh-water ecosystems, and most species of algae are not harmful.
- Algal blooms can deplete the oxygen and block the sunlight, and some can produce toxins that are harmful to the health of the environment, plants, animals, and people.
- When algal blooms block sunlight from reaching underwater plants, an ecosystem can be impacted. As with most plants, sunlight is vital for the growth of aquatic vegetation, which provides food and a place to live and grow for fish and animals.
- Blooms can last for months at a time, which can harm a waterway's ecosystem by causing declines in dissolved oxygen, underwater plant growth and fish populations.
- The more common problem associated with harmful algal blooms is the impact recreational activities and commerce due to the unsightly green scum and accompanying unpleasant odor.
- Cyanobacteria (blue-green algae) and red tides are examples of algae that can bloom and produce toxins that may be harmful to human and animal health.

What are the Causes of Algal Blooms?

Blooms can be due to a number of reasons. Two common causes are nutrient enrichment and warm waters.

Nutrient enrichment of water, especially phosphates and nitrogen, is often the result of pollution from nonpoint sources and can cause algal blooms.

Water temperature has also been related to the occurrence of algal blooms, with unusually warm water being conducive to blooms. Algal blooms on the river can be dramatic and are a result of excess nutrients from fertilizer, wastewater and storm water runoff, coinciding with lots of sunlight, warm temperatures and a wide, shallow river.

How are HABs Dangerous to Fish and Humans?

Harmful Algal Blooms (HABs) are dangerous to fish because they can deplete oxygen in water. When oxygen levels become too low, fish suffocate and die. Some algae species in blooms produce toxins that can kill fish and cause illness in human.

How do Fish get Exposed to HAB Toxins?

Most illness associated with HAB exposure is the result of consuming toxins that are present in shellfish or finfish. Some HAB toxins can become airborne during a bloom and people can become ill by inhaling toxins.

Harmful Algal Blooms

The important algal blooms which cause serious problems are:

1. Cyanobacteria (blue-green algae)
2. Harmful Marine Algae
3. Red Tide

Cyanobacteria

- ☆ Cyanobacterial blooms occur when organisms that are normally present grow extremly. Within a few days, an bloom of cyanobacteria can cause clear water to become cloudy.
- ☆ The blooms usually float to the surface and can be many inches thick, especially near the shoreline.
- ☆ Blooms can occur at any time, but most often occur in late summer or early fall.

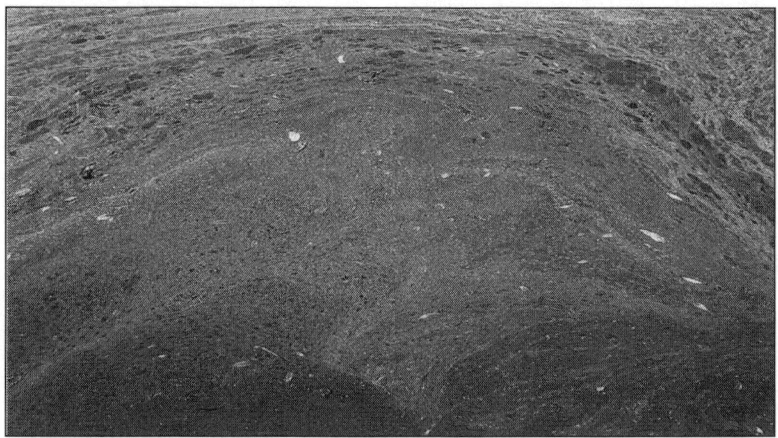

Harmful Marine Algae

- ☆ Harmful marine algae, such as those associated with red tides, occur in the ocean and can produce toxins that may harm or kill fish and marine animals.
- ☆ NSP is caused by eating shellfish contaminated with brevetoxins, which are produced by Karenia brevis, the marine algae associated with Florida red tides.
- ☆ NSP is a short-term illness with neurologic symptoms (such as tingling fingers or toes) and gastrointestinal symptoms.

Red Tide

- Under certain environmental conditions, microscopic marine algae called Karenia brevis (K. brevis) grow quickly, creating blooms that can make the ocean appear red or brown. People often call these blooms "red tide."
- K. brevis produces powerful toxins called brevetoxins, which have killed millions of fish and other marine organisms.
- Scientists know little environmental exposure of brevetoxin–such as breathing the air near red tides or swimming in red tides–may affect fish.

Algal Toxins

Some dinoflagellates and cyanobacteria produce toxins that can affect domestic animals and humans. Some of these toxins such as domoic acid, saxitoxin (paralytic shellfish poisoning or PSP toxin), brevetoxin, and cyanobacterial toxins (including anatoxins, microcystins, and nodularins) have been suspected.Marine algal toxins such as domoic acid, saxitoxin, and brevetoxin that bioaccumulate or are magnified in the food chain by fish and shellfish, and anatoxins from freshwater cyanobacteria, affect the nervous system; cyanobacteria that contain microcystins or nodularin cause liver damage.

Toxin Producing Algae

1. Cyanobacteria (blue green algae)
2. Golden algae (*Prymnesium parvum*)
3. Euglenoids
4. Dinoflagellates
5. Diatom bloom
6. Filamentous algae (blanket weed)

Some Toxic Producing Species

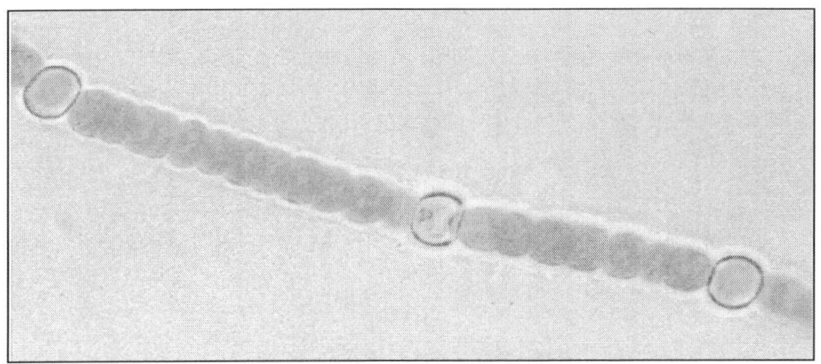

Nodularia **sp. (Cyanobacteria)**

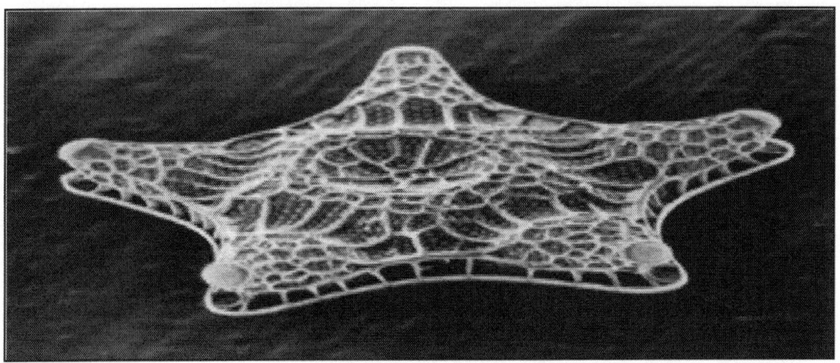

Aureoumbra lagunensis **(Golden algae)**

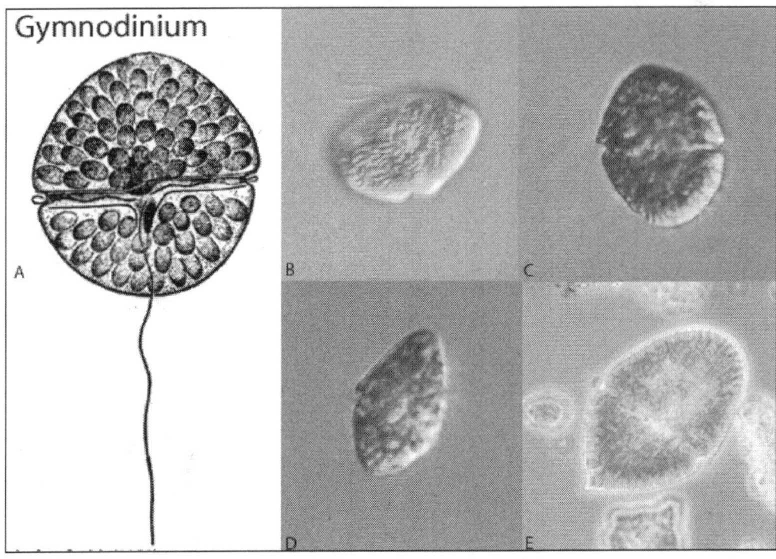

Gymnodinium **sp. (Dinoflagellates)**

Different Types of Fish Poisoning Caused by Algae

Poisoning	*Fish Species*	*Location*
Ciguartera Fish Poisoning	Reef fish such as barracuda, snapper, grouper, kingfish, amberjack (jack).	Caribbean, Florida, Hawaii.
Diarrhetic Shellfish Poisoning	Shellfish such as molluscs, oysters, and scallops.	West Coast (California, Oregon), and Northeast Coast (Maine).
Neurotoxic Shell-fish Poisoning	Oysters, fish, and clams. Exposure from breathing in fumes from sea foam during a bloom will also cause symptoms to occur. These HABs usually result in red tides.	Gulf of Mexico (Texas, Louisiana, Mississippi, Alabama, Florida). Has been associated with fish kills.

Remedy of Common Algae

Types	*Erradication Process*
Hair Algae	Adding any fertilizers to stop it from growing or add more potassium so the plants will use the phosphates quicker. It is easy to pull off so you can just pull it off until you get rid of it.
Green Spot Algae	If it is on the glass however most sucker fish like a pleco will get it off. For plants that keep getting it make sure to move them so no direct sunlight hits them.
Green water algae	Filtering with micron cartridges or diatom filters will successfully remove these algal blooms.
Brown Algae	However you could just buy either a group of SAE (siamese algae eaters) or otocinclus (commonly referred to as oto's). They will make a feast out of it. It is best to buy them in groups of 3 or larger.
Red Brush Algae	The only good fish to get rid of it is the SAE (siamese algae eater). Get a group of them and it should be gone shortly. The common plec has been known to also eat it (very rarely though).
Blue-green algae	No known cure. No fish will eat it because it releases toxins. Some people have tried using antibiotics before but make sure you know what you are doing first.
Fuzzy green en-crustations	Addition of grazing fish
Filamentous algae	Easily removed with a toothbrush.
Branched algae	Change water, from one fifth to two thirds of it per day

Algal Bloom Control

1. Change water, from one fifth to two thirds of it per day, depending on how turbid it is, and whether odour is included.
2. Also try feeding your fish less.
3. If your tank does not contain plants, or only contains low-light plants such as java moss, try lessening the time the tank stays lit.
4. If your tank does not contain plants, or if it is sparsely planted, try adding a plant or two.
5. If the water stays green, wrap your tank in black cloth or paper and turn off the light for a few days.
6. Avoid using algaecide.

Controlling Algae in the Aquarium

Physical Control

Filamentous algae can be reduced by gently combing and winding the strands around a small brush (*e.g.* toothbrush) in physical method.

1. Dense Planting

Planting your aquarium densely from the onset will help to deter the establishment of algae.

2. Water Changes

Regular, large water changes help to eliminate the nitrogenous waste and phosphates that algae need to thrive. Replace half the water every week (or at least every second week).

3. Light Starvation

Decreasing the amount of penetrating light that an aquarium receives can sometimes control algae. Blackwater extracts or wood (tannins) can help to darken the water and starve algae of the light needed for photosynthesis. Floating plants may help, however they will also reduce the amount of light that penetrates down to your plants.

Biological Method

Introduce Algae Eating Fish

Suitable fish include: *Otocinclus* spp. dwarf plecos, whiptail catfish, and members of the *Loricariidae* genus and most live bearers (*e.g.* Mollies). Members of the *Corydoras* spp. help to stir up the substrate and help deter algae from settling and taking hold.

Chemical Control

A weak solution of copper sulphate (bringing the copper concentration up to around 0.3-0.5ppm) will kill algae but may kill some aquarium plants too. Overdosing will result in plant and fish deaths.

Ultraviolet Sterilisers

These lights are set up after the filter unit so that all water passes within a specific distance to kill any microscopic plant or animal life within the water.

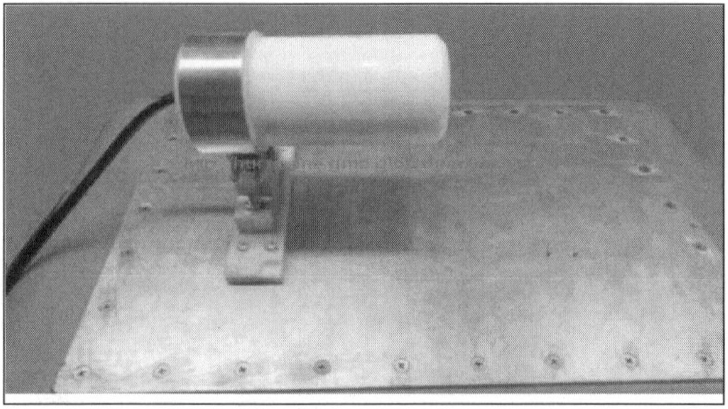

Other Methods of Control

Some others methods of algal control include: ozonization, reverse osmosis and resin exchange filters. Bottled or distilled water for water changes can help exclude the addition of tap water contaminants.

Conclusion

Marine biotoxins, produced by phytoplankton usually during harmful algal blooms, are some of the most potent toxins in the world and extremely dangerous. For some toxins, doses at the microgram per kilogram level are more than sufficient to kill. When enough toxin is accumulated in fish or shellfish, small amounts of cooked or raw tissue can kill a human. For example, recorded harmful algal blooms have produced enough paralytic shellfish poisoning PSP toxin in mussels that the consumption of one or two small mussels could have killed a normal, healthy adult human. While some toxins are very potent, *i.e.*, requiring only small amounts to produce illness or death, other less potent toxins may accumulate to such high levels that they still can cause harm. So its our duty to eradicate them to save ourselves.

Chapter 16
Problems of Flooding in Ponds and its Solution

Introduction

Flood can be described as a form of natural disaster when there is more water than the lakes, rivers, oceans, or ground can hold.Flooding is an extreme weather condition naturally caused by rising global temperature, resulting in heavy downpour, thermal expansion of the ocean, and glacier melt, which can also lead to rise in sea level, thereby causing salt water to inundate coastal lands. It is one of the most common environmental hazards and it regularly claims over 20,000 lives in a year and adversely affects around 75 million people world widely. Floods are often deadly, damaging and devastating. They kill lots of people, damage houses and crops, and cause extensive destruction. It also causes death of fishponds and rivers. In broader terms, floods are of two types;

1. Natural floods and
2. Catastrophic floods.

"Any flooding around the pond is purely due to the tide".Pond flooding generally occurs due to heavy rainfall, river flooding, no inlet and outlet system or by erosion of embankment. Flood majorly destroys pondsfollowed by pollution of water (31.3 per cent), flood affects the water pH (13.8 per cent) as well as oxygen depletion in the pond.

Classification of Flood

Floods can be classified by the factors and conditions of their generation. Thus, river floods comprise one large category, and inundations of shores of lakes and seas are another. Large scale floods that originate from, and are exacerbated by, human activity can be put into a specific group.The speed of a flood wave is slowed by obstacles, constrictions and dams in the river channel.

1. Coastal Floods

Coastal Floods usually occur along coastal areas. When there are hurricanes and tropical storms which will produce heavy rains, or giant tidal waves created by volcanoes or earthquakes, ocean water may be driven onto the coastal areas and cause coastal floods.

2. Arroyos Floods

An arroyo is river which is normally dry. When there are storms approaching these areas, fast-moving River will normally form along the gully and cause damages.

3. River Floods

This is the most common type of flooding. When the actual amount of river flow is larger than the amount that the channel can hold, river will overflow its banks and flood the areas alongside the river and this may cause by reasons like snow melt or heavy spring rain.

Factors and Conditions of Flood Generation

Various factors, processes, and their combinations can induce floods. Generally, two major factors usually combine to generate floods and determine their specific characteristics. The first is the physical process that generates the change in the interaction of the lithosphere, atmosphere and water masses. The second depends on the geographic situation in the area where the flood takes place, and this determines the scale of a flood, such as the area and depth of inundation, and its duration. Regional specificity, including climatic conditions, determines the most characteristic features of floods and their recurrence. The role of the geographic situation in giving rise to floods is very important. For instance, heavy rains can be the cause of inundation in river valleys, but they never createfloods on sea coasts. In the same way, storms generate surges on low-lying shores of seas and lakes, but never create surges on shores with elevated and steep coasts.

In specific conditions, the effects of a flood depend on a large number of additional circumstances, such as depth and duration of inundation, velocity of the flood's wave movement along the river, height and velocity of tsunami wave travel, and so forth. Accordingly, the protection and mitigation measures for floods vary greatly too.

Impact of Flood on Aquatic Life

Flood is a very devastating problem in aquaculture and its impact is very serious for future periods. However, sometimes it has some beneficial impact.

We can categorized the impact of flood into 4 types:

1. Structural change
2. Biodiversity change
3. Changes in physico-chemical parameters
4. Economic damage/loss

1. Structural Damage/Loss

A. Breakdown of Pond Dykes

The flood may destroy the pond dykes due to hydraulic pressure of the water if the pond is fully loaded. The water of the pond and surrounding will mix up and create problem in culture units and ponds.

B. Topography of the Aquatic System Changed

A devastating flood can completely wiped out several ponds, situated at that area. In a management practice, there are several ponds of several size and depth for stocking fry, fingerling or brood fish. But flood completely wipe out the entire barrier between these pond, mixing or vanishing the entire stock. Due to flood, topography of the aquatic system completely changed.

C. Accumulation of Debris

Due to flood, several unwanted weeds, debris accumulated in the pond. Their decomposition deteriorates the water quality causing a big problem in fish pond management.

D. Reduction of Pond Depth

to introduction of excess sediment and nutrients (river flooding) the pond depth is reduced. Accumulation of heavy sediment in the inlet and outlet system will also hamper the proper drainage system. It actually hampers the fish culture, and also the removing of the extra sediment costs a lot for a farmer.

Benefit

Sometimes it may be beneficial due to introduction of several important nutrients and make the soil organically rich.

2. Biodiversity Change

Due to river flooding in pond, the complete biodiversity can be wiped out. The cultivable fishes escaped and there is a chance of damages to the habitat, food chain, species diversity and stability of the pond ecosystem. Several weed and predatory fishes may get entry to the culture units. The total plankton and fish population may be destroyed due to entry of saline water from flooding of river in coastal areas.

Benefit

Sometimes due to rain flooding the soil and water become productive in future and thereafter it is a chance of more production in culture units.

3. Changes in Physico-Chemical Parameters

After flooding the physico-chemical parameters of soil and water are changed according to the type of flood.

- ✰ The temperature of the water may be changed.
- ✰ If the river is flooded and the water enters to the freshwater pond, then the salinity, ph, hardness and other parameters are changed.

- ☆ The Calcium and Phosphorus level may be changed in any type of flooding.
- ☆ Turbid water is a common matter in case of river flooding.

4. Economical Damage/Loss

A. Production Loss

The water coming from flood may bring several bacteria or parasites with them and when this water mixes with the pond water, all the fish stock may be infected and leads a very economic loss to the farmer. If the pond is adjacent to agricultural or industrial area, the waste product and other toxic chemical disposal will mix with the pond water during flooding and these poisonous chemical can lead up to mass mortality and leads a very economic loss to the farmer.

B. Reconstruction

High siltation totally damages the inlet-outlet system and very often the flood may also break the dyke. So, a farmer has to again reconstruct his pond and drainage system which really costs a lot.

General Solution for Flooding in Pond

Mitigation Process

To avoid severe economic loss during flooding some steps should be followed by the farmers:

1. Hydrographic charts should be followed at the time of pond construction. Flood prone areas should be avoided.
2. Pond dyke should be constructed properly by maintaining height and other measurement.
3. The fish farm should maintain a proper drainage system and must contain pumping arrangement.
4. The total fish population should be harvested if very possibilities of flood.
5. The entire pond area should be covered by net.
6. On pond dyke plantation is very much scientific to prevent soil erosion of dikes.

After flooding a farmer have to maintain several steps-

1. Check the Physico-chemical Parameters

After flooding, conditioning is very much important for the good health of fish.

- ☆ If pH is changed lime or organic manure should be added in proper dose as required.
- ☆ If pond becomes turbid it should give time to settle the soil particles.
- ☆ If the salinity is changed the whole water should be drained of.

2. Check the Health of Fish

If the fish is affected by diseases treatment should be taken.

- ☆ External fungi- Malachite green(0.1 ppm flash for 1 hour)
- ☆ External protozoa- formalin+malachite green
- ☆ Parasitic copepod- Chlorophus(0.25-0.5 ppm. 3-5 times at weekly interval)
- ☆ Ulcer disease- Oxytetracycline @ 25-75 mg/kg per day for 6 days
- ☆ Columnaris disease- Chloramphenicol orchlorotetracycline @ 10-20 ppm.

Conclusion

Flood is a large amount of water covering an area that is usually dry. It is an overflowing of a great body of water over land not usually submerged.Floods are among the most devastating natural disasters in the world, claiming more lives and causing more property damage than any other natural phenomena. Pond flooding is a natural or artificial disaster that can affect the pond ecosystem as well as the production of pond. It also deteriorates the soil and water quality and fish health also. So the farmer should construct the farm scientifically using hydrographic charts and keep the proper mitigation measures and also proper disaster management tools and practices so that, they can face a minimum loss.

Chapter 17
Comparative Study of Drugs

Introduction

"The increasing problems associated with infectious diseases in fish, the limited number of drugs available for treatment and prevention of these diseases, and the rapid increase in resistance to these antibiotics represent major challenges for this source of food production worldwide."

All aquaculture operations will have a demand for drugs, biologics, and other chemicals, collectively referred to as "regulated products". This may include:

1. Disinfectants as part of biosecurity protocols,
2. Herbicides and pesticides used in pond maintenance,
3. Spawning aids,
4. Vaccines used in disease prevention,
5. Marking agents used in resource management.

Despite the best efforts of fish culturists to avoid pathogen introductions, therapeutic drugs are also occasionally needed to control mortality, infestations, or infections. It is critical that culturists have access to regulated products that are safe and effective and apply them in a manner that is consistent with their intended use, best management practices, and relevant rules and regulations.

Uses of Chemicals and Drugs

1. Soil and water quality of the pond is improved by using different types of chemicals and drugs, as a result fish culture and fish production increase.
2. With the application of antibacterial drugs fish diseases can be controlled.

3. Phytoplankton production may be increased by using different types of chemicals which increase the oxygen concentration in the pond and fishes get sufficient nutrition.
4. Nutritional chemical compounds are supplied through feed to the fish, so as to increase the growth of the fish.
5. Some chemicals help for colouration of fish.
6. During transportation different types of anaesthetic are use.
7. By using different types of pesticides and herbicides, undesirable aquatic weeds and harmful insects are controlled.

Different Types of Chemicals and Drugs

For Soil and Water Treatment

Lime

Lime is a major chemical used for soil and water treatment in aquaculture. It is used to correct pond bottom and stabilize water pH. It is also reported to ensure a healthy plankton bloom.

There are at least four types of lime use:

- ☆ Agricultural lime/lime stone or crushed shell ($CaCO_3$)
- ☆ Hydrate lime or slake lime ($Ca(OH)_2$)
- ☆ Quicklime/burnt lime or burnt shell lime (CaO)
- ☆ Dolomite or dolomite lime ($CaMg(CO_3)_2$)

Use: It is used @ 3kg/bigha – 150kg/bigha.

Organic and Inorganic Fertilizers

Disodium Ethylene Diamine Tetraacetate (EDTA)

EDTA is a chemical used to improve water quality by reducing heavy metal concentrations. In shrimp larval rearing, it is applied at 10 mg/L prior to stocking of nauplii.

Dosage: EDTA is normally applied at 1-5 ppm to remove organic substances in the water.

Zeolite

Zeolite is applied to shrimp ponds to remove hydrogen sulphide, carbon dioxide and ammonia, as it has a strong capacity to absorb molecules. Shrimp farmers use zeolite to clean pond bottoms. Zeolite is available in the market under various brand names, and it is supplied as fine grains in bags of 20 kg.

Dosage: The recommended dose is 180-350 kg/ha.

Single Super Phosphate

It is a fertilizer used in aquaculture to increase the productivity of the pond.

Dosage: It is used @ 30kg/bigha.

Iodine

The commercial solution of iodine, as used for disinfection of wounds, contains 10 per cent iodine, dissolved in alcohol. The solution has a dark brown Colour.

Method of use: One part by volume of commercial I2 is mixed with nine parts by volume of water. This solution is used directly for touching sore spot, wounds or fungus. This is done by brush or tuft cotton wools, dipped into the solution. I2 is never used for preparing baths.

Copper Sulphate

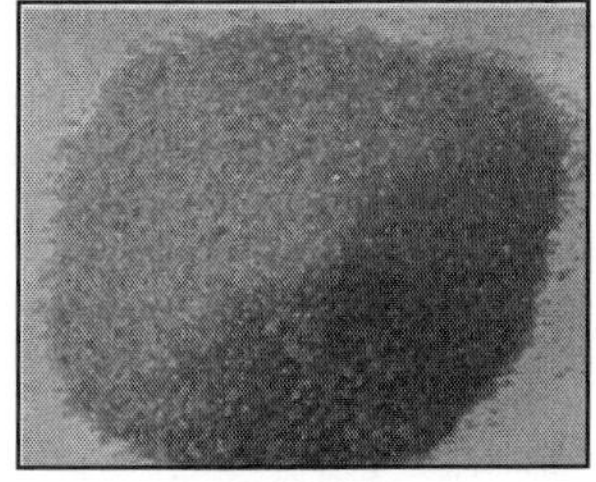

Copper sulphate is vigorously used in aquaculture sector in W.B. It has been reported very useful against bacterial gill diseases. It is active against protozoa and has a particular indication for Amyloodinium and Cryptocaryon infections in marine aquaria where the target species tolerance is much better than in freshwater. It is also toxic to fish which have been exposed to a reduced osmotic pressure without time to acclimate; the loss of sodium ions exacerbates copper toxicity.

Use: it may be used in case of ulcer disease by dip treatment at a conc. of 500ppm for 1min.

Hydrogen Peroxide

Hydrogen peroxide is used for treatment of cotton wool disease. Its effect on bacteria due to its decomposition to hydroxyredicals and on protozoa, and possibly on flukes, due to intracellular oxidation.

Method of use: Hydrogen peroxide may be used only in fresh tap water, in containers in which no plants, sands or other materials are present; otherwise the effective activity of the chemical is considerably decreased. 3 per cent commercial solution of hydrogen peroxide is added per 17.5cu.cm. of freshwater. In this solution fishes are bathed for 15-20 min. Treatment must be repeated for several times.

Fumagillin

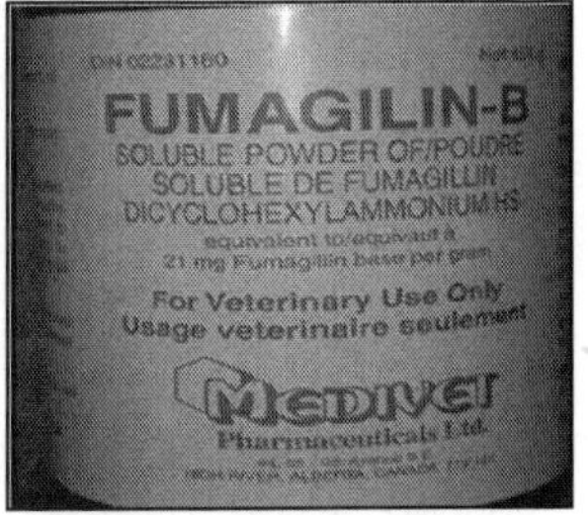

Fumagillin is a antibiotic produced by parasitic fungus, *Aspergillusfumigatus*, and acts by inhibiting RNA synthesis. It is acidic and is normally presented as dicyclohexylamine salt. This salt is sparingly soluble in water but moderately soluble in ethanol. Fumagillin is heat-labile so it cannot be incorporated in feed before pelleting. Fumagillin has low anti-bacterial and anti-fungal activity and is primarily active against protozoa. It has also been shown to be active in control of proliferative kidney disease, a condition of freshwater salmonids for which a putative protozoan pathogen has been hystologically but never isolated and cultured.

Use: A diet containing 0.1 per cent Fumagillin effectively prevented the development of disease.

Chloramin-T

Chloramine-T is used as a disinfectant and as a treatment for bacterial gill disease and occasionally fin-rot. Bacterial gill disease is a multi-factorial infection in which the bacterium, *Flexibacterbranchiophilum*, is usually present, and protozoa, Ichthyophthirius, Chilodonella and Tricodina may all be involved. It is traditionally treated with Chloramine-T. It has been shown to prevent water-borne *Aeromonassalmonicida* infection.

Dosage: For dip treatment concentration of 8.5 – 10ppm are recommended, with exposure for one hour daily for three days.

Therapeutants

Acriflavine

This is a deep orange, granular powder. This chemical is used to treat fish eggs and aquarium fish for infections by bacteria and external protozoans. It is recommended as a treatment for *Flexibactercolumnaris* in seabass and against bacteria in walking catfish.

Use: It is used @ 130ml/350kg fish. 100 ppm for dipping eggs for 3-5 sec and 25 ppm for prolonged treatment of *F. columnaris* in seabass. For catfish, only 5 ppm is recommended. For aquarium fish, 5-10 ppm bath for 24 hr is recommended.

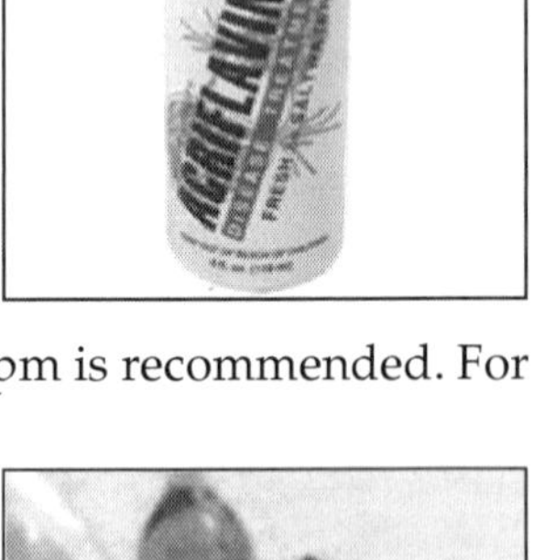

Potassiuum Di-Chromate

This is an orange-red crystallization chemical, but it also be obtained in powder form. For treatment of wounds, fungal infections, sore spots or moldy places.

Use: stock solution is prepared by dissolving 1gm in 99cu.cm of water. It is used at conc. of 1:25000 or 1:20000 for bathing the fishes for several days, but not more than 10 days.

Dipterex

Dipterex is widely used to treat for crustacean, monogenean, and protozoan parasites in pondcultured fish. Recently, Dipterex has been applied to water storage canals and reservoirs to eliminate wild crustacean vectors of shrimp viruses that enter ponds through the water supply.

Use: A rate of 0.25-0.3 ppm is used for prolonged treatment; however, it is necessary to repeat treatment 2-3 times at 3-day intervals.

Trichlorphon

It has been reported to be effective as a cure for parasitic disease specially argulus.

Use: Fishes may be treated by using gradually increasing strengths (0.2mg/lit. for 24 hrs, 0.4mg/lit. for 30 min.)

Trypaflavine

Trypaflavine may be used in treatment of Myxobacterial disease by bathing fishes in a solution of 3-6gm/m^3 for 6-12 hrs for consecutive 1-3 days.

Antibacterial Drugs

Penicillin (Benzylpenicillin Sodium)

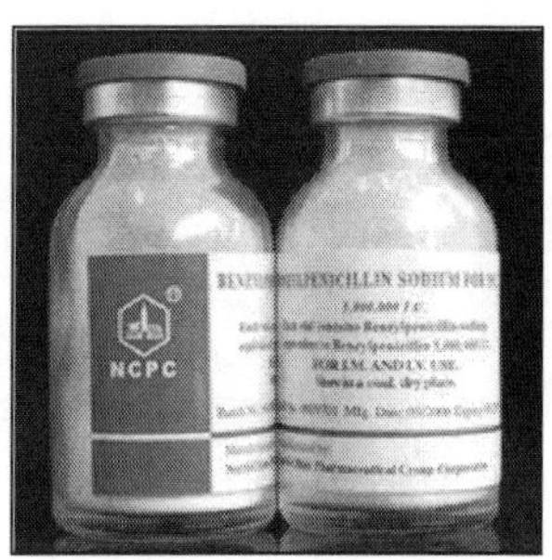

Penicillin is the most well-known antibiotic. It is produced by the fungus, *Penicilliumnotatum*. It has a narrow spectrum of action. It is active against a number of gram- positive bacteria, but inefficient against most gram-negative bacteria. Since most bacteria pathogenic to fish are gram- negative, so, Penicillin is without red value for treatment of bacterial infections in fish. Penicillin is effective against *Ichthyophthirius*sp, Furanculosis, Edwardsiellosis disease.

Use: The dose rate in the range 40-80mg/Kg/day have been recommended.

Sulfonamides

The Sulfonamides are a large range of synthetic compound. They are amphoteric but most of them are more soluble in alkalis than in acids. The Sulfonamides have a broad spectrum of activity. One advantage which sulfonamides have over other antibacterial drugs is that they are absorbed through the gills. This may be used in treatment of Vibriosis, caused by *Vibrio anguillarum.*

Use: It is used through feed at the rate of 200mg/Kg fish/day initially, later maintained at 100 mg/Kg fish/day for minimum 10 days.

Sulphamethazone

It is used in treatment of red mouth disease caused by *Yersinia ruckeri*, through feed @ 200 mg/Kg/day for 5 days.

Sulphadizine

One of the modern sulpha drugs, experimentally used for treatment of bacterial infections in fish. Sulphadiazine is a white crystalline powder.

Use: It is only slightly soluble in acidic water(130 mg/lit. at pH 5.5 at a temperature of 37°c) much more soluble in alkaline water(2 gm/lit pH 7.5 at 37°c).

Other Chemicals and Drugs

Ammonia

Ammonia may not be used in planted tanks or ponds since it has very bad effect on plant growth. It has strong prickling smell.

Method of use: It is mixed with water and used in ratio of 1:2000. But when fishes have strong slimy skin then it is used in 1:1000 ratio.

Ammonium Sulphate

It is used in pond for oxygen production from bloom starved of nutrients. It is spread over the pond surface for the treatment of algalblooms caused by *Prymnecimparvum*.

Method of use: It is used @ 10 ppm, pH is 9 and temperature is 20°c.

Aureomycin

This is an antibiotic isolated from cultures of *Streptomycesaureofaciens*. It is a golden yellow crystalised substance with weak basic reaction.

Use: It is effective against many gram positive and gram negative bacteria, certain protozoa (Icthyphthirius, Oodinium) and some viruses. Food containing aureomycin increases growth rate of young fishes and intensification of colours.

Method of use: Aureomycin treatment should be extended over a longer period than is absolutely necessary to cure fish; 4 days will be maximum for bacteriostatic lower concentration and not more than 2 days for strong, bactericidal concentration.

CIFAX

It is a drug which is used for Ulcers. It is used @ 100-200ml/bigha.

Cleaner

It is a medicine which is used for Argulus treatment. It is used @ 350ml/bigha.

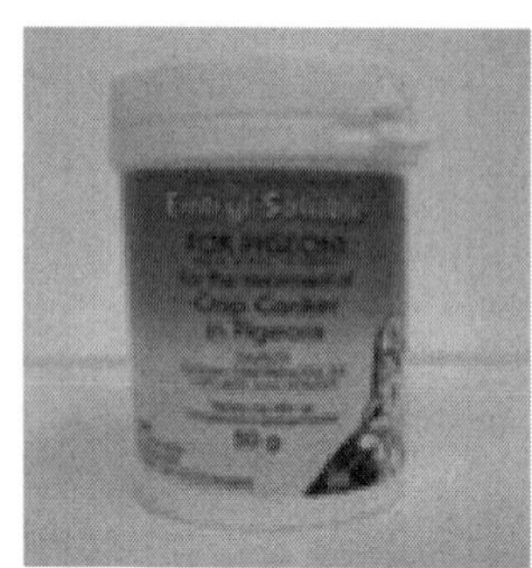

Furnace

It may be used in treatment of bacterial disease by bathing fishes in a solution of 1mg/lit. for 1 hr.

Emtryl

It is a medicine for endoparasitic diseases. It is given through feed @ 40mg/Kg fish/day for 3-5 days.

Furazolidone

It is used during soil treatment with lime to kill spores. It is a nutrofuran antibiotic. Only slightly soluble in water. It has not toxicity and secondary side effects.

Use: It is used as feed treatment @152-194 mg/Kg fish/day.

Gammexane

This is a crystalline substance with a persistent acid smell. It is used as insecticide.

Method of use: A 1 per cent stock solution in absolute ethanol is prepared and 1cu.cm. of this is mixed with 10lit. of water, with through stirring. The treatment must not be prolonged more than 2-3days.

Gentamicin

It is used for treatment of bacterial disease of ornamental fishes. It is injected through intrapectorial region of fish @ 20-40mg/Kg fish.

Hemisulphate

It is used for bacterial diseases by bathing or flowing it at the conc. of 20ppm for ½ hr.

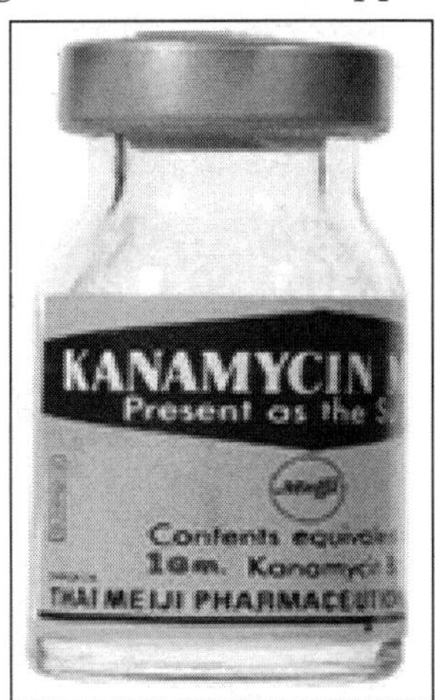

Kanamycin Sulphate

It may be used in treatment of bacterial disease caused by *Micobacterium* sp. through feed @ 10mg/Kg feed for 4 days.

Magnesium Sulphate

White crystal of magnesium sulphate is used as a laxative in human medicine, has also been recommended for treatment of constipation of fish. The fish should be fasted for 2 days and then a pinch of crystals should be dropped into the water in the hope that the fish will swallow them.

Methylene Blue

It is a dye. The solid substance appears brownish red, but solution colour is blue.

Use: It may be used for treatment of Ichthyophthiriasis, skin and gill flukes and as a palliative medicine in all cases of diseases of gills, where fishes suffer from difficulty in breathing.

Method of use: Stock solution is prepared by dissolving 1 gm in 100 cucm. of hot water.

Metronidazole

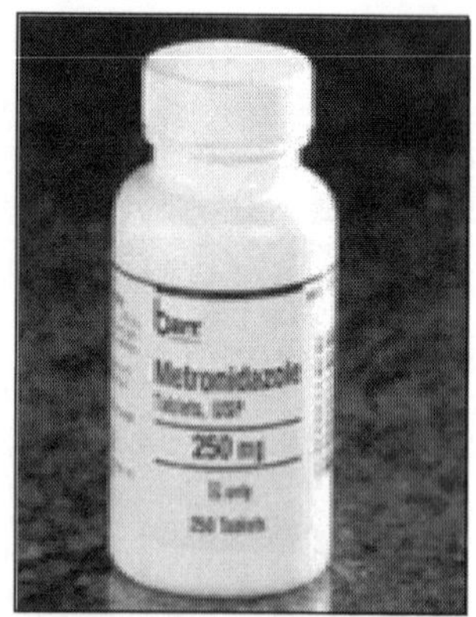

It has been reported to be effective as a cure for parasitic diseases specially Trypanosoma and Hexamita.

Use: It is given through feed @ 50mg/Kg fish/day for 6 days.

Pentamycin

It is a well-known antibiotic. It is commercial available in tablet form. It is used in case of bacterial diseases.

Use: It is used @ 5 tablets/bigha.

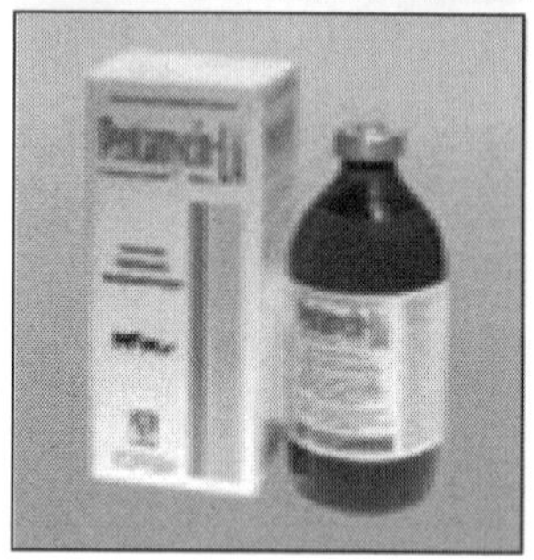

Quinine Hydrochloride

It is a white, crystallized chemical of several quinine compounds. It is used for treatment of white spot disease and some case of sliminess of skin.

Method of use: Stock solution is prepared by dissolving 3 gm. of Quinine Hydrochloride in 300cu.cm. of water. 95 cu.cm.of stock solution is added per cu.ft. water. This quantity should not be added at once, but in 3 approx. equal parts at intervals of half to one day.

Quicklime

It kills all free living leeches. It is used atthe concentration of 1gm/2lit. of water.

Salt

Salt may be used in treatment of sliminess of skin, in early stages of skin flukes and as an additional treatment in several cases of bacterial diseases. Fishes may be treated by bathing them in a strong solution for a short period or treatment may be permanent by gradually increasing strengths.

Use: 15-20gm salt is used per 1 lit. water. Salt treatment should not be done in presence of galvanized iron or zinc, otherwise zinc poisoning could occur.

Hazard and Averse Impacts

Excessive use chemicals and drugs brings adverse effect to the fish

- ☆ The use of synthetic drugs in hatcheries, especially chloramphenicol and oxolinic acid, may adversely affect larval growth and inhibit defence mechanisms. They may reduce the activity of acid or alkaline phosphate acid, which are considered to play a major role in the protein metabolism of shrimp.

- Ionics incorporated with steroids usually have good appearance and uniform size. Once they leave the hatcheries and reach the grow-out ponds, however, they become sensitive to external stresses and pathogens, resulting in growth retardation and/or mass mortality. Shrimp broodstock fed diets containing nitrofurans for a long period may produce eggs of lower quality.
- Some chemicals, such as a mixture of formalin and malachite green, give a very effective treatment against "Ich," but malachite green has teratogenic and toxic potential.
- Insecticides such as Dipterex have been widely used as effective treatments for crustacean parasites in fish ponds. Since their introduction into shrimp culture systems to treat small shrimp and other viral disease vectors such as crabs, the amount of these chemicals applied by aquaculturists has increased considerably, causing a great concern for the environment, as very little is known about their residues in aquatic animals and in the environment.

Recommendation

To Farmers

- Do not use chemicals to attempt to overcome poor management.
- Do not let aggressive promotion by suppliers of chemical products lead to the overuse of these drugs.
- Do not use chemotherapeutants due to their availability; use them only as a last resort.
- Treat diseases based on accurate diagnoses, and treat them as early as possible.
- Remember that high stocking density leads to a greater risk of disease and an increase in the need for chemical use.

To Producers and Suppliers of Chemicals

The industry must be responsible for giving accurate information on the specificity of chemicals and must clearly exhibit the ingredients of each product.

Conclusions

Clearly, chemicals and drugs play an important role in present aquaculture systems, whether they are intensive, semi-intensive, semi-closed or closed systems. However, it is also obvious that good management can considerably reduce the chemical use in aquaculture. Certain types of chemicals, if used inappropriately, can cause damage to animals and the environment. Overuse of chemicals, especially antibiotics, not only increases production costs but also intensifies adverse consequences. The adverse impacts of chemical use in aquaculture may be outweighed by their advantages; however, there has been increasing concern about their uses, and they must be used with great caution. Chemical residues in food products originating from aquaculture and the effects of chemicals on the aquatic environment should be monitored.

Chapter 18

Comparative Study of Common Fish Food and its Use in Aquaculture

Introduction

Feeding of artificial diet balanced in all nutrients has assumed foremost importance in aquaculture industry. Artificial feeding is an essential practice in an aquaculture operation accounting for over 60 percentage in total input cost. Consequent to researches done in India and abroad, a strong database has been developed about nutrition of the great majority of these cultivable fish and shellfish species,and several feed formulation have been successfully undertaken.

Common Types of Feed

- ☆ Groundnut oil cake
- ☆ Coconut oil cake
- ☆ Soya bean meal
- ☆ Cotton seed meal
- ☆ Rice bran
- ☆ Sorghum
- ☆ Wheat bran
- ☆ Maize
- ☆ Sun flower seed cake

Groundnut Oil Cake

Groundnut oilcake is a safe feed for fish. It has been demonstrated in India that diets consisting of 50 per cent of the material as the principal protein source can be used in complete feeding of carps. Its high polyunsaturated fatty acid (PUFA) content also makes further addition of fats to such diets unnecessary. Because groundnut protein is especially low in methionine, the oilcake should be used with methionine-rich protein supplements or with synthetic methionine to achieve a proper balance of essential amino acids in the diet.

Quality in groundnut oilcake depends on whether the material is made from decorticated nuts. Removal of the fibrous hulls yields a better quality product with higher protein content. This will also depend on the amount of residual oil in the cake. In a few countries there are two varieties of groundnut oilcake: the mill produced and the 'country' produced. The former is an industrial product with a generally lower residual oil content (less than 7 per cent), whereas the latter is produced on often a very small in villages and may contain as much as 13 per cent oil.

Contaminant of Groundnut Cake

The principal (and often very serious) contaminant of groundnut cake is aflatoxin. Aflatoxin is a group of highly toxic substances produced by the mould

Aspergillusflavus. The most prominent of these are Aflatoxins B_1, B_2 and G_1. The toxins are produced only when the mould exists as a pure culture. Aflatoxin is a hepatotoxin and mortality among afflicted animals and fish invariably results from severe liver damage. A small dose over an extended period of time produces cancer of the liver in humans. The half lethal-dosage, for trout weighing 100 g is 0.5 ppm in the diet. Improper postharvest handling of nuts is usually blamed for the presence of aflatoxin in groundnut although pre-harvest infestation has also been documented. Aflatoxin is not destroyed by heat.

Coconut Oil Cake (*Cocos nucifera*)

Coconut oilcake is widely available as a feed for fish and livestock in countries such as Sri Lanka, Indonesia and the Philippines. On the average, 1000 nuts yield 60 kg oilcake (called poonac in South India and Sri Lanka). Compared with residues of oil seed extraction, coconut oilcake is low in protein and high in fibre. It is difficult to make a water-stable pelleted feed if the formulation contains too much coconut oilcake, as the material tends to absorb large amounts of water, causing the pellet to disintegrate. Poonac is commonly used as a supplementary feed in the pond culture of carp in Sri Lanka. It has also been included in the diet of milkfish in Indonesia.

One of the main problems in its use is the tendency of coconut fat to turn rancid in storage. The effects of feeding rancid coconut oilcake to fish have not been studied, although short-chain free fatty acids in the diet have been shown to suppress growth in at least two species, the trout and channel catfish. However, rancid soya bean oil does not appear to affect the growth of carp.

Soya Bean Meal

Soya bean is one of the world's most important oil seed crops. A legume, like the groundnut, soya bean has relatively low oil content. Because of this, and because the seeds can be easily flaked by crushing, oil removal of flaked soya bean is carried out by solvent extraction. The residue is very low in fat (less than 1 per cent) and high in protein. Decorticated soya bean meal may contain up to 55 per cent protein.

Soya bean meal is a good source of essential amino acids (EAA) and is one of very few plant sources rich in lysine. In recent years, soya bean meal has been increasingly used as a substitute for more expensive fish meal in compound fish feeds.

Soyabean meal has a relatively low methionine content which can be corrected by addition of synthetic forms of the amino acid in the feed. Anti-nutritive factors, principally urease and trypsin inhibitor present in the raw bean are usually destroyed by heat during the oil extraction process. Full-fat soyabean processed under high pressure has recently been successfully used as a complete substitute for fish meal in the diets of both warm-water (catfish) as well as cold-water sp. Considerable work has been carried out on the extent to which processed soya bean products can replace fish meal as a protein source in fish diets. While such replacement can be total for carp without effect on growth similar replacement in diets for trout resulted in slower growth and poorer feed conversion (trout) species.

Cotton Seed Meal

Cotton seed is rarely used as feed. The seed is processed by first removing the lint and then cutting the tough hull to release the kernel. These operations are carried out by machines. The kernels are then crushed and the oil removed by mechanical means or by solvent extraction. In developing countries the former process is most common. Although cotton seed cake has high protein content, it is rather lacking in lysine. For fish, therefore, it should be used in combination with other materials such as soyabean meal.

Gossypol Content

An important factor limiting usage of cottonseed cake or meal in fish feed is its gossypol content. Gossypol is a toxic phenolic compound confined to the genus of cotton, Gossypium. It is the yellow pigment which constitutes 20 to 40 percent of the substances inside the glands of the seed kernel.

The amount of gossypol is proportional to the number of glands present as it does not occur elsewhere in the seed. Cottonseed usually contains 0.4 to 1.7 percent gossypol. There have been few studies on the toxic effects of gossypol in fish although from animal studies, it has been shown to inhibit digestive enzymes. Studies with trout have shown that the toxin persisted in body tissues 12 months after its feeding to the fish. Much of the retained gossypol was found in the liver, kidneys and spleen. On the other hand, glandless (gossypol-free) cottonseed meal has been used at up to 40 percent in diets for salmon without deleterious effect.

Binding of gossypol in the oilcake or meal renders the product non-toxic. This can be achieved by treating with water and cooking with steam. The gossypol effect may be counteracted by the addition of ferrous sulphate, although the level of additions required has not been established for fish. Only free gossypol which can be extracted with aqueous acetone is physiologically harmful.

Rice Bran

Freshly produced rice bran has high oil content (14-18 per cent). The oil is sometimes recovered by solvent extraction because of its high commercial value, especially in countries where demand for cooking oil exceeds supply. The oil present in the bran is rich in polyunsaturated fatty acids which undergo rapid oxidation under normal storage conditions. Rice bran that has turned rancid has markedly reduced feed value. Rice bran has higher protein content than the grain. It is also fairly high in fibre, thus limiting its use in fish feeds. Often, rice bran is mixed with rice polishing at the mill and sold as rice bran. Adulteration with hulls lowers its feeding value considerably. Good quality pellets can be made from feeds containing unextracted rice bran.

Oil extracted rice bran is almost totally devoid of oil. It is the type of rice bran favoured by feed mills because of its longer shelf-life. Its protein content is somewhat higher than fresh rice bran while its energy content is lower. Pelleted feed containing a high proportion of de-oiled rice bran are not very water-stable. This is because of the water absorption characteristics of fibre, which will be fairly high in such feed. Rice bran has been successfully used in compound fish feeds for both warm as well as cold water species.

Sorghum

Sorghum is grown in warm areas where rainfall is inadequate for maize cropping. Its feed value is comparable to maize for most species of livestock. In Mexico where maize is normally reserved for human consumption, sorghum is used as a substitute in compound feed for trout and carp. Sorghum contains tannic acid which decreases the availability of methionine in the diet. The amount of tannic acid in sorghum depends not only on the variety of sorghum but also on the conditions under which the crop is cultivated.

Wheat Bran

Wheat bran is the primary coat of the wheat grain. It has fairly high protein content. Because it is also high in fibre, it has a laxative effect when fed at excessive levels to animals. Such an effect, however, will be more difficult to determine in fish. Wheat bran has been successfully fed at fairly high levels to various species of fish without adverse effects on growth. Too much wheat bran in a formulation results in pellets with poor water stability due to the water absorption characteristics of fibre.

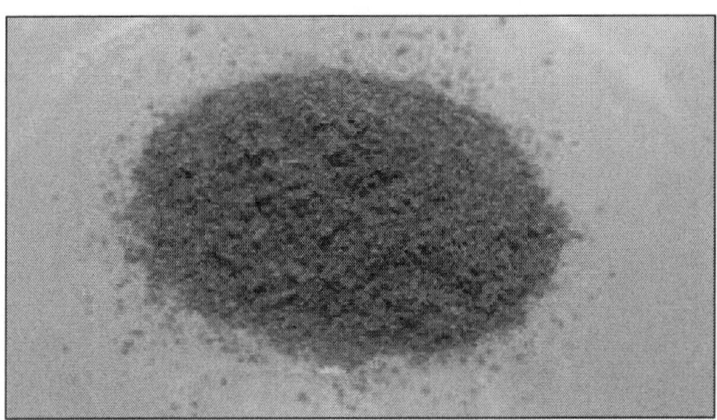

Maize

Maize is grown throughout the tropics and is the principal cereal grain employed in the feeding of livestock. Maize has also been fed successfully to fish. Whole grain has been used as a supplementary feed for tilapia and carp. However, maize is usually used in finely ground form as an energy component in compound feeds. While there is little published data on the digestibility of starchy components of cereal grains in general, studies on maize have indicated enhanced utilization after cooking. Normal pelleting processes only partially gelatinize the starch granules whereas total starch gelatinization occurs when the cereal is subject to high temperatures under moist conditions as in extrusion cooking. For channel catfish, such processing methods greatly increase the digestible energy value of maize, indicating the relatively low digestibility of raw starch for that species.

Wet milling of maize is an industrial process for production of valuable components of the maize kernel. The main products of the process are: corn oil and starch (from which dextrin and glucose are obtained after further processing). The by-products include: maize gluten meal, maize bran and maize germ meal. Often all three are mixed together and sold as maize gluten feed. The protein content of maize gluten meal is comparable to that of soyabean meal. However, maize protein has very low lysine content, and maize gluten meal or maize gluten feed should be used along with other lysine-rich protein sources (animal protein, or soyabean meal).

Sunflower Seed Cake

Sunflower is a fast-growing cash crop grown primarily for its oil. The leaves and stems of the plant, at two months, are rich in protein and low in fibre and are suitable as feed for fish. There are no known phytotoxinspresent in the sunflower plant. Sunflower seed cake made from dehulled seeds has fairly high protein content. It is also richer in the sulphur-amino acids than soyabean meal although its lysine content is much lower. Sunflower seed meal has been shown to be a good substitute for fish meal in diets for tilapia.

Fish Meal

Trash fish used as raw material for fish meal. There are two principal ways of making fish meal. The first involves direct oven drying of whole fish and/or residues (the latter usually from canning operations). The product is then ground to produce what is known commercially as white fish meal. The second method involves the cooking of whole fish followed by pressing to remove the oil and finally grinding up of the press cake after it has been dried in a steam jacket. This process produces a dark coloured fish meal. Peruvian and Thai fish meal belong to the latter category. There is a notable difference between these two varieties. Whereas Peruvian fish meal is made from the main catch, usually of a single species, anchovy, Thai fish meal is derived from by-catch consisting of many species. The Peruvian product is therefore more consistent with regard to quality. Protein content of Peruvian fish meal is also higher - 65 per cent compared to 50-60 percent for Thai fish meal.

Fish meal quality is also very much affected by the drying process. Some backyard fish meal factories in West Malaysia dry the processed fish over fire in

concrete tubs. Although the meal is mechanically stirred during this operation, without proper supervision, over-cooking frequently occurs. The fish meal becomes very dark coloured, a sign of scorching of the product. On the east coast of India and in Sri Lanka, fish meal is usually made from sun-dried by-catch consisting mainly of silver belly (*Leiognathus splendens*). Often the drying is not complete with the result that the product becomes infested with Salmonella bacteria. This contaminant poses a serious risk for users of sun-dried fish meal. Another contaminant of fish meal is urea, sometimes added by unscrupulous merchants to give otherwise inferior product deceptively high crude protein content. Good quality fish meal should also contain no more than 2 per cent salt added as a preservative before processing, and, if kept for long periods before use, should contain an anti-oxidant.

Blood Meal

Animal blood is obtainable in most areas from abattoirs. In many village abattoirs, it is merely discarded. Animal blood may be used fresh, or processed into blood meal before using. Although dried blood has high-protein content and is highly digestible, its amino acid content is not as well balanced as that of muscle tissue.Fresh blood may be used to enrich commonly available feedstuffs such as rice bran and wheat bran in fish diets. The bran also soaks up the excess moisture in the production of moist pelleted feed. In combination with rumen contents, fresh bovine blood has successfully substituted for fish meal in channel catfish diets.

Chemical Composition of some Feed Ingredients in Per cent

Ingredients	*Moisture (Per cent)*	*Crude Protein (Per cent)*	*Ether Extract (Per cent Dry Matter)*	*Fibre (Per cent)*	*Nitrogen-Free Extract (Per cent)*
Ground nut oil cake	7-10	40-43	4-8	6-7	30-33
Soyabean meal	8-10	50-55	1-2	4-5	30-32
Rice bran	7-10	12-16	12-14	8-12	40-45
Mustard oil cake	8-10	25-38	6-9	10-16	35-40
Maize flour	10-12	5-6	3-4	2-3	70-80
Fish meal	8-10	45-65	8-12	-	-
Blood meal	10-15	85-90	1-2	-	-

Bone Meals

Bone meals are made by heat processing of animal bones followed by grinding to produce a fine powder. Contrary to common belief, bone meal is not all calcium and phosphorus. Depending on the processing method employed, bone meal can contain up to 36 per cent protein. Much of this protein is, however, of low quality since it is mainly collagen, the substance extracted industrially from bones to produce gelatin and glue. With the development of fish diets containing ever-decreasing content of animal protein such as fish meal and meat meal, bone meal may become more prominant as a dietary source of calcium and phosphorus.

Chemical Composition of some Feed Ingredients in Dry Matter (DM)

Use of Feed in Aquaculture

- ☆ The feed is formulated to meet the nutritional requirement of the fish at every life stage to ensure rapid growth and optimum health.
- ☆ Various types of probiotics and antibiotics mixed with fish feed which help in good health of fish.

- To ensure that the marine and freshwater ecosystems are protected, measures are in place to monitor aquaculture activities to minimize environmental impacts:

  1. Using feed that is easier to digest with less fecal waste from the fish.
  2. Using underwater video cameras to monitor the amount of uneaten feed settling beneath the open pond. Feed levels are adjusted to eliminate uneaten feed resulting in very efficient feed conversion.
  3. Using electronic devices to control feed.Feed conversion ratios determine the amount of food required to produce healthy farmed fish. The aquaculture industry strives to keep these conversion ratios as low or efficient as possible.

- Feeds and feeding methods have become more efficient with more nutrients being used for fish growth rather than being wasted as fecal matter or uneaten feed.

Conclusion

Several agro-based ingredients have been identified and analyzed for their use in fish feeds.The ingredients that contain less than 20 per cent protein and 18 per cent fibers are classified as energy supplements. Research work done in India,from which nutrition of fish and some feed formulation technique is given in aquaculture industry. With optimum pond management as well as feeding of formulated balanced diets, it has been possible to increase aquaculture production.

Chapter 19

Herbal Uses and its Limitations in Aquaculture

Introoudction

Many herbs and plants have been used in many cultures around the world for human and animals as well as fish. Some of these remedies have anti-viral properties as well as anti-bacterial and anti-fungal properties. These natural plant products have been reported to have various properties such as anti-stress, growth promoters, appetizers, tonic and immunostimulants.Moreover, these substances also possess other valuable properties; they are nontoxic, biodegradable and biocompatible. The properties of herbs and plants are used in herbal medicine around the world.

Herbal Use in Aquaculture

Neem Leaf

- ☆ Neem (*Azadirachtaindica*) is a highly esteemed "wonder" tree of India that is widely dispersed throughout the country.
- ☆ From the practical side these compounds also exhibit a wide variety of biological activity, for example, pesticides, antifeedants, and cytotoxic properties.

Parts Used: leaves tied up in bundles are dipped in the fish ponds infected. They will work best when placed at the access of the water flow or at the 4 corners of the pond.

Dosage: 150-200 twigs for a 1000 square-meter pond.

Effect of Neem against Bacteria

- Neem has significant effects on several bacterial strains.
- Among some of the more prominent strains studied were staphylococcus aureus, streptococcus pyogenes, cornebacterium, E. coli, and salmonella typhosa. Neem's ability to exert significant effects over the above mentioned bacterial strains indicates it's ability to resolve the aforementioned conditions

Effect of Neem against Viruses

- There are certain compounds in neem that demonstrate a unique ability to surround viruses, which prevents them from causing infection.
- Neem also inhibits viral multiplication by interacting with the surface of the cells to prevent the cell from becoming infected by the virus.

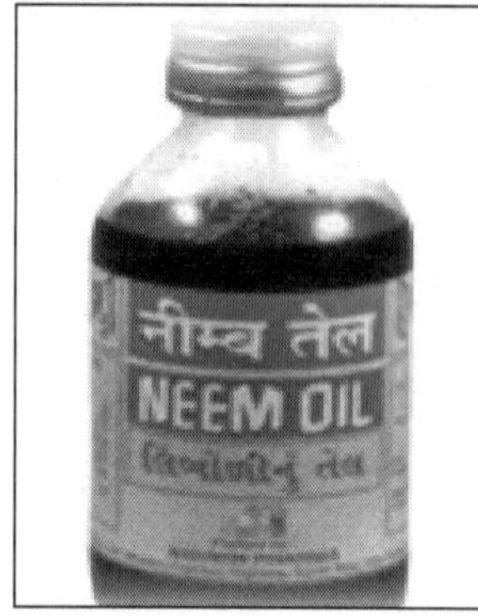

Effect of Neem on the Immune System

- Modern clinical studies have identified a number of compounds in the neem tree that effectively regulate immune system functions.
- There are immunomodulatory polysaccharide compounds, especially present in neem bark, that apparently increase antibody production.

Use of Neem Seed Oil

- Neem seed oil (NSO) is composed of a complex mixture of biologically active compounds, azadirachtin is most active of these. Azadirachtin-A is most abundant.
- There are more than 25 other active compounds found in neem, including meliantriol, vepol and salannin.
- Neem oil can also suffocate mites, whiteflies, aphids and other types of soft bodied insects on contact.

- The principle active ingredient, azadirachtin, acts as an insect's growth regulator (IGR) preventing exoskeleton development and impeding the molting process.

Tulsi (*Ocimum tenuiflorum*)

- The plant extract *O. sancutum* treated fish group showed no mortality and more weight gain.
- The antibody titre of the plant extract treated fish exhibited earlier and more antibody titre and the phagocytic activity was also more.
- Therefore, *O. sancutum* induces both non-specific and specific immunity.
- During an outbreak of disease among Indian major carps the causative agent was found to be *A. hydrophila, Ocimum sanctum* may act at various levels in the immune mechanism.

Parts used: leaves, stem, flower, root, seeds and even whole plant.

Dosage: It is concluded that 10 ppm *O. sancutum* is more effective.

Kalmegh (*Andrographis paniculata*)

- Kalmegh contains bitter principles andrographolide, a bicyclic diterpenoid lactone and kalmeghin.
- Columnaris, a fish disease caused by the bacterium *Flavobacterium columnare*. Of all tested herbs shows that kalmegh (*Andrographis paniculata*) have astrongest antimicrobial activity against *F. columnare*.

Parts used: Mostly the leaves.

Dosage: The minimal inhibitory concentration was 31.25 µg/mL.

Garlic

- It contains antimicrobial compound.
- It is use for control of EUS disease.

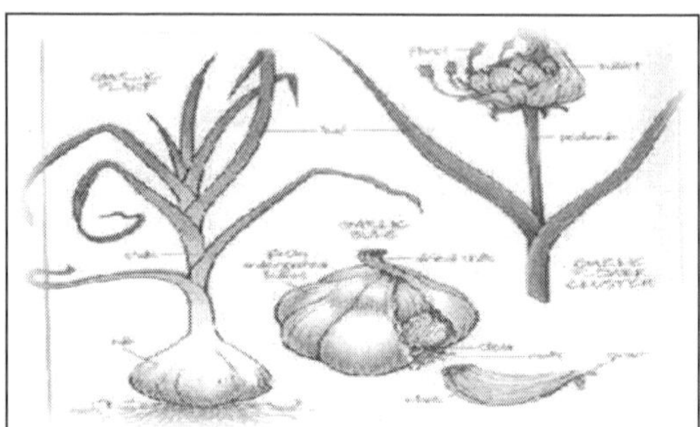

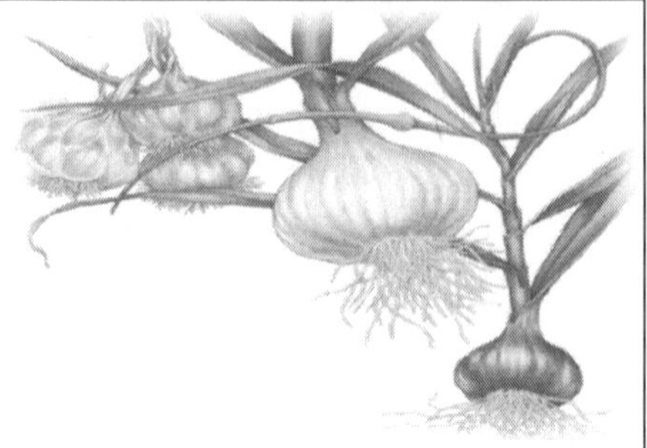

Application: Garlic (2 kg)+ Salt (2 kg)+CuS04(20 gms)+ KMn04 (20 gms).

Dosage: Ingredients made into paste, is mixed in 30-50 liter of water and sprayed over pond water of 0.133ha pond.

Use of Banana Logs in Pond

Uses: Use for to minimize Protozoan disease andworm of fish. Banana plant cell helps in increasing pH of the pondwater through their alkaline secretion. The pores on the surface on banana trunk are the opening to the pockets of oxygen, which ultimately helps to increasing the oxygen content.

Method: Banana logs are cut into pieces andimmersed into pond water.

Broken Catechu Nut Extract

Uses: Use for to obtain higher Hatching rate.It helps in shell hardening and prevents immature release of hatchlings

Method: In the hatching pool, the extract of water-immersed catechu nut is used in proper dose to make the eggshell hard.

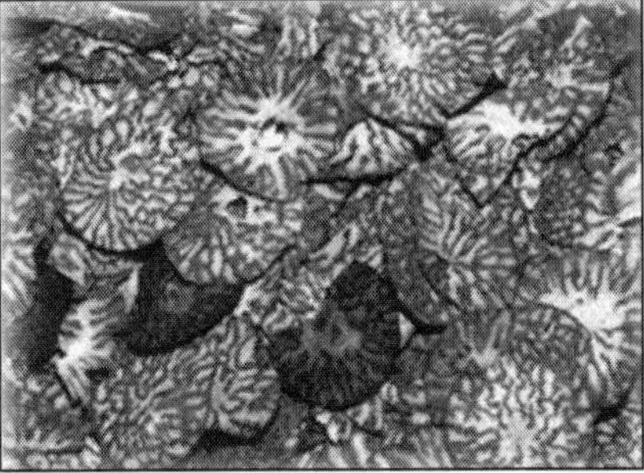

Castor Oil Plant (*Ricinus communis*)

Uses: Use for control Epizootic Ulcerative Syndrome (Red-spot Disease).

Dosage: 5-30kg for a 1000 square meters in width and 1.5 - 2 meters in depth pond.

Method: The leaves of this plant are tied up in bundles are dipped in the fish ponds infected.

Little Hog Weed (*Portulaca oleracea*)

Uses: Use for control the bacterial effect.

Dosage: 1.5 - 3 kg/100kg of fishes

Method: Wash the leaves with freshwater and then with 3 per cent salt water. Feed the fish once/day during 5-7 days.

Bhringaraj (*Eclipta alba*)

Uses: Use for parasites control.

Dosage: 10g of *Eclipta alba* leaves, 1 - 3 times/day

Method: 10g of *Ecliptaalba* (L.) leaves, combined with 10g of Betel leaves, mixed with 3g of squid liver oil to create a mixture. Put this mixture into 1kg fish-feed, then feed the fish 1 – 3 times/day.

Water-Pepper

Uses: Use forEpizootic Ulcerative Syndrome control.

Dosage: 3kg Water-pepper leaves/100kg fishes, or 1-2kg water pepper powder/100kg fishes

Method: Water pepper trunk and leaf boiled, mixed with fish-feed, feed the fish continuously 3-6days

Water Hyacinth

Besides rectifying the foul odor of the water, the plant takes care of absorbing the heavy metal particularly Mercury (Hg) Lead (Pb) and Arsenic (As).

Noduar Plants

The nodular plants are mainly (Family- *Leguminaceae*) used to increase the nitrogen content of the pond by action of nitrifying bacteria

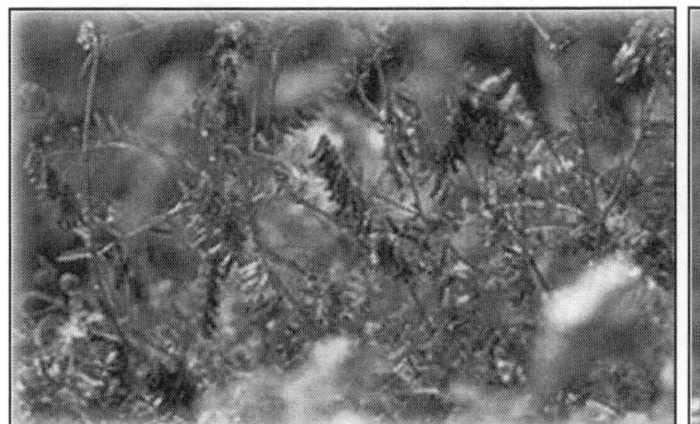

Bamboo Sticks

- Farmers place bamboo sticks vertically in rural fish ponds so that,
- Fishes rub their body against the bamboo sticks. This helps in removing a large number of ecto-parasite mainly *Argulus* from their body.
- Also helps to prevent poaching.

Fig.1 Use of banana logs in pond

Fig.2 Use of banana logs in pond

Fig.3 Bamboo logs are placed on the pond

Fig. 4 'Pala' are placed on the pond for resting place

Fig.5 Mixture of Garlic, Salt, $CuSO_4$, $KMnO_4$ as a prophylactic measure for fish diseases

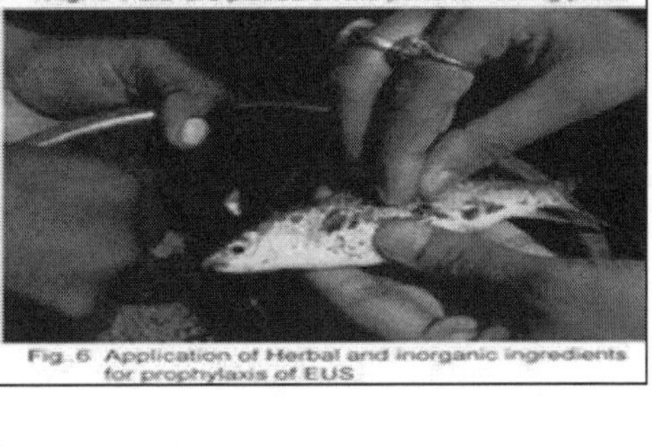

Fig. 6 Application of Herbal and inorganic ingredients for prophylaxis of EUS

Spice/Herb	Microorganisms
Garlic	*Salmonella typhymurium, Escherichia coli, Staphylococcus aureus, Bacillus sp.*, Mycotoxigenic *Aspergillus*
Onion	*Aspergillus flavis, Aspergillus parasiticus*
Cinnamon	Mycotoxigenic *Aspergillus, Aspergillus parasiticus*
Cloves	Mycotoxigenic *Aspergillus*
Mustard	Mycotoxigenic *Aspergillus*
Allspice	Mycotoxigenic *Aspergillus*
Oregano	Mycotoxigenic *Aspergillus, Salmonella* spp., *Vibrio* sp.
Rosemary	*Bacillus cereus, Staphylococcus aureus, Vibrio parahaemolyticus*
Bay leaf	*Clostridium botulinum*
Sage	*Bacillus cereus, Staphylococcus aureus, Vibrio parahaemolyticus*
Thyme	*Vibrio parahaemolyticus*

Use of Herbal Extracts in Shrimp Broodstock Diet

Inhibitory Effects of Spices and Herbs

Botanical Name	Family	Distribution	Useful Parts	Effect in Aquaculture
Cinnamonum zeylanicum	Lauraceae	India, Sri Lanka	Bark	Endocrine system, Growth promoter
Elettaria cardomomum	Scitaminaceae	India, Sri Lanka	Dried ripe seeds	Endocrine system, Growth promoter
Eugenia caryophyllata	Myrtaceae	India, Sri Lanka	Fruits and dried flowerbuds	Endocrine system, Growth promoter
Asparagus racemous	Liliaceae	India	Leaves and Root	Endocrine system
Mucuna pruriens	Papilionaceae	Tropics	Seeds, roots and legumes	Endocrine system
Witania somnifera	Solanaceae	India	Root and leaves	Endocrine system
Cinnamonum zeylanicum	Lauraceae	India, Sri Lanka	Bark	Endocrine system, Growth promoter
Elettaria cardomomum	Scitaminaceae	India, Sri Lanka	Dried ripe seeds	Endocrine system, Growth promoter

Botanical Name	*Family*	*Distribution*	*Useful Parts*	*Effect in Aquaculture*
Eugenia caryophyllata	Myrtaceae	India, Sri Lanka	Fruits and dried flowerbuds	Endocrine system, Growth promoter
Asparagus racemous	Liliaceae	India	Leaves and Root	Endocrine system
Mucuna pruriens	Papilionaceae	Tropics	Seeds, roots and legumes	Endocrine system
Witania somnifera	Solanaceae	India	Root and leaves	Endocrine system

Limitations of Herbal Plants

- **Lack of knowledge**: though the herbal plants are used in many places but the farmers don't know the actual dose and method of use. This causes many problems.
- Cure using herbal medicine and supplements would take some time.
- **Low availability:** the herbal plants are not sufficiently available in every place. So these can't be used everywhere as the availability is restricted.
- **Restricted in small scale operation:** the herbal plants can't be used in a large scale. These are useful only in small scale operation.
- **More amounts:** though the cost of herbal plants is less than the chemicals means but the amount to be used is very high. So it causes various disturbances.
- **Decomposition**: as the herbal plants are organic in nature if they are used in excessive amount they can cause various problem by decomposing in the water body like obnoxious gas formation, growth of microbes, increase ammonia level, acidosis *etc.*
- **Water quality:** by spreading over the water body and also by decomposing, the herbal plants may cause water quality deterioration.
- Remember, the government does not regulate herbal medicine industry. Hence, there is no quality assurance for herbal products.

Conclusion

Herbal compounds have the ability to inhibit the generation of oxygen anions and scavenge free radical, hence reducing stress effects. The herbal compounds have some advantage and some disadvantage. But it will be good if we use herbal plant extract instead of chemicals.

Chapter 20
Turbidity Problems and its Permanent Solutions

Introduction

Turbidity is an important indicator of the amount of suspended sediment in water, which can have many negative effects on aquatic life. The more total suspended solids in the water, the murkier it seems and the higher the turbidity. The suspended particles absorb heat from the sunlight, making turbid waters become warmer, and so reducing the concentration of oxygen in the water. As a consequence of the particles settling to the bottom, shallow lakes fill in faster, fish eggs and insect larvae are covered and suffocated, gill structures get clogged or damaged.Long-term turbidity monitoring is very important because it helps us measure the amount of soil erosion due to construction and agriculture, and helps to determine increasing urban development on water quality.

What is Turbidity?

The term turbidity generally describes about the "cloudiness" or "muddiness" of water. Turbidity can be caused by many substances, including microscopic algae (phytoplankton), bacteria, dissolved organic substances that stain water, suspended clay particles, and colloidal solids.Turbidity caused by suspended clay tends to occur most often in soft, poorly buffered (low alkalinity) waters. Turbidity is a measure of how clear the water is. High turbidity means that there are a lot of particles suspended in the water and light cannot get through. Low turbidity means that there are fewer particles in the water and it is clearer.

Sources of Turbidity

Turbidity can result from several different sources:

1 Due to **suspended clay particles**: Clay particles found in many soil types are small and repel each other because they carry the same electrical charge. When these particles enter a pond; they do not settle out, and the water stays turbid.

2. Due to **mechanical activities**: Feeding activities of fish such as carp, bullhead cause bottom sediment to become suspended. Channel catfish can be a source of turbidity. Bottom sediment can also be disturbed by crayfish and aquatic insects such as burrowing mayflies. Another mechanical cause of muddy water is shoreline erosion by wind and wave action against shorelines lacking vegetation or other soil holding cover.

Other Sources of Turbidity

1. Runoff from clear-cut or overgrazed,
2. Watersheds,
3. Road or building construction,
4. Pond bank erosion from wave action, and
5. Excessive aeration.

Types of Turbidity

Water turbidity is caused by the presence of these suspended particles in varying quantities. There are three types of turbidity:

(a Mineral Turbidity

It is caused by a high content of silt and/or clay particles, which turn the water a light brown, sometimes reddish colour. It may occur when the water supply is turbid or a bottom feeding fish, such as the common carp, stirs up the bottom mud.

(b) Plankton Turbidity

It is caused by a high content of minute plants and animals which colour the water in various shades of brown, green, blue-green or yellow-green, depending on which plankton species is dominant.

(c) Humic Turbidity

It is caused by the presence of humus which turns the water a dark brownish colour. Its origin is usually the water supply, although it can also be caused by an excess of organic matter entering the pond. With 10 per cent turbidity, light will reach the bottom of a pond.

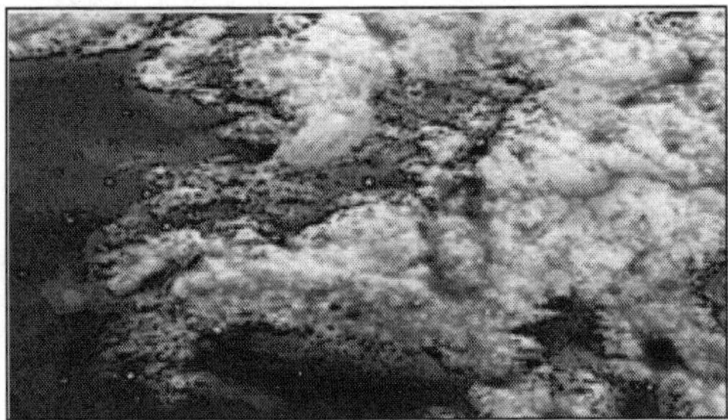

Problems in the Aquaculture Related with Turbidity

- ☆ Turbidity effects on aquatic organisms, both plant and animal, and the general health of the pond.
- ☆ Turbidity mainly occurs due to the biological factors and watershed factors.
- ☆ Turbidity reduces sunlight penetration into the water, inhibits plant growth, decreases oxygen production and limits primary food production.
- ☆ Aquatic organisms still demand the same amount of oxygen, but less oxygen is being produced.
- ☆ If the water becomes too muddy, sight feeding fish like largemouth bass and bluegill will have a difficult time finding and capturing food. This could result in poor growth rates and a reduction in the numbers and sizes of these fish.

- ✰ Excessive silt can irritate fish gills, smoother fish eggs, and suffocate bottom dwelling aquatic insect communities.
- ✰ In addition, aquatic herbicides used to control excessive aquatic vegetation may be ineffective when applied to turbid water.

Herbicide Treated Ponds Before and After.

Turbidity Measurement

- ✰ Turbidity of pond water varies from almost zero to highly turbid, depending on the amount of suspended particles.
- ✰ The method used for its measurement varies according to the kind of turbidity present.
- ✰ If it is a mineral turbidity (brownish water), you will need the help of a laboratory to determine the weight of material suspended in a given volume of water.

☆ When taking samples, be careful not to disturb the water too much, as you can increase the TSS very easily. Also, do not take the water only from the surface, as it is often much less turbid. Amount of total suspended solids (TSS) present in pond water:

Mineral turbidity.

Less than 25cm: Low turbidity.

25-100 cm: Medium turbidity.

Over 100cm: High turbidity.

If it is a **plankton turbidity** (greenish water), we can measure by the following process:

(a) Slowly wade into the shallow part of your pond, trying not to disturb the pond bottom too much.

(b Stretch one arm, and immerse it vertically into the water until your hand disappears from sight.

(c) Note the water level along your arm:

1. If it is well below your elbow, plankton turbidity is very high.
2. If it reaches to about your elbow, plankton turbidity is medium.
3. If it reaches well above your elbow, plankton turbidity is low.

Measuring Turbidity with Secchi Disc

The Secchi disc is a very simple tool which can be used to give a better estimate of turbidity. It is particularly useful in green-coloured ponds to estimate plankton turbidity.

This measurement is then called the Secchi disc transparency.

Measuring the Secchi Disc Transparency

To measure the Secchi disc transparency, we have to proceed as follows.

(a) Slowly lower the disc into the water.

(b) Stop when it just disappears from sight.

(c) Note at which point the line breaks the water surface. Mark this point A.

(d) After noting at which point along the line the disc just disappears, lower the disc a little and then raise it until it just reappears. Mark this point B.

(e) Mark point C, midway between points A and B.

(f) Measure the transparency of the water as equal to the distance from the top of the disc to this point C, counting the knots along the line.

To Obtain the Best Measurement with the Help of Secchi Disc, take note of the following points

- ☆ Measure transparency between 09.00 hours and 15.00 hours on calm days.
- ☆ Whenever possible, make the readings when the sun is out, not behind a cloud.
- ☆ Look at the sinking disc from directly above, if possible with the sun behind you.
- ☆ Keep the disc clean, particularly the two white quadrants. If necessary, repaint the disc black and white.

If the transparency is:

1. Less than 40 cm, there is too much plankton and your fish are in danger during the night when oxygen is not produced by photosynthesis and when too much oxygen is consumed by the respiration of this plankton.
2. 40 to 60 cm, the fish production will be the best.
3. Over 60 cm, there is too little plankton, and your fish do not have enough natural food to eat.

Preventive Measures

1. Do not stock and prevent the introduction of carp, goldfish, and bullheads into the pond. A major source of these nuisance fish species is the bait bucket. It is not unusual to see carp and bullhead fingerlings in buckets of minnows.

2. Many pond owners prohibit fishing with minnows or any bait fish.
3. Maintain dense populations of largemouth bass. This predator species can control nuisance fish species in ponds.
4. Try to keep domestic geese and ducks away from the pond. One pair per surface acre is tolerable.
5. Try to maintain vegetative cover throughout the watershed. If the watershed is largely agricultural, turbidity of water can be prevented by maintaining most of the watershed in hay crops rather than field crops.

6. If hay crops are not an option or the pond owner does not control the entire watershed, maintain a wide buffer strip of vegetation around the pond.

   Typically, buffer strips of 100-150 feet are suggested.
7. Establish and encourage moderate vegetative growth around the pond edge. Shorelines that have areas of rushes, sedges, and cattails are better protected from wave action and therefore less likely to add clay particles to the pond.
8. It is important to fence livestock away from the pond to prevent erosion of the pond bank and disturbance of shallow water soils.

Remedies

Some important remedies for the permanent solution from turbidity are as follows:

- ☆ If nuisance fish species are already present and causing muddy water, they need to be removed. Either drain the pond and remove them or treat the pond with a fish toxicant to eliminate the fish community.
- ☆ Restock with desirable species. If nuisance species are present but water clarity is acceptable, be sure to maintain a dense population of largemouth bass to ensure future control of the nuisance fish species.
- ☆ Keeping geese away from ponds is becoming increasingly important as their population increases. If geese have caused a muddy water problem, harassment tactics are an option used by many. They range from air cannons to scare the birds away to dogs that continually harass the birds until they leave. There are also products on the market that make the vegetation and lawns around the pond unpalatable to geese. If they do not like the taste, they will eventually leave to find food elsewhere.

Preventive Measures with Chemicals

Gypsum

- ☆ Agricultural gypsum is another material for removing suspended clay and does not cause the concern of a fish kill associated with adding hay.
- ☆ It does not cause pH problems.
- ☆ Typical application rates are from 1,000-1,500 pounds per surface acre of water, depending on the severity of the clay suspension.
- ☆ Again, it is wise to add the gypsum at a conservative rate of 250-500 pounds per surface acre of water, wait several days, and determine if additional gypsum is needed.(This prevents excessive application and therefore helps keep costs down).
- ☆ Dissolve the gypsum in clean water and spray over the surface on a calm day.

- Late evening is often an ideal time to make the application as most nights are wind-free. Water movement from the wind prevents the suspended clay from quickly settling out, reducing the effectiveness of gypsum.

Alum (Aluminum Sulfate)

- Alum is the most effective material for clearing clay turbidity from a pond, often within a few hours.
- Application rates are typically 100-450 pounds per surface acre.
- After application, there is a chemical reaction that impacts the pH (acidity) of the water. The reaction produces small amounts of sulfuric acid which can decrease pH significantly in some waters to levels harmful to aquatic life. Therefore, alkalinity and pH should be tested prior to application. Alkalinity should exceed 100 mg/l and pH should be greater than 7.0. If not, hydrated lime needs to be added simultaneously to buffer the effects of the acid produced by the alum addition. Application rate for lime is 50 pounds per acre-foot.
- Hydrated lime also removes suspended clay.

Limestone (Calcium Carbonate)

- Agricultural limestone is a material commonly used to remove suspended clay from the water.
- Application rates of 500-1,000 pounds per surface acre are typically used. Limestone can be added in the same manner described for gypsum.

Conclusion

Turbidity reduces sunlight penetration into the water, inhibits plant growth, decreases oxygen production and limits primary food production.Most problematic to the aquatic animals, when DO is reduced in pond. Excessive silt can irritate fish gills, smoother fish eggs, and suffocate bottom dwelling aquatic insect communities. In addition, aquatic herbicides used to control excessive aquatic vegetation may be ineffective when applied to turbid water. Chemical method is best to prevent the turbidity in the pond.

Chapter 21

Problems of Pesticide and its Solutions in Aquaculture

Introduction

The aquaculture industry in India as well as all over the world is growing rapidly, especially the production of freshwater aquaculture fish.The uses of pesticides in fish farming are common practices to avoid the overgrowth of unwanted plants and fish diseases besides promoting the fast growth of the fish. Generally pesticides residues used at farm level can be accumulated in fish and could cause chronic health effects to consumers. They have the potential to gradually accumulate in the body and cause certain organ damage.Among the health problems caused by chemical residues are cancer, nerve problems and immunological problems.

Pesticide

A pesticide is any substance or mixture of substances intended for preventing, destroying, repelling, or mitigating any pest. Pests can be insects, mice and other animals, unwanted plants (weeds), fungi, or microorganisms like bacteria and viruses. Though often misunderstood to refer only to *insecticides,* the term pesticide also applies to herbicides, fungicides, and various other substances used to control pests. Under United States law, a pesticide is also any substance or mixture of substances intended for use as a plant regulator, defoliant, or desiccant.

Types of Pesticides

Pesticides are categorized according to their target use. The major groups of pesticides are:

1. Herbicides (weed control).
2. Insecticides (insect control).

3. Fungicides (disease control).
4. Nematicides are pesticides used to control stem-dwelling nematodes (round worms).
5. Acaricide is a pesticide that controls mites and ticks.
6. Piscicides for eradicating fish vary widely.
7. Molluscacide(aquatic snail control)

Herbicides

Herbicides are the most commonly used pesticides. They are directly applied to lakes and ponds to control nuisance growths of algae (colonial, filamentous, and single cells), submersed water grasses (coontail, milfoil, naiad, pondweed), floating water plants (water lily, spatterdock, duckweed), and emergent water plants (cattails, rushes, reeds). Herbicides generally are less toxic to fish and aquatic life than insecticides. Many are short-lived and do not accumulate in the environment. However, some are highly toxic to aquatic animals and should be avoided or used with extreme caution near water ways and aquatic environments.

Chelated Copper Complexes and Copper Sulphate

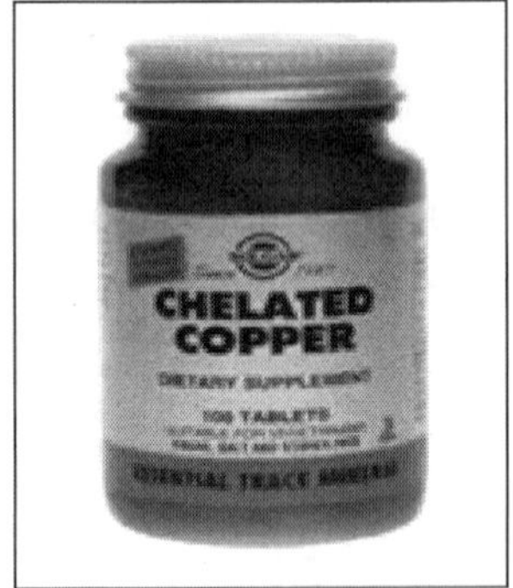

Use: It is used to control algae, not rooted aquatic plants. Most algae species are effectively controlled by these herbicides.

Disadvantage: copper is a toxic metal that is long-lived (persistent) in the environment. Copper sulfate can be toxic to fish and aquatic animals at concentrations near levels used to control algae, especially in soft water. Copper toxicity increases as water hardness decreases. Copper sulfate is not as safe to use as chelated copper compounds, and it should not be used in soft waters (alkalinity values less than 50 mg/L.

Solution: high air or water temperatures (thermal degradation). Use Safe/ Sensible Application Methods.

2,4-D

Use: 2,4-D (Aquacide, Aqua-Cleer, Weedar, Weed-Rhap, Weedestroy, Weedtrine) is effective for controlling submersed aquatic plants.

These compounds rapidly and completely decompose in about 3 weeks. Toxicity of these herbicides increases as pH decreases.

Disadvantages: more toxic in acidic waters (pH<6).

2,4-D can be highly toxic to rainbow trout.

It causes Weight Loss, Low Diseases Resistance, Sterility, Reduced Egg Production, Loss of attention and Low Predator Avoidance.

Solution: it is very much toxic to aquatic animals, it should be avoided or used with extreme caution near water ways and aquatic environments.

2,4-D should not be used in water for irrigation, livestock, or domestic purposes.

Fluridone

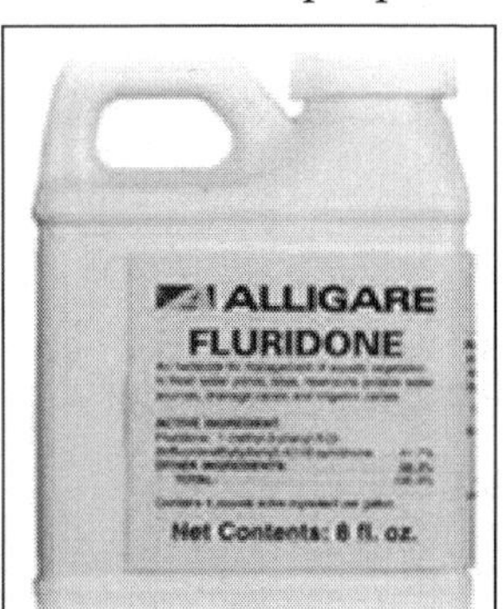

Use: Fluridone (Sonar) is perhaps the safest of the registered herbicides to use in fish ponds.

Advantages: It is a persistent, slow-acting herbicide.

Disadvantages: It is expensive. Require high maintenance while applying to fish pond.

Solution: There are no restrictions for fishing, swimming, or livestock.

Rodeo

Use: It is sold as solid and it is affective in emergent and shore line plants.It is a broad spectrum herbicide and used to control cattail, grasses around pond margin. Application is affective when the weed is in flowering or fruiting stage.

Disadvantage: Rainfall occurs within 6hr of application it reduces effectiveness of rodeo.

Solution: through monitor while application.Do not give excess amount otherwise it creates adverse effect in aquatic animal.

Insecticide

Use: Insecticides are used to control unwanted insects in aquatic environment. The predominant insecticides used were synthetic chemicals called organochlorine insecticides (OCs).The most infamous OC is DDT (dichlorodiphenyl-trichloroethane).

Disadvantages: Its effect on fish, wildlife, and natural environments was devastating.

Solution: they are banned from use. The ban on many OC insecticides in the United States has been important in the survival of fish and aquatic species and the protection of water quality.

Pyrethroids (PYS)

Use: PYs, especially synthetic ones, are the most toxic group of insecticides.

Advantages: (1) they are strongly absorbed to bottom muds, (2) they are short lived and usually last only days, (3) they rapidly decompose in 1 to 10 days when exposed to sunlight, and (4) they usually are applied at lower rates compared to the other insecticides.

Disadvantages: PYs are highly toxic to aquatic animals, they seldom cause fish kills.

Solution: They should be used with extreme caution near waterways

Organophosphates (OPS)

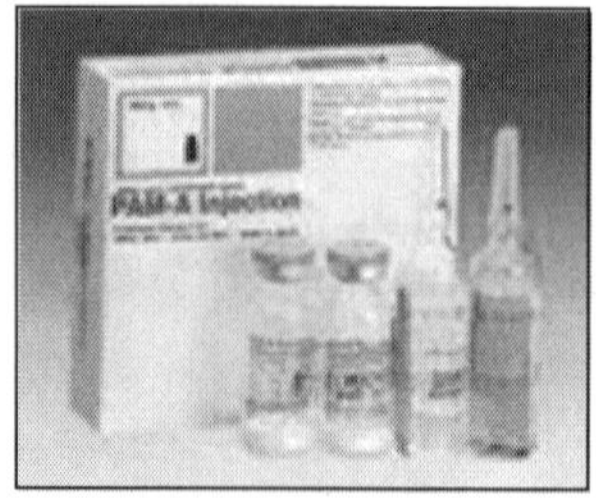

Use: It is used as an insecticide.

Advantages: insecticides are water soluble and metabolized quickly. They generally have short persistence (half-lives of days to months), and their residues do not pose long-term problems for aquatic animals.

Disadvantages: Insecticides are extremely hazardous to fish. Fish kills involving these insecticides have been documented. OP insecticides can bio concentrate in fish, frogs, tadpoles, and toads to levels that pose hazards to their predators.

Solution: They should be used with extreme caution near waterways.

Biological Insecticides (BIS)

Advantages: Some BI insecticides are less hazardous to fish and other aquatic animals, because many target specific insects (narrow spectrum). BIs include microbial and insect growth. Inhibits the formation of an insect's hard exoskeleton (outer shell).

Solution: as it is not hazardous to aquatic animals, precaution is not necessary.

Conclusion

Aquaculture is not just a matter of producing fish – it is part of a complex value chain that is itself influenced by a range of environmental, societal and governmental factors that make the difference between a successful or failed initiative.There is also a move to improve the environmental performance of aquaculture through the development of systems such as 'multitrophic aquaculture' that use low-trophic-level species to utilise the nutrient-rich by-products from high-trophic-level species; for example, farming shellfish in fish farm drainage canals, 'polyculture' (*e.g.* combining carps with different feeding niches in the same pond) and rice/ fish culture (where fish can be farmed in rice paddies, raising overall yields and predating on pests).

Chapter 22
Problems of Heavy Metals and their Remedy

Introduction

Heavy metals are normal components of aquatic environment brought about by the natural weathering of rocks. However, the increased discharge of wastes from industrial, agricultural and mining activities have pushed the levels of heavy metals well beyond normal trace amounts.As metals can interact with a variety of biochemical pathways and cellular macromolecules, the increased metal load in aquatic systems would most certainly impact the inhabitants at all trophic levels.

Definition of Heavy Metal

- ☆ A heavy metal is a member of a loosely-defined subset of elements that exhibit metallic properties. It mainly include Arsenic, cadmium, chromium, lead, manganese, mercury, nickel, zinc, copper, selenium.
- ☆ There is an alternative term *toxic metal*, for which no consensus of exact definition exists either. Heavy metal can include elements lighter than carbon and can exclude some of the heaviest metals.
- ☆ Heavy metals occur naturally in the ecosystem with large variations in concentration. In modern times, anthropogenic sources of heavy metals, *i.e.* pollution, have been introduced to the ecosystem.

Types of the Heavy Metals

Heavy metals are grouped into two categories:

1. **Traditional metals** (*e.g.* copper, iron, manganese *etc.*) include those elements essential for metabolic function of organisms at low concentration but may be toxic at high concentration.

2. **Metalloids** (*e.g.* arsenic, cadmium, lead, mercury, chromium, zinc, selenium, nickel *etc.*) are gradually not required for metabolic activity but may be toxic at low concentration.

Arsenic **Cadmium**

Copper **Chromium**

Iron **Lead**

Selenium

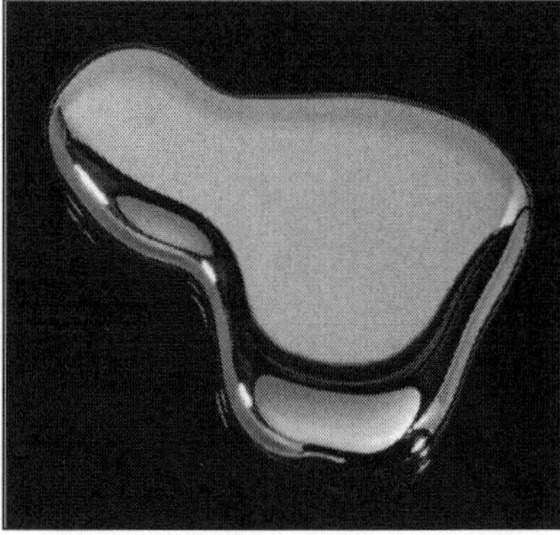

Mercury

Sources and Concentration of the Heavy Metsls

Metals	*Sources*	*Concentration Range (mg/l)*
Arsenic	Insecticide manufacturing; ore extraction, wood products preservation, arsenic trioxide manufacturing *etc.*	10-1000
Cadmium	Metallurgical alloying; ceramic and inorganic pigments; electroplating; textile printing, batteries.	4-1000
Chromium	Corrosion control additives; inks, pigments and dyes, tanning, metal cleaning, preservatives.	1-1300
Copper	Metal cleaning, pickling and plating, copper sulphate production.	10-900
Lead	Battery manufacturing, electroplating, television, paint pigment, tetra ethyl lead production.	<1-400
Mercury	Chlor-alkali plants, paper mills, smelting, fertilizer mills, water based paint.	<0.3-20
Zinc	Steel works(galvanizing), zinc and brass metal works, zinc smelting, mine drainage.	<1-2000
Iron	Mine drainage, coal storage pile drainage, steel manufacturing and rolling, titanium oxide, paint and ink manufacturing.	40-5000
Nickel	Pickling and planting nickel alloys, basic steel works.	<1-900
Selenium	Paint, pigment and dye production, electronics and glass manufacturing.	<1-10

Environmental and Health Risk from the Heavy Metals

Heavy metals are natural components of the Earth's crust. They cannot be degraded. To a small extent they enter organism bodies via food, drinking water and air. They are dangerous because they tend to bioaccumulation. Bioaccumulation means an increase in the concentration of a chemical in a biological organism over time, compared to the chemical's concentration in the environment. The capacity of the aquatic organisms to take up, store remove, or detoxify heavy metals differ considerably. Many intrinsic and extrinsic factors influence heavy metal uptake:

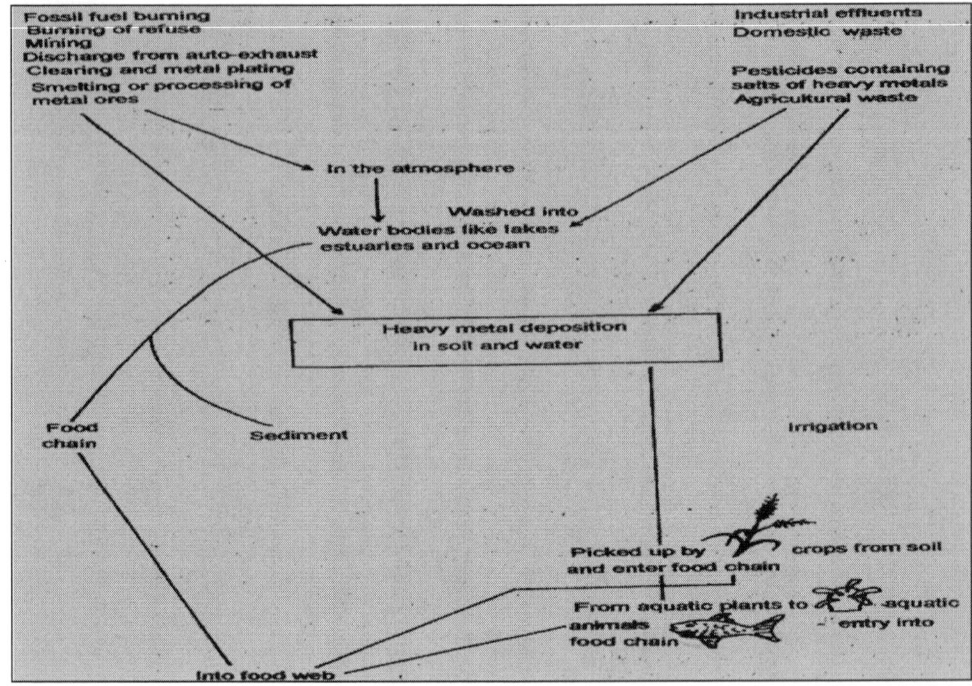

Entry of Heavy Metal into the Ecosystem

1. Intra and interspecifically variable intrinsic factors, such as surface impermeability, nutritional state, stage of molt cycle.
2. Extrinsic physico-chemical factors, such as dissolved metal concentration, temperature, salinity, presence or absence of chelating agents.
3. At toxic levels, heavy metals act as enzyme inhibitors in aquatic organisms.
4. They also adversely affect cell membranes.
5. The species often damage reproductive and central nervous system.
6. Tissue inflammation and degeneration, genetic derangement may also occur.
7. Feeding behavior, digestive efficiency and respiratory metabolism are adversely affected.

Impact of Cadmium on the Fish, Aquatic Organsims and Environment

Cadmium is bio persistent and once absorbed by an organism, remains resident for many years although it is eventually excreted. Brown trout living in cadmium waters have been reported to demonstrate evidence of chronic stimulation of the hypothalamo-pituitary- inter- renal axis of the brain, disrupting the neuroendocrine control mechanisms and resulting in both hypertrophy and hyperplasia of the inter-renal cells in the kidney. Cadmium alters the appearance and composition of the population of sensory dendrites on the olfactory lamellae in carp. Cd may also produce bone defects, bone softening, gill damage, high rise to blood pressure and other heart diseases in humans and animals. The most serious effects of Cd poisoning are damages to the kidney, particularly the renal tubes.

Impact of Mercury

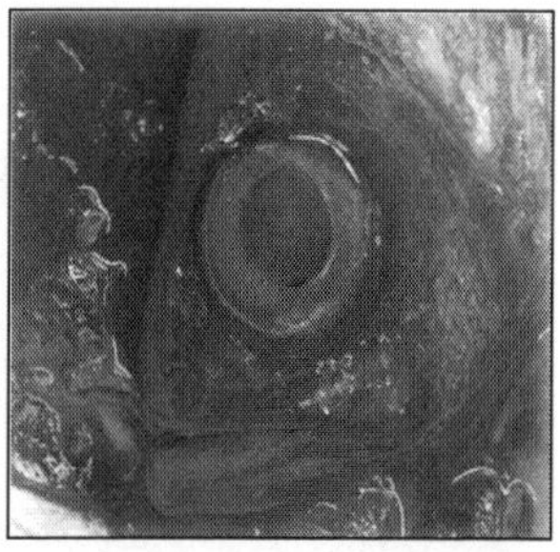

Generally mercury and its derivatives are bound effectively to sulfur and disulfide groups, inactivating critical proteins and replacing the trace element zinc. Mercury is highly toxic to the fish and impacts the fish nervous system in various areas. Consequently, structural alteration of proteins, inhibition of enzymes, binding to nucleic acids, and disruption of cellular membranes observed in mercury toxicity. $HgCl_2$ has been demonstrated to disrupt binding to olfactory receptor. $HgCl_2$ inhabits monoamine oxidase in the brain of snakehead fish, potentially altering catecholamine in the fish brain. $HgCl_2$ inhabits sodium ion, potassium ion in the snakehead fish. Calcium ion in the brain of the channel catfish is also inhibited by $HgCl_2$. The calcium ion is very important in regulating cellular homeostasis in neurons.$HgCl_2$has also been reported to lower AChE (Acetyl cholinesterase) in the brain, gill, kidney, intestine, liver and muscles of freshwater carp.

Impact of Lead

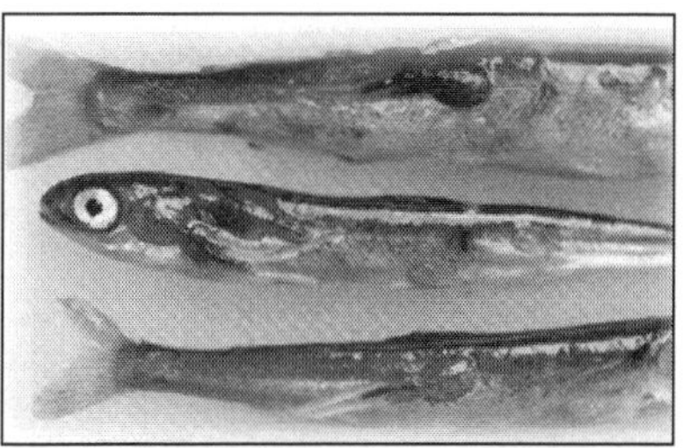

Lead is ubiquitous and is found in almost all components of the environment. Lead can activate the calcium ion dependent second messenger protein kinase c in the brain, resulting in disruption of the maintenance of normal cellular metabolism. It is presumed that lead is not very toxic to fish because of limited uptake resulting from low solubility across and limited access to cell membranes. High levels of exposure may result in toxic biochemical effects in humans which in turn cause problems in the synthesis of hemoglobin, effects on the kidneys, gastrointestinal tract, joints and reproductive system and acute or chronic damage to the nervous system.

Imapct of Copper

Copper has frequently been used therapeutically for the control of parasites in fish. High level of copper is toxic to fish. Copper affects nervous tissues in fish and causes behavioral alterations. In the brain of non-teleost fish, copper has been shown to cause mitochondrial degradation and decreased the amount of Golgi apparatus. This type of damage could disrupt the metabolic process in the brain and could result in behavioral alterations and possibly mortality. Copper also damages fish tissue of gills and sensory organs including those associated with sight, smell and taste. It also affects the growth rate of fish.

Imapct of Zinc

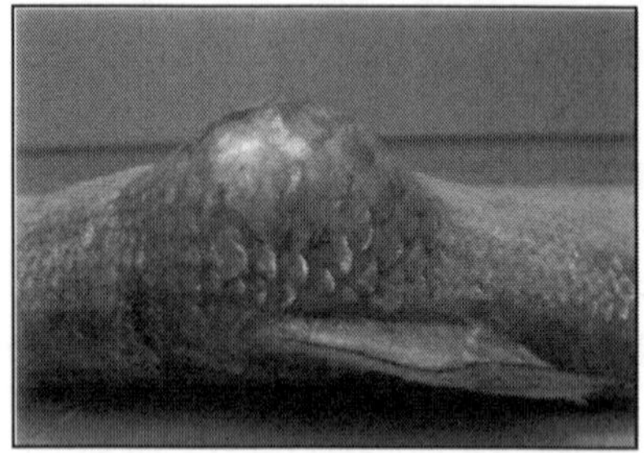

Zinc is naturally present in the teleost CNS. Exposure to excessive concentrations of zinc has caused behavioral alterations in fish. In the brain of brown trout living in zinc contaminated waters, there is evidence of chronic stimulation of the hypothalamo-pituitary-inter renal axis. In teleost's zinc affects the sensory tissue function. Short-term exposure of catfish olfactory mucosa to zinc induced extensive degradation of olfactory receptor cells. Zinc have been shown to reduce the growth rate and fecundity of the fishes.

Imapct of Arsenic

Arsenic is nonspecific growth stimulation for the aquatic organism. Long-term exposure over years to higher levels of arsenic in drinking water may also cause: thickening and discoloration of the skin; nausea and diarrhoea; decreased production of blood cells; abnormal heart rhythm and blood vessel damage; and numbness in the hands and feet and may also occur skin cancer and tumours of the bladder, kidney, liver and lung. Depending on the amount, arsenic can be toxic to freshwater fish, invertebrates, and aquatic plants. At low amounts it can slow aquatic plant life, and at higher concentrations it is lethal.

Imapct of Chromium

Chromium regulate the metabolism of glucose and cholesterol. Low concentrations of hexavalent chromium cause sub lethal toxic effects in aquatic plants and animals. Low level exposure can irritate the skin and causes ulceration. Long term exposure can cause damage kidney, liver, circulatory system, nervous tissue, respiratory system. Lactic acid content of blood and muscle increase in fish. Enzymatic activity decrease in the liver. The glycogen content increased in liver but decreased in muscle of fish.

Imapct of Iron

In freshwaters, iron is also an important nutrient for algae and other organisms. Excessive intake of iron damage the gills, digestive tract of the fish. High amount of iron, present in the cells, reduce the oxygen uptake and inhabit the ion regulation in fish. The DNA and membrane damage occur for iron toxicity. In vertebrates it is observed that high cellular concentration of iron may cause cell degeneration. The toxic effect of iron on rainbow trout, at natural pH is due to the precipitation of ferric hydroxide on the eggs and gill surfaces.

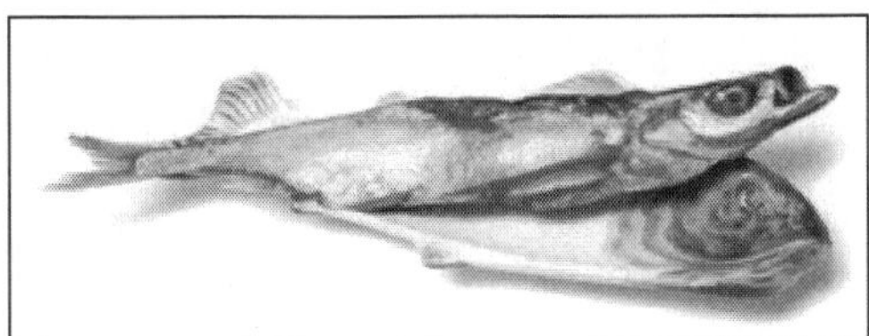

Disease which Occurs by Mercury Pollution

Minamata Disease

In Japan in 1952 mercury poisoning occurred due to consumption of fishes which had died of Hg poising. Mercury had contaminated the water where it

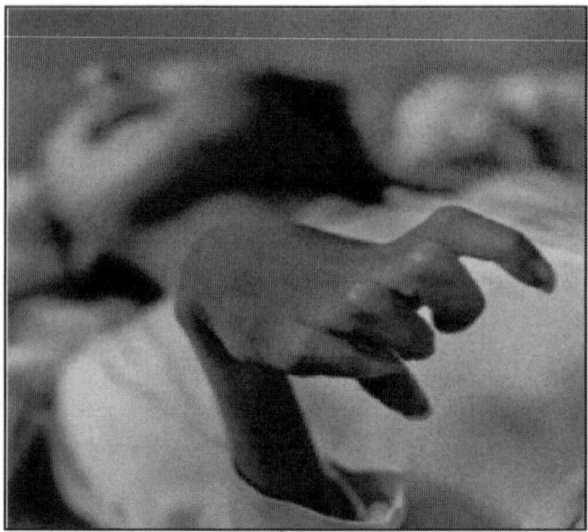

had reached as effluent of a vinyle chloride factory. The disease effected people are suffered from Minamata disease, whose symptom were weakened muscles, impaired vision, mental retardation, paralysis and death.

Disease which Occurs by Arsenic Pollution

Arsenicosis

Drinking water rich in arsenic over a long period leads to arsenic poisoning or arsenicosis. The health effects are generally delayed and the most effective preventive measure is supply of drinking water low in arsenic concentration.

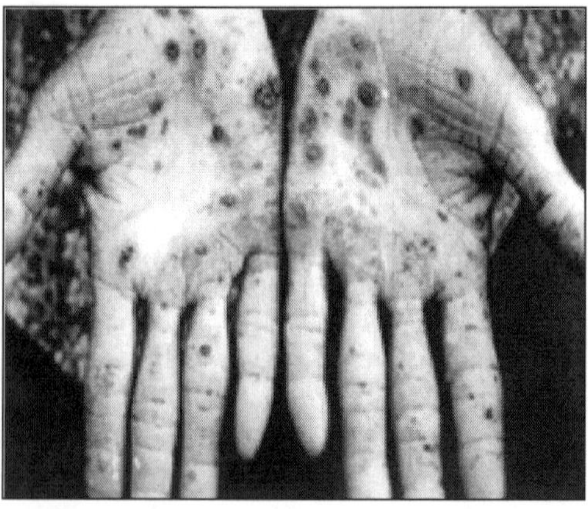

Disease which Occurs by Cadmium Pollution

Itai-itai Disease

Itai-itai disease was the documented case of mass cadmium poisoning in Toyama Prefecture, Japan, starting around 1912. The cadmium poisoning caused softening of the bones and kidney failure. The disease is named for the severe pains caused in the joints and spine. The cadmium was released into rivers by mining companies in the mountains. Itai-itai disease is known as one of the Four Big Pollution Diseases of Japan.

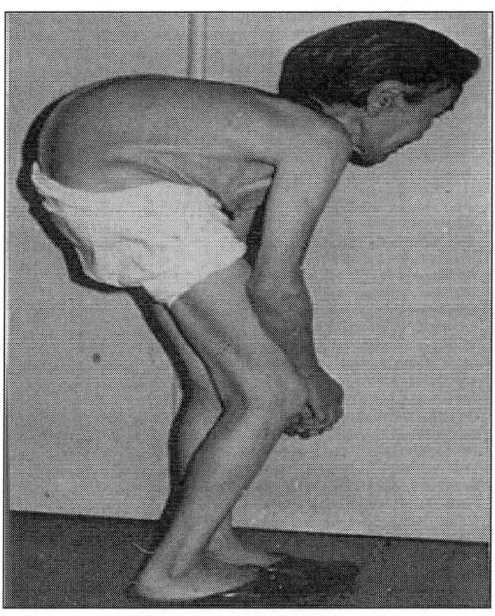

How to Prevent Heavy Metal Pollution

To save the human lives and other aquatic organisms from disaster occurring from toxic effect of heavy metals, preventive need to be enforced. Govt. of India has set up in January 1985, a separate Ministry of Environment and Forest which look into the issues of environment. For prevent of contaminated by heavy metals, the following has been suggested. The layers of uncontaminated clean soil over contaminated sediment have to be placed, so metal containing sediment does not washed out by rivers. Treat with calcium carbonate which increases pH of the sediment and immobilizes heavy metal. Limestone, gypsum, iron sulphate and activated charcoal can be used as detoxifying agents. Some water plants such as *pistia* and *hydrilla* pick up mercury from water bodies.

Conclusion

Every pollution in the aquatic environment which impacts physiology, development, growth or survival of fish, affects human that, at the top of the food chain, consume fish. The accumulation of heavy metals in the tissues of organisms can result in chronic illness and cause potential damage to the population. The development of biological monitoring techniques based on fish offers the possibility of checking water pollution with fast responses on low concentrations of direct acting toxicants.

Chapter 23

Biofertilizers: Its Role and Problems

Introduction

Aquaculture is developing, growing and intensifying in almost all regions of the world. Global production from capture fisheries has leveled off and most of the main fishing areas have reached their maximum potential. Capture fisheries will therefore not be able to meet the growing global demand for aquatic food. Aquaculture has the potential to make a significant contribution to this increasing demand. To meet this ever-increasing demand the aquaculture sector should have a sufficient and a healthy production. To get a higher production most of the farmers nowadays use different types of chemical means like fertilizers, pesticides, antibiotics and several other things. But these chemical means cause so many serious problems. But by using the organic means like manures, biopesticides, biofertilizers, *etc.* we can minimize those problems and get a good and healthy production.Biofertilizers can serve a significant role in the development of sustainable aquaculture.

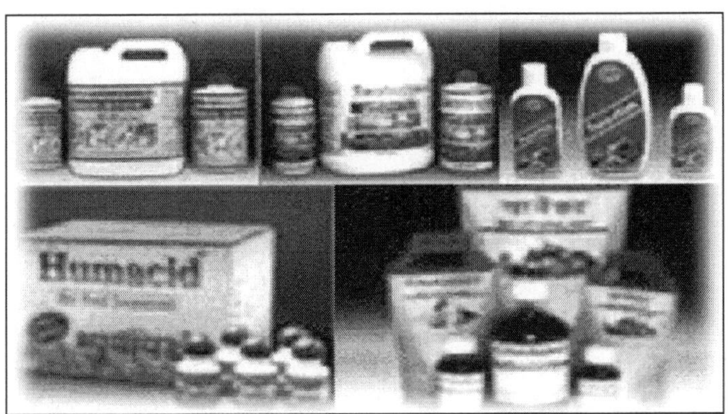

Biofertilizers

Biofertilizers can be defined as "*microbial inoculants which contain live or latent cells of selected strains of nitrogen fixing, phosphate solubilizing microorganisms used for application to seed, soil or composting areas to accelerate certain microbial processes, thus augmenting the availability of nutrients in an easily assimilable forms to plants.* "

- Use of biofertilizers in the soil, makes the fishes healthy as well as protect them from getting any diseases.
- Biofertilizers contains living microorganisms which colonizes the rhizosphere or the interior of the plant and promotes growth by increasing the supply or availability of primary nutrient and/or growth stimulus to the target crop, when applied to seed, plant surfaces, or soil.
- They are environment friendly, renewable, low cost and do not cause the pollution of any sort.

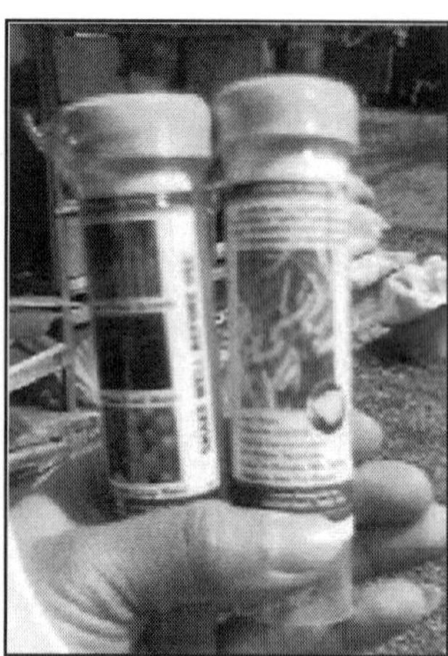

How Biofertilizers Work

- Biofertilizers fix atmospheric nitrogen in the soil.
- They solubilize the insoluble forms of phosphates like tricalcium, iron, and aluminum phosphates into available forms.
- They scavenge phosphate from soil layers.
- They produce hormones and anti-metabolites which promote root growth.
- They decompose organic matter and help in mineralization in soil.
- When applied to seed or soil, Biofertilizers increase the fertility.
- UsingBiofertilizers, healthy plants can be grown while enhancing the sustainability and the health of soil.

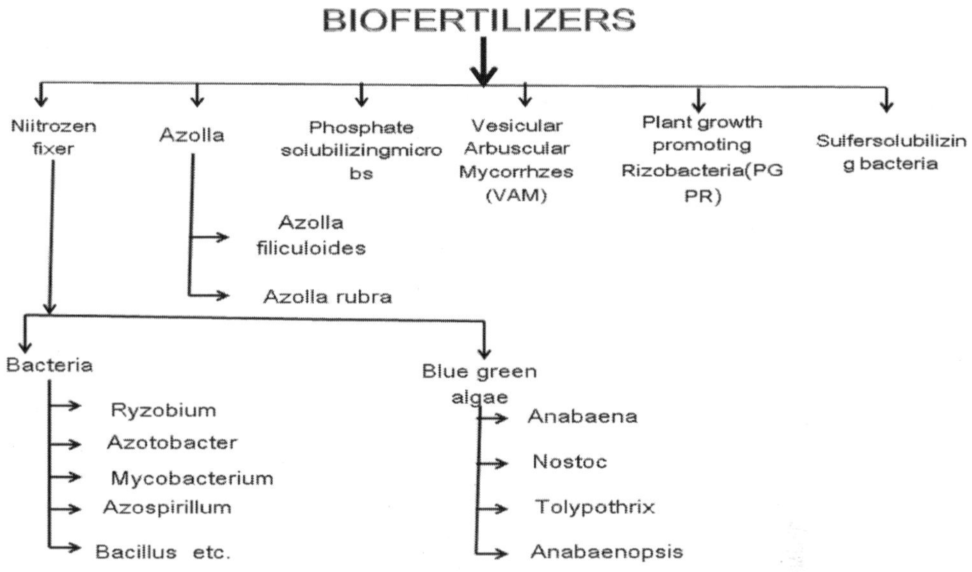

Sl.No.	Groups	Examples
		Nitrogen Fixing Biofertilizers
1	Free-living	*Azotobacterial, Beijerinkia, Clostridium, Klebsiella, Anabaena, Nostoc*
2	Symbiotic	*Rhizobium, Frankia, Anabaena azollae*
3	Associative Symbiotic	*Azospirillum*
		Phosphate Solubilizing Biofertilizers
1	Bacteria	*Bacillus megaterium var. phosphaticum, Bacillus subtilis*
		Bacillus circulans, Pseudomonas striata
2	Fungi	*Penicillium*sp, *Aspergillus awamori*
		Phosphate Mobilizing Biofertilizers
1	Arbuscular mycorrhiza	*Glomus* sp.,*Gigaspora*sp,*Acaulospora*sp., *Scutellospora*sp. *and Sclerocystis*sp.
2	Ectomycorrhiza	*Laccaria*sp., *Pisolithus*sp., *Boletus* sp., *Amanita* sp.
3	Ericoid mycorrhizae	*Pezizellaericae*
4	Orchid mycorrhiza	*Rhizoctoniasolani*
		Biofertilizers for Micro nutrients
1	Silicate and Zinc solubilizes	*Bacillus* sp.
		Plant Growth Promoting Rhizobacteria
1	Pseudomonas	*Pseudomonas fluorescents*

Groups of Biofertilizers Based on their Nature and Function

Biofertiizers Generally Used in Aquaculture

- ☆ *Azolla*
- ☆ Phosphate solubilizing bacteria
- ☆ Nitrogen fixing bacteria
- ☆ BGA (Blue Green Algae)

Azolla

- ☆ *Azolla* is a free-floating water fern that floats in water and fixes atmospheric nitrogen in association with nitrogen fixing blue green alga *Anabaena azollae*.
- ☆ *Azolla* fronds consist of sporophyte with a floating rhizome and small overlapping bi-lobed leaves and roots.
- ☆ The BGA is supplied with carbohydrate by the fern and the fern in turn receives inorganic N_2 that is fixed by the BGA and it acts as a store house of carbon and nitrogen which can be directly fed to the fish as green manure.

Phosphate Solubilizing Bacteria

- ☆ Phosphate solubilizing micro-organisms (PSM) secrete organic acids which enhance the uptake of phosphorus.

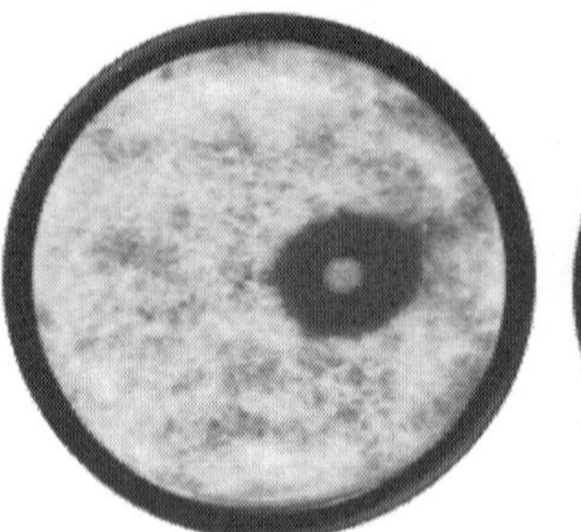

- ☆ PSM (*e.g.* **Phosphatika**) are very important in aquaculture ponds because they can regulate substantially the mineralization of phosphorus which is considered limiting nutrient in the aquatic condition.

Nitrogen Fixing Bacteria

Rhizobium

- ☆ Rhizobium is a soil habitat bacterium, which can able to colonize the legume roots and fixes the atmospheric nitrogen symbiotically.
- ☆ The morphology and physiology of Rhizobium will vary from free-living condition to the bacteroide of nodules.
- ☆ They are the most efficient Biofertilizers as per the quantity of nitrogen fixed concerned.

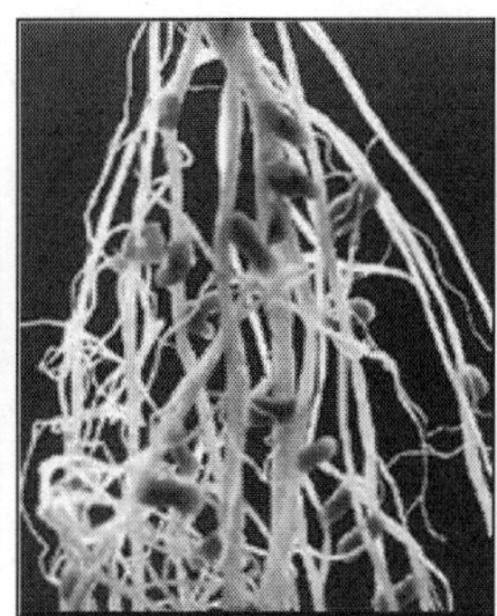

Azotobacter

- ☆ Of the several species of *Azotobacter, A. chroococcum* happens to be the dominant inhabitant in soils capable of fixing N_2 (2-15 mg N_2 fixed/g of carbon source) in culture media.
- ☆ The bacterium produces abundant slime which helps in soil aggregation.

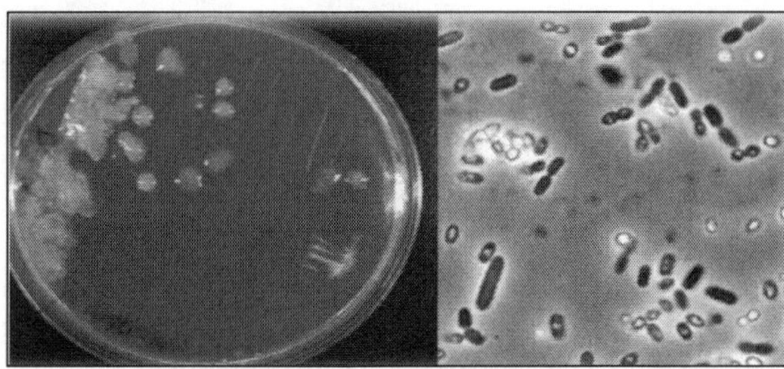

Azospirillum

- ☆ *Azospirillum lipoferum* and *A. brasilense* (*Spirillum lipoferum* in earlier literature) are primary inhabitants of soil, the rhizosphere and intercellular spaces of root cortex of graminaceous plants.

- ✰ They perform the associative symbiotic relation with the graminaceous plants.
- ✰ The bacteria of Genus *Azospirillum* are N_2 fixing organisms isolated from the root and above ground parts of a variety of crop plants.
- ✰ It has a good disease resistance capacity and drought tolerance

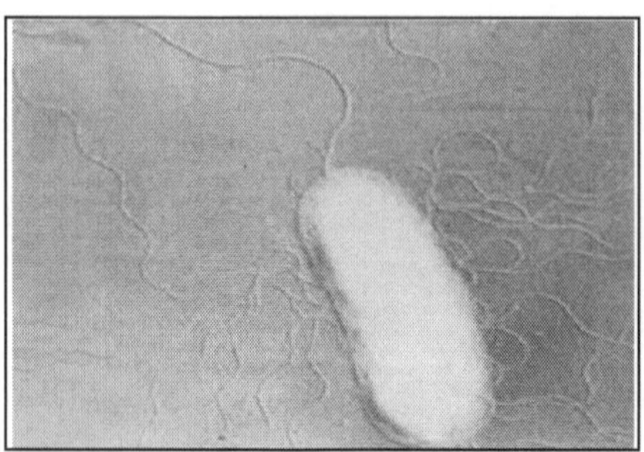

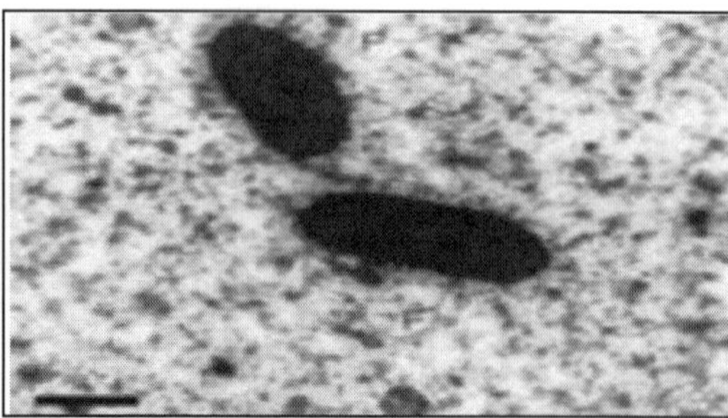

BGA (Blue Green Algae)

- ✰ The blue-green algae (Cyanobacteria) are another source of biological nitrogen.
- ✰ They are symbiosis with the water fern Azolla. They reduce molecular atmospheric nitrogen to ammonium which can then be utilized for amino acid and protein biosynthesis.
- ✰ Nitrogen compounds which can be assimilated by other plants are ultimately released to the soil after death of the algal cells and subsequent mineralization and nitrification.

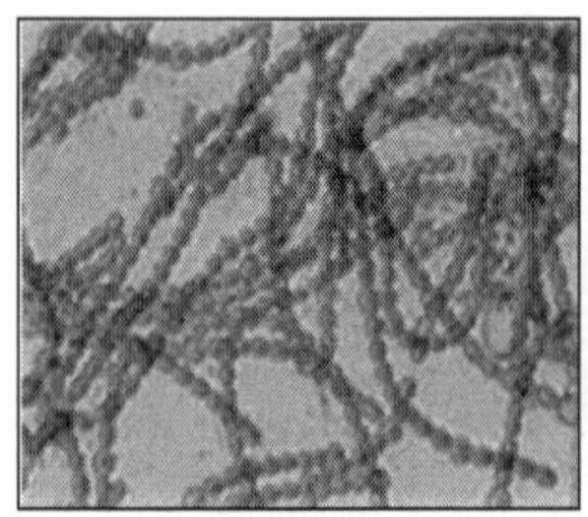

Cyanobacteria

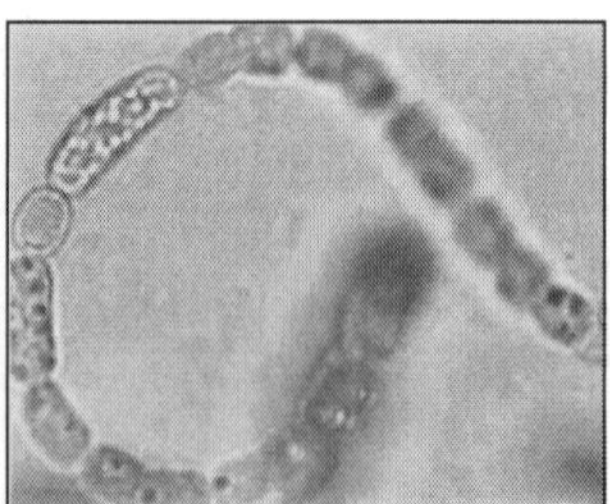

Anabaena

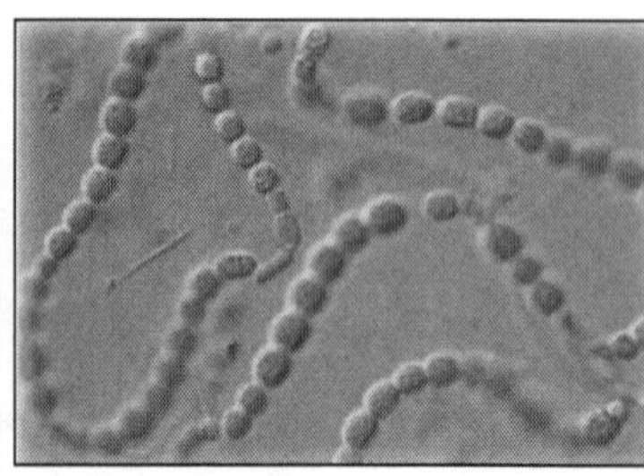

Nostoc

Comparative Study of Common Biofertilizers

Name of Biofertilizers	*Contribution*
Rhizobium {Symbiotic}	Fixes 50-30 kg N/ha
	Leaves residual nitrogen
	Increase yield by 10 –30 per cent
	Maintains soil fertility

Name of Biofertilizers	Contribution
Azotobacter	Supplies 20-40mg N/g of carbon source
	Promotion of growth substances like vitamins, B Group, IAA and Gibberellic acid
	10-15 per cent increase in yield
	Maintains soil fertility
	Biological control of diseases.
Azospirillum	Fixes 20-40 kg Nitrogen
	Results in increase mineral and water uptake.
Blue Green Algae {BGA}	20-30 kg N/ha in submerged fields.
	Production of growth substances like auxins, IAA, giberellic acid
Azolla	Fixes 40-80 kg N/ha
	Used as green manure because of large bio-mass

Composting Biofertilizers

- Composting Biofertilizers are used for hastening the process of composting and for enriching its nutrient value.
- Composting process is enhanced by enzymes secreted by microorganisms (Cellulolytic fungal culture) to hydrolyzepectins, xylans, hemicellulose, cellulose releasing beneficial micronutrients.

Exotic Biofertilizers

- Seventeen Rhizobacteria isolated from different ecological regions, *i.e.* Brazil, Indonesia, Mongolia and Pakistan were studied to develop inoculants for making Biofertilizers.
- In addition, Vesicular ArbuscularMycorhizae (VAM fungi) are free-living soil forms that increase nutrient uptake (especially by converting organic phosphorus into inorganic phosphorus), nodulation and nitrogen fixation in soils.

Role of Biofertilizers

- They have high nitrogen-fixing capabilities.
- They can solubilize phosphates.
- They can also supply Vitamin Bl2, Auxin and Ascorbic acid.
- They involve less energy, low cost and less effort in their production.
- Microbial processes are quick.
- No pollution hazards by using biofertilizers.
- They save non-renewable form of energy.
- They can be stored for long duration.
- They need little space for storage.

- They are environment eco-friendly.
- Their use leads to soil enrichment and the quality of the soil improves with time.
- Though they do not show immediate results, the results shown over time are extremely spectacular.
- Microorganisms convert complex organic material into simple compounds, so that the plant can easily take up the nutrients.
- These fertilizers harness atmospheric nitrogen and make it directly available to the plants.
- They increase the phosphorous content of the soil by solubilizing and releasing unavailable phosphorous.
- Bio-fertilizers improve root proliferation due to the release of growth promoting hormones.
- They help in increasing the crop yield by 10-25 per cent.

Problems of Biofertilizers

- Use of improper strains for production.
- Lack of qualified technical personnel in production units.
- Unavailability of good quality carrier material.
- Non-availability of essential equipment, laboratory and storage facility.
- Non-availability of sufficient fund.
- Seasonal variability in demand for biofertilizers.
- Short- shelf life of innoculance.
- Non-availability of quality specification and quality control measures.
- Unawareness of the benefits of the technology.
- Non-availability of right inoculants at right place and in right time.
- Narrow genetic base of mother cultures and lack of efficient and virulent strains suitable to various agro-environments.
- Not employing properly trained microbiologist.
- Biofertilizers are live product and require care in storage.

Conclusion

Despite having certain limitations, Biofertilizers are potential alternative to chemical fertilizers for improvement of soil fertility for sustainable crop productivity and revegetating the mine spoils in cost effective and eco-friendly manner. Biofertilizers also restore the natural habitat of the soil and to some extent it reduces the negative impact of global warming. So to meet the required demand and to maintain a sustainable aquaculture environment it is desirable to use biofertilizers.

Chapter 24

Jute Retting and its Problems in Aquaculture

Jute and Jute Retting

- ☆ **Jute** is a long, soft, shiny vegetable fibre that can be spun into coarse, strong threads. It is produced from plants in the genus ***Corchorus***, which was once classified with the family **Tiliaceae**.
- ☆ Jute is one of the most affordable natural fibres and is second only to cotton in amount produced and variety of uses of vegetable fibres.

Jute retting is a biological process by which the jute plant are soaked in water to extract the fibre from the main plant.

Jute and Aquaculture

- ☆ Jute is a most economic crop in West Bengal as also India. In West Bengal several water bodies of Murshidabad, Hooghly, Nadia, North 24 parganas, Malda district are used for retting of jute.

- But it is noticed that most of the water bodies are not used in such type of integrated aquaculture.
- If proper precaution is taken then a great number of aquaculture activities especially fish culture can be done from the water bodies.
- The most advantageous fact is that these water bodies contain plenty of nutrients and foods for fish are coming from the process of jute retting.

Problems of Jute Retting in Aquaculture

Jute retting causes several problems in aquaculture particularly it changes the water quality in terms of the following:

- It increases the BOD and COD of the water as a result decreases the DO level.
- Fluctuation of plankton level during retting and post retting period.
- Increases aerobic bacterial count deteriorating the water quality.
- It decreases the pH of the water thus resulting in acidity of the water.
- Causes blackening of water.
- Producing toxic gases.

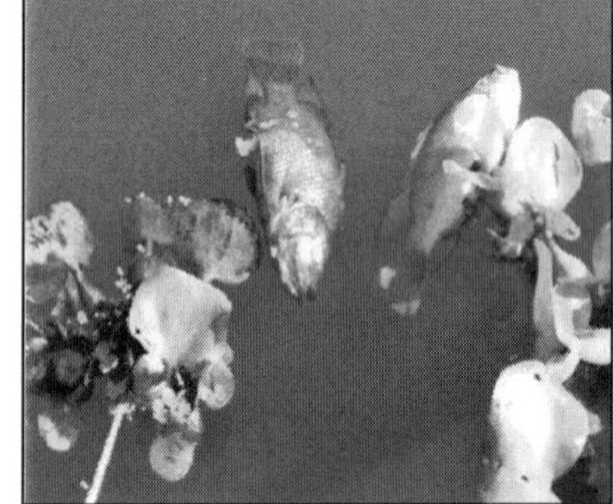

Problems Related to do

- BOD and COD increases due to microbial decomposition of the jute plant for extraction of fibre.
- Increase in BOD which may cause hypoxia conditions with consequent adverse effects on aquatic biota.
- High amounts of suspended particles has detrimental effects on aquatic flora and fauna and reduce the diversity of life in aquatic system and promote depletion of oxygen.
- This indicates that the effluents is unsuitable for the existence of aquatic organisms due to the reduction of DO content.
- DO is less that cause anoxic condition in pond.
- This cause respiratory distress in fish and fish show erratic movement.

Fluctuation during Retting

During the process of retting the plankton level decreases drastically due to low oxygen and action of different aerobic bacteria thus decreasing the fish food level making unsuitable for fish culture. But post retting due to presence of sufficient amount of nutrients plankton bloom may occur which is also detrimental to fish culture.

Problems Related to Microbial Growth

- Utilizing the carbohydrate of the jute plant many aerobic and anaerobic bacteria which are necessary for the retting process are produced but they are detrimental to the fish culture as many of them act as pathogen.
- From a study it has been found that the microbial count increases as follows:

	Pre-Retting	*Post-Retting*
Aerobic bacteria	$2.9X10^8$	$4.2X10^8$
Anaerobic bacteria	$1.0X10^8$	$4.2X10^8$
Fungus	Nil	$1.5X10^8$

Chemical Propeties of Post-Retting Water

Parameters	*Amount*
Dry residue (g)	10
Temporary hardness (ppm)	7.20
Permanent hardness (ppm)	70.00
O_2 (ppm)	3.00
CO_2 (ppm)	14.00
Bicarbonate (ppm)	105.00
BOD (mg/l)	5.10
COD (mg/l)	940.00

Decreases pH

Due to the process of jute retting the microbes utilizes the carbohydrate contain in the plant and produces organic acids which reduces the pH of the water causing fish mortality.

Blackening of Water

The retting process involves production of many organic and inorganic materials which causes blackening of water as a result the fish which grow in the water have less demand and unhealthy for human consumption.

Production of Toxic Gases

Different types of obnoxious, toxic gases are produced from jute retting which make the pond side unpleasant to visit and also these gases may cause different diseases to the aquatic biota.

Solution/Post Reting Remedy to make the Water Suitable for Aquaculture

1. Eradiction of Floating Weeds

After the jute retting process, for making favorable condition for fish culture the unwanted materials of retting and different types of floating weeds need to be eradicated by mechanical process. Water hyacinth is also used as a water cleaner.

2. Application of Lime

Assessing the water condition 1-1.5 kg/kata $CaCO_3$ can beapplied by spraying. Prior to that monthly 1kg/kata lime can be applied.

3. Application of Chemicals

A. $KMnO_4$ @ 300 – 400 ppm/bigha for increase of oxygen as well as a disinfectant. It also help in mineralization resulting in increase of NPC in the pond.

B. $CuSO_4$ @ 1-2 ppm for microbial disinfection.

Conclusion

- ☆ Investigation suggests there was no significant variation in the physico-chemical parameters except for the increase of COD and BOD for a short period when fish mortality was observed.
- ☆ There was no qualitative variation in the plankton population. However, plankton population decreased during the jute retting period and increased in the post jute retting period. Hence jute retting has both detrimental and favorable effect on the aquatic ecosystem.
- ☆ For a short period it creates ananaerobic condition and there may be fish mortality if the retting is done in a small closed water body.
- ☆ But with the passage of time the nutrients released from the decaying jute, help the growth of the aquatic life and hence fish culture.

Chapter 25
Problems Related to Fish and Shrimp Seed and an Insight to it

Introduction

When the Food and Agriculture Organisation first compiled production statistics on shrimp and fish in 1950, production came solely from wild catches. In Asia, fish and shrimp seeds had for centuries been traditionally grown in low-density monocultures, in polyculture with fish, or in rotation with rice in the Bheries of WB and Pokkali of Kerala in India. With improved technologies and the introduction of commercial formulated feeds, the industry boomed during the 1980s.

But unfortunately, recently the aquaculture industry has been severely criticised for the environmental and socio-economic problems generated in seed production field. The adoption of better farming practices can to a large extent be self-regulated by the aquaculture industry to direct their support towards sustainable seed production.

The sole reliance on wild catches of shrimps to meet the increasing Nigerian demand can cause a number of problems including over fishing, destruction of shrimp fishery increased by catch problem and habitat destruction. Freshwater shrimps (Prawn) of the genus *Macrobrachium*is found in West African waters and is widely distributed in Nigerian waters. These shrimps are more manageable than their marine relatives.The economic viability of culturing these shrimps should be investigated and research on the hatchery production of their post larvae should be intensified.

What is a Seed?

Actually Fish seed denotes spawn to fry and to advanced fingerlings. It is a collective name of hatchlings, post larvae, fry and fingerlings.Advanced fingerlings are used in beel and reservoir management.

Present Status of Aqua Sector: Responsible for Creating Problem

- ✰ Competition among fish seed producers to meet demand sometimes leads to poor management of brood fish and fish seed which may negatively affect seed quality. Substandard quality seed are frequently observed as a result of high stocking density in nurseries.
- ✰ Most fish hatcheries in India are concerned more about the quantity rather than the quality of fish seed and produce them without following any selection norms. Consequently, seed suffer from high rates of mortality, poor growth and high susceptibility to diseases and parasites.

Details about Problems Related Fish and Shrimp Seed Quality Problems Due to Improper Brooder Quality

To obtain more production farmers are busy in rat race. They are not concerned about quality brooder.

While selecting brood fish for spawning, small size and young fish should be avoided. Although most carps attain first maturity in their 1 to 2 years, there is an optimum age and weight at which they should be selected for induced breeding.

After selection, the young brood fishes are kept under optimal conditions (stress free environment) before choosing the most appropriate female and male individuals for hatchery operation.

But farmers are not maintaining that, they give much more emphasis on quantity of seed production; so they breed **underage brooder**. This is a great problem of our aquaculture sector.

Poor Seed Quality

Poor quality seed, perceived as a major constraint to expansion of fish and shrimp culture, can have deleterious effect on fish production and brood stock development.

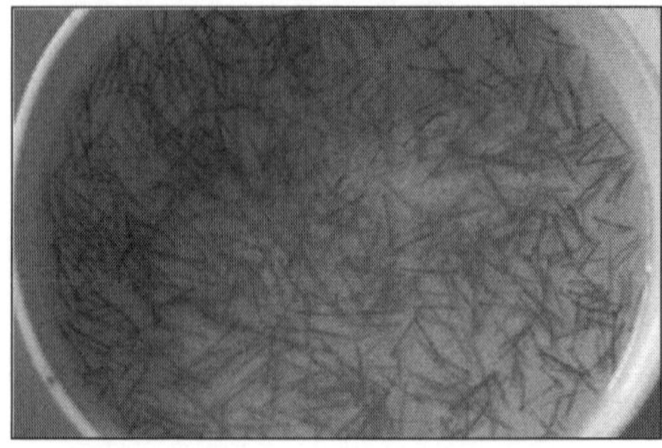

As per "Aquatic Seed Management-Ministry of Agriculture", fish breeders should obtain the quality seed from one of the centres established under the "National Aquatic Breed and Wild Seed System" (NABWSS), which includes "Genetic breeding centre" (GBC), Wild variety collection centre (WVCC), and "seed quality inspection centres" (SQIC).

Some farmers have experienced sudden spawn deaths and occurrence of deformed larvae/fry, more common for spawn produced during late breeding season, thus, farmers tend to avoid breeding during late spawning season.

Problems due to Physico-chemical Incompatibility

A. Dissolved Oxygen

A fish's metabolism increases with feeding and digestion, resulting in increased oxygen demand. An increase in respiration and uptake of water following handling can lead to ion imbalance and result in mortalities. Depletion of DO is one of the cause of poor seed production.

B. Temperature

Water temperatures greater than 28°C accompanied by declining dissolved oxygen and increasing ammonia, create a hostile environment. It is mostly problematic in tropical country.

C. Carbon-di-oxide

High levels of carbon dioxide lowers the affinity of oxygen to bind with hemoglobin in the blood of the fish. In water, carbon dioxide forms carbonic acid (H_2CO_3). As the levels of carbon dioxide and carbonic acid increase, the pH of water decreases. Due to increases in CO_2, pH often drops while fish are in transit.

Overcrowding: High Stocking Density

Overcrowding is an unpleasant environment for fishes.

- ✰ Ammonia levels can rise if there are too many fish seed in a tank, and ammonia is deadly to fish.
- ✰ They compete for their proper nutrition.
- ✰ DO level decreases fast.
- ✰ More contamination.

A good rule of thumb to use when it comes to stocking freshwater tropical fish is an inch of fish per gallon of water. This means only 10 small fish in a 10-gallon tank. So, Tank should have good filtration.

Problems during Seed Transportation

The seed collectors normally sell their seed to the seed traders. The seed normally passes through a few tiers of middlemen before it reaches thee farms. Head-loading, country boats, motorized boats, bicycles and buses are employed for seed transportation. Generally, **Aluminium vessels of 20–30 litre** capacity are

the most popular containers for long distance transportation. At that time direct sunlightpenetratesthe container; so delicate skin of seed gets burn which is known as **"Sun cancer"**. It is reported that due to "**sun cancer 10–50 percent seed mortality occurs".**

How to Overcome it?

- ☆ Weak and sick one should be removed before packaging.
- ☆ Feeding should be stopped 24-48 hrs prior to transportation.
- ☆ The seed transporters sometimes put table salt into the container water with the belief that it will keep the seed healthy and strong.
- ☆ Harvest fish for transporting at the coolest part of the day (very early of the morning).
- ☆ Some chemicals are also used to transportation
- ☆ The sedatives are used in transportation can decrease the rate of oxygen consumption and reduce the rate of excretion of CO_2,NH_3, and other toxic wastes.
- ☆ Tranquilization: The fish seed are first put in a full strength solution of MS 222 (*i.e.*, 5 g MS 222 in 100 lit water). It very soon puts them to "sleep".

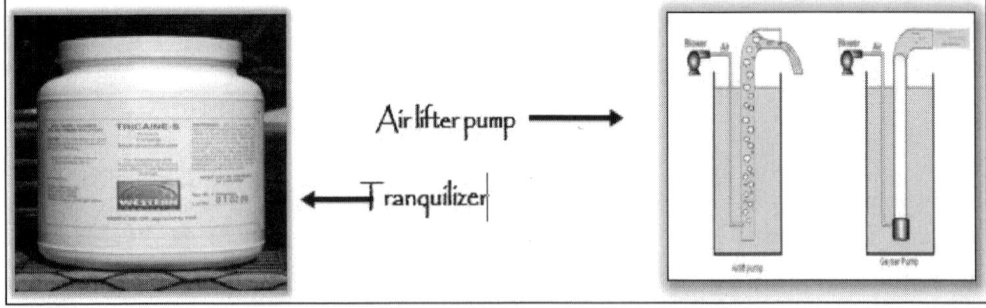

Quality Degradation due to Inbreeding Depression

Any fishery operation that results in a limited number of fish being available to produce progeny for use as brood stock in the next generation may lead to a constriction in the gene pool of that population, either in a hatchery or in a natural fishery. As a result seed is not at per quality.

Problems of Inbreeding in Fish Population

- ☆ Fish growth is retarded.
- ☆ Reproductive capability is reduced (eg: fecundity, egg size, hatchability)
- ☆ Offspring may show low survivability, deformities, poorer growth, less food conversion efficiency or may even suffer from various diseases.

Poor Handling

- ☆ Handling Stress during handling and packing may be so severe as to cause chronic and acute mortalities. Poor handling and packing procedure may also cause osmoregulatory and metabolic dysfunctions. Therefore it is important to proceed gently and quickly.
- ☆ Live-transport of fish seed and brooders entail handling often involving chasing with or without net, catching and transfer from the nursery/farm ponds to the conditioning system. The stress of this pre-transport handling can itself be high, sometimes causing injury and immediate death.

Problems Associated with Emerging Diseases

Emerging disease epizootics frequently cause substantial, often explosive, losses among populations of fish and shrimp seed, resulting in large economic losses in commercial aquaculture and threats to valuable seed stocks.

- ☆ Ignorance about "Fish seed quality assurance" through health management is considered as one of the important problematic factors in the seed industry.
- ☆ Arasitic (both ectoparasite and endoparasite) diseases in nursery are one of the most important limiting factors for growth and survival of fry and fingerling.

Mostly occurred carp seed diseases include:

Ichthyopthirius sp. (ich disease), *Trichodina* spp., *Ichthyobodo* spp., *Lernaea* spp., *Myxobolus* spp. and *Dactylogyrus* spp. *Myxobolus* and *Henneguya* affect the gills and causes heavy mortalities incarp seed.

In case of shrimp:

Common diseases are: Cotton Shrimp Diseases, Faulingin prawn due to *Epistylis* sp *etc.*

By Cach Problems Associated with Seed Inputs

The fisheries for wild shrimp and fish seed as well as brood stock are fraught with by catch problems. The seed are caught in coastal waters using fine-meshed nets that inevitably trap large quantities of other invertebrates and fish seeds, which are usually sorted out and discarded on land. In India and Bangladesh, where the

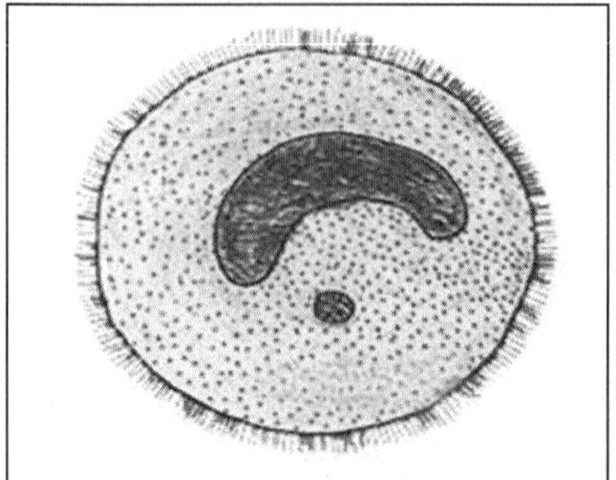

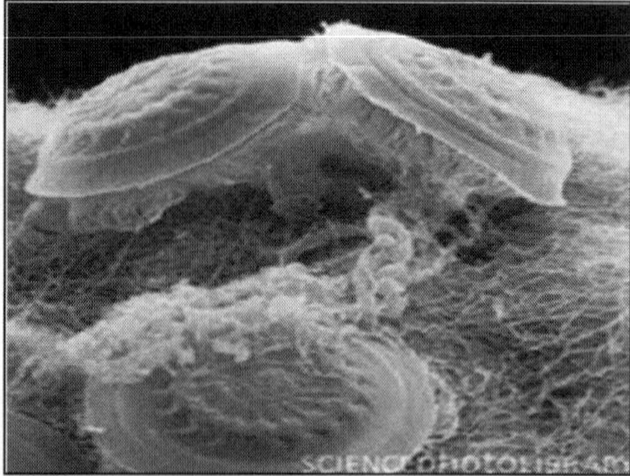

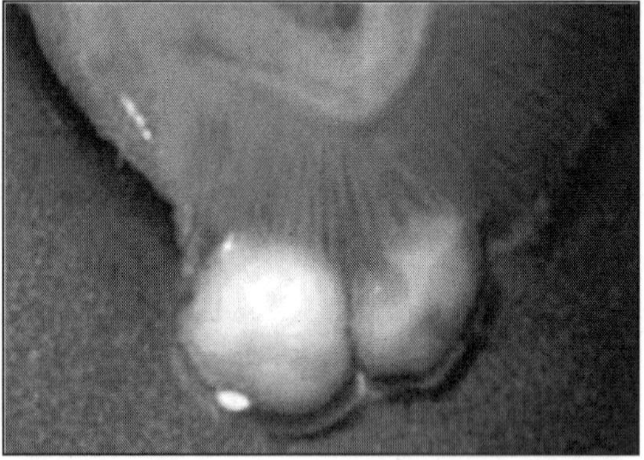

***Ichthyophthirius multifiliis, Henneguya* affects the gill, *Trichodina* sp.**

collection of wild *Penaeus monodon seed supports* major fishery operations, up to 1000 fish and other shrimp fry are discarded for every penaeid shrimp collected. The amount of by catch destroyed globally is staggering and could have major consequences for biodiversity and capture fisheries Production.

Seed Certification

Fisheries and aquaculture are oftentimes both governed by fisheries policy. In recent past, there is progress in the recognizing aquaculture in government policies. Like mandatory registration of all hatcheries with the Department of Fisheries;

- Mixed spawning of carps is an illegal act;
- Growing of only pure strains of cultivable varieties of carp seed;
- Certification of seed by the competent authority;

But this proper fish seed certification process does not exist in many countries in the region.

Introductin to Alien Species

- Introductions and transfers of seed from outside region (known as Alien sp) may lead to competition with endemic fauna, genetic introgression with local fauna, and introduction of pathogens and parasites.

- ☆ Improved control and regulation of the worldwide transfer of shrimp post larvae and spawners are a vital step needed to reduce the spread of disease between regions, countries and continents. Only native species should be cultured to avoid the introduction of alien species and minimise the spread of disease.
- ☆ Introductions of fish seed, including any "Genetically modified organisms" (GMO) and Exotic Species, have a role in the supply of fish seed for aquaculture. Any movement of fish seed between natural ecological boundaries/watersheds may involve risk associated with disease spread, biodiversity and management of genetic resources.

Huge Amount Usages of Chemical and Antibiotic

- ☆ Huge amount of Chemicals are used in fish and shrimp seed culture which may be classified as therapeutants, disinfectants, water and soil treatment compounds, algicides and pesticides, plankton growth inducers (fertilisers and minerals) and feed additives.
- ☆ Excessive and unwanted use of such chemicals results in problems related to toxicity to non-target species (cultured species, human consumers and wild biota).
- ☆ Prophylactic agents that can be used to limit disease prevalence include immunostimulants and probiotics.

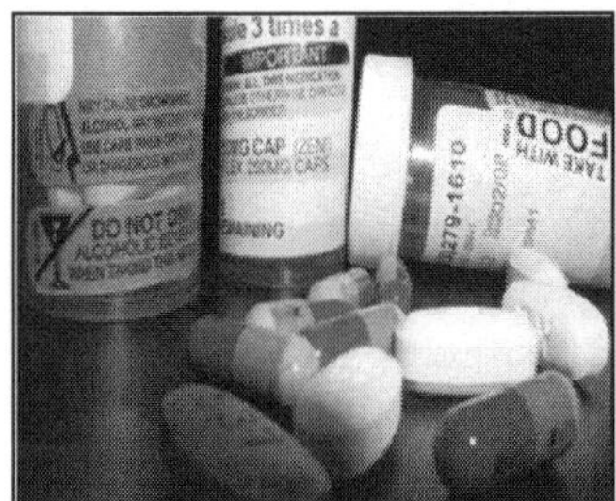

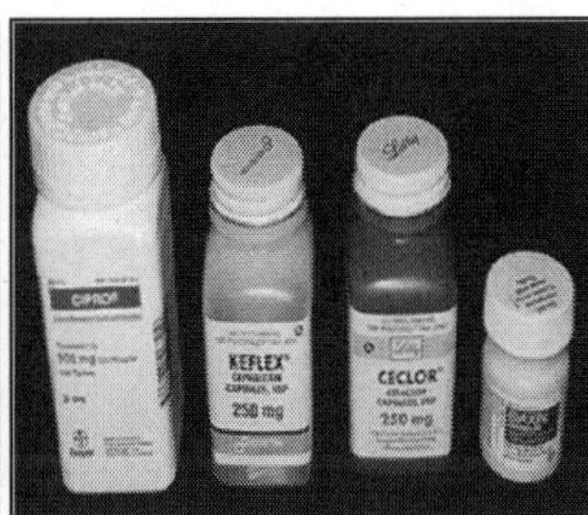

Various Types of Drygs and Chemicals Used in Aqua Farm

Metomidate,Etomidate, Propoxate, Ketamine Hydrochloride, quinaldine Sulfate - are generally used to prevent diseases. But they can be dangerous if dosages are not proper. It can leads to mortality of seed. Most of the rural farmers are unaware of the proper dosages.

- Higher doses of this Propoxat can cause respiratory arrest after 15 minutes at 64 mg/L, and after 1 hour at 16 mg/L.
- Over dosages of Quinaldine can cause an integumental and ocular irritant.

Consideration (Solution)

Development of Ancillary Units

- It is always necessary to make available some of the ancillary items like packing material, oxygen cylinder, breeding and hatching equipmentsetc at the time of need. These markets are to be developed at the nearest accessible areas for the benefit of seed producers and growers.

Creation of Trained Marketing Groups

- Unscientific and careless handling of fish seed often lead to high mortality in transportation and selling. *Training may be organized for the benefit of seed producers and unemployed youth in handling of fish seed, transportation etc.*

Transquilization

- It **must** be done during seed transportation.

Research and Development (R&D) Proposal to Overcome Poor Seed Quality Problem

- Research institutes should build an 'Institute-industry Research Partnership' with hatchery and nursery operators to improve quality brood stock and produce quality seed.
- Undertake research studies to determine the cause/s of poor quality seed.
- Investigate ecological innovations to use rice fields for fingerling production.
- Research on the use of wild-caught fish as feed as basis for support to policy decisions for efficient resource use.

Conclusion

- Inspite recent advancements in selective breeding and seed production technologies for aquaculture organisms and the availability of genetically improved aquaculture strains, there is still a pressing need to improve seed quality and yield for sustainable aquaculture in the region.
- Majority of the member countries share the same concerns and therefore issues as presented must be addressed to further boost aquaculture production in the next decade.

Chapter 26
Problems in Quality Seed Production and Broodstock Health Management in Aquaculture

Goals of a Fish Health Management Plan

Fish Health Management Plan aims to:

1. Prevent the introduction of exotic diseases or disease causing agents.
2. Reduce the occurrence of diseases in fish held in the culture facility.
3. Minimize the spread of disease to stocks within and outside the facility.
4. Maintain an environment that promotes the health and productivity of cultured fish and reduces the susceptibility of fish to disease.
5. Protect public health and minimize disease risks to cultured and wild fish through judicious use of drugs and chemicals.
6. Provide culturists and regulators with appropriate information from which rational, evidence-based fish health management decisions can be made.

Required Elements for Fish Health Management Plan

A. Characterizing the Health Status of Fish at a Culture Facility

1. Fish Health Records

Facilities must have an information management system that provides culturists with timely information to identify and assess changes in fish health to allow them to make fish health management decisions.

For individual groups of fish in the facility, operators must:

a) Keep up-to-date fish health records including:
   - Disease history and management;
   - Pattern of morbidity and mortality, sampling and diagnosis of disease;
   - Actions taken to prevent, control, and treat disease;
   - Records of movements of fish within the facility; and
   - Health risk factors specific to the site and/or the affected fish group;

b) Keep records for a minimum, the entire time that the fish are being cultured (*i.e.* until they have been released to the wild or harvested); and

c) Be able to link fish health records to other production records (*i.e.* feed, environment, transfers).

2 Monitoring Disease and Infection

Operators must have a plan for routine assessment of fish to determine their disease status and an action plan to prevent the spread of disease that will minimize the impact on fish and other organisms.

Operators must:

a) Regularly and systematically inspect fish and fish holding units for signs of disease;

b) Increase monitoring efforts for groups of fish showing unusual mortality rates, signs of morbidity, or subjected to stressful events that could predispose them to disease;

c) Routinely evaluate fish health and other production records; and

d) Develop an action plan to prevent, control or treat disease.

B. Identifying and Managing Risks to Fish Health

Risk factors must be identified and managed to minimize their effect on fish health and their role in predisposing fish to disease.

1. Water Quality

Maintaining good water quality is vital to good fish health.

Operators must have:

a) Regular programme for monitoring and recording water quality (*i.e.* temperature, oxygen, effluent);

b) Contingency plan to restore water quality.

2 Factors that Predispose Fish to Disease

Certain activities can have a negative impact on fish, the effectiveness of their immune system and their ability to combat infections and disease.

Operators will:

a) Minimize the time fish are exposed to stressful events such as anaesthesia/ sedation, crowding, and out-of-water events (*i.e.* handling, counting, grading, tagging, injecting);

b) Minimize predator interactions;

c) Provide fish with suitable rearing conditions and appropriate nutrition; and

d) Ensure equipment and methods used to handle fish will not result in significant injury or predispose fish to disease.

3 Vaccination

Vaccination is not a substitute for other aspects of fish health management, but part of an integrated disease prevention program.

Operators must use:

a) Vaccination procedures that minimize injury, secondary disease or losses to fish; and

b Vaccination programs based on local disease/infection conditions and information on the safety and efficacy of vaccines.

4 Brood Stock Health Management

Brood stock should be reared in a manner that recognizes and manages disease risks specific to this life stage and their progeny.

Operators must:

a) Use a system to identify individual or select groups of brood fish and their respective gametes from other production fish;

b Address the hygienic, nutritional and handling concerns unique to brood stock and

c) Use protocols for hygienic handling of brood stock or their gametes to prevent the transmission of disease causing agents to other fish.

C. Reducing Exposure to or Spread of Disease Causing Agents

Minimizing the exposure of fish to disease causing agents will reduce the spread of pathogens and disease.

1. Outbreak Investigation and Management

Detecting and managing outbreaks will help reduce the spread of disease or disease causing agents.

Operators must:

a) Have access to the resources and qualified personnel needed to detect and manage a disease outbreak;

b) Develop a rapid response plan to reduce the spread of disease and initiate it when a disease outbreak is detected;

c) Detail monitoring activities during and after an outbreak to establish the distribution of the disease and monitor the effectiveness of control and treatment measures;

d) Keep details of investigations and verification of all outbreaks (this must be under the supervision of a qualified fish health professional); and

e) Notify Provincial and Federal authorities in the event of outbreaks in accordance with existing regulations or surveillance agreements.

2. Management of Dead Fish

Operators must:

a) Regularly remove dead fish from holding units and dispose of the fish in a manner that will not facilitate the spread of disease; and

b) Plan for the removal and disposal of increased levels/numbers of mortalities during unexpected disease outbreaks or loss of fish.

3. Bio-security

Maintaining a clean, safe work environment reduces the potential for spread and exposure of fish to infectious or parasitic disease causing agents.

Operators must:

a) Reduce the potential movement of infectious or parasitic agents within and between facilities by:

- Using good hygiene and sanitation protocols that manage the movement and disinfection of staff, contractors, visitors, vessels or vehicles; and
- Routinely cleaning and disinfecting equipment and holding units;

b) Safely handle and dispose of disinfectants (In accordance with occupational safety waste management and pollution regulations).

4. Release or Escape of Fish from a Culture Facility

Operators must:

a) Minimize the risk of escape or release of sick or infectious fish from fish culture facilities; and

b) Consider disease and treatment status when planning deliberate release or relocation of fish stocks where escapes can occur.

5. Movement of fish

Fish Health status must be considered when evaluating the risks of moving fish stocks.

Operators must:

a) Ensure all equipment used to transport fish:

- Safeguards the health of the fish being moved;

- Minimizes the risk factors that may predispose the fish to disease;
- Minimizes transfer of disease causing agents; and
- Reduces the risk of accidental loss of fish and gametes during transport activities;

b) Assess and mitigate the disease risks from cultured fish that are known to be, or are suspected to have, a disease or infection before they are sold, or given away;

c) Assess and mitigate the risks associated with intentional release of fish to the wild;

d) Minimize the transfer of disease agents between stocks within and in the vicinity of the facility; and

e) Ensure optimal water quality during transport events.

D. Use of Drugs and Chemicals in Fish Health Management

1. Diagnostic Support

Operators must have a veterinary-client relationship with a licensed veterinarian to ensure:

a) Reliable diagnoses are obtained and

b) Prescriptions for treatments can be obtained in an effective manner.

2. Drug, Chemical and Biological Use for Disease Treatment and Prevention

Operators must:

a) Ensure staffs have access to information on the drugs, chemicals and biologics that are used on site;

b) Ensure safe handling and storage of drugs and chemicals;

c) Keep records of the amounts of drugs, chemicals, biologics and medicated feeds purchased or moved into a fish culture facility and/or used during treatment;

d) Ensure groups of fish that are treated can be identified during treatment and subsequent withdrawal times and

e) Keep records of treatment for the entire time that the fish are being cultured or until they have be released or harvested.

Common Disease Problems in Carp Hatcheries and Early Rearing Systems

Diseases and Parasites in Carp Hatcheries

The nursery and rearing systems of carps are often very rich organically and provide an ideal environment for many of the pathogens. Important problems in hatcheries and early rearing systems are caused by some of the following pathogens:

1. Protozoan ciliates like *Ichthyophthirius multifilis* (white spot disease), Trichodina complex and ecto-dermal ciliates like Epistylis, Vorticella. All these ectoparasites can cause mass mortality of younger stages of carp very quickly and the situation becomes worst in waters with low oxygen and high organic matter.
2. Disease caused by Myxosporidians are serious threat to the fish seed farms. These sporozoan spores present in the pond soil are normally ingested by the developing fry. Once inside the target tissue the sporozoans cause massive destruction of their target tissue and produce large scale mortality.
3. Worm parasites like Dactylogyrus (gill fluke) and Gyrodactylus (skin fluke) with their well developed attachment haptor and feeding apparatus can cause mortality in early developmental stages of carps.
4. Opportunistic secondary bacteria (Aeromonas) and fungi (Saprolegnia) can become serious problem in fish larvae which are heavily parasitized. In carp rearing ponds major problems are because of larger ectoparasites, secondary invaders, systemic bacterial pathogens and viruses. Ectoparasites like Dactylogyrus and Argulus (fish lice) and endoparasites like sporozoans are very important. Bacterial problems like surface ulcerative conditions and acute systemic diseases are common in carp rearing systems. Many a times mortalities seen in carp culture systems are a result of ectoparasitic and systemic bacterial diseases. External fungal problems are normally associated with fish which are poorly handled. The possibility of viral diseases causing mortalities in carp nurseries cannot be ruled out.

Bacterial and Fungal Diseases in Carp Rearing Systems

- ☆ Bacteria can cause diseases either as secondary invaders or as primary pathogens.
- ☆ Bacterial diseases in larvae can be broadly classsified as surface ulcerative, acute systemic and chronic granulomatous type.
- ☆ Surface ulcerative types of diseases are characterized by haemorrhagic surface ulcers and are normally caused by *Aeromonas, Pseudomanas, Vibrios, Flexibacteria, Myxobacteria, etc.* Surface ulcerative disease conditions at times develop to acute systemic disease.
- ☆ These are characterized by the presence and proliferation of bacteria in internal organs like kidney, heart, spleen, blood and other visceral organs.
- ☆ These diseases produce significant necrotic changes in all the affected organs and can cause mortality in a short time scale.
- ☆ Bacterial haemorrhagic septicaemia caused by numerous serotypes of *Aeromonas hydrophila* is a major problem.

Treatment Chart for Common Disease Conditions of Carp Larval Rearing Systems

Disease Agent	*Chemical*	*Method*	*Concentration/Time*
1. Ectoparasitic protozoans			
a. *Ichthyophthirius* sp.	Formalin	Short bath	60-100 ppm for 30 minutes
b. *Trichodina* sp.	Formalin	Long bath	20-30 ppm
c. *Epistylis* and *Vorticella*	Formalin	Dip	200-300 ppm for 1 minute
d. *Ichthyobodo* sp.	Formalin	Dip	20-30 pm for 30 minutes
2. Monogenetic worms			
a. Dactylogyrus	Wormid	Pond treatment	50gms per ton body wt of Fishes up to 5 days
b. Gyrodactylus	Wormid	Pond treatment	50gms per ton body wt of Fishes up to 5 days
3. Crustaceans			
a. *Lernaea*	Licex, Arguklin	Pond treatment Through feed	50gms per ton body wt of Fishes up to 5 days 250gms/ton of fish feed for 1st day and then 125gms for 15 days.
b. *Argulus*	Licex, Arguklin	Pond treatment Through feed	50gms per ton body wt of Fishes up to 5 days 250gms/ton of fish feed for 1st day and then 125gms for 15 days.
4. Endoparasites			
a. Sporozoans	Licid	Pond treatment	100gm/ton body wt of fishes continue up to 5 days
b. Cercaria and metacercaria of digenetic trematodes	Licid	Pond treatment	100gm/ton body wt of fishes continue up to 5 days
5. External Mycosis Saprolegnia	Malachite green Formalin	Dip Bath Dip Bath	60 ppm for < 1 minute 1-2 ppm for 1 hr 50-75 ppm for 30 minutes 100-200 ppm for 1-3 minutes
6. Surface bacterial diseases	Proflavine Oxytetracycline and Furnace	Bath Short bath	20 ppm 1-5 ppm
7. Systemic bacterial diseases	Enrofloxacin Oxonilic acid Chloramphenicol	In feed In feed In feed	50mg/kg fish/day 10mg/kg fish/day 50 mg/kg fish/day

Gas Disease

Heavy mortality of fish seed due to 'gas disease' has been observed on several occasions. The symptoms of the disease are the presence of gas bubbles in gill filaments, heart and blood vessels and also in the gut. The air bladder gets highly distended. The 'gas disease' also referred to as 'gas embolism' occurs due to super saturation of water with either oxygen or nitrogen and is generally encountered in fish nurseries

Maintenance of Brood Stock

Raising of properly maintained brood stock of both sexes to prime mature condition is a pre-requisite for successful induced breeding. Potential brood fishes of over 2 years age, need be stocked in ponds @ 1000-2000 kg/ha by November in Assam and January in other parts of the country and reared for a period of about 6 months. For a target production of one crore spawn (60 lakh of Indian major carps and 40 lakh of Chinese carps), during a season 750 kg of brood stock (300 kg of Indian major carps+450 kg of Chinese carps) comprising males and females in an approximate ratio of 1: 1 by weight and 2: 1 by number would be required. The management of the pond includes maintenance of sanitary condition in the predator free pond devoid of weeds, enriching the nutrient status by application of manures and fertilizers as in stock ponds (CIFRI Aquaculture Extension Manual, New Series No.2) and providing supplementary feed mixture of groundnut/mustard oilcake and rice bran in 1: 1 ratio by weight @ 1 - 2 per cent (generally 1 per cent) of total body weight of fish per day.

Practical Handling and Care of Fish

This is the very first step in fish production especially through aquaculture source because there cannot be harvest without recruitment. It entails all the precautionary steps and techniques to protect the fish especially from the seed stage to adult stage when it appears on the table for consumption. The fish seeds (fry and fingerling) are usually 'nursed' rather than reared to ensure proper recruitment or else the harvest will be low. All processes *e.g.* the contact materials with fish (fish seed), water treatments, fastness during operations *etc.* are referred to as the practical handling and care of fish to ensure the success of each venture into fisheries management.

For example, as a rule, gear that is used for collecting fish especially fish seed is constructed of fine mesh (<10mm) netting materials and should inflict minimal physical injury on the body. These gears include drag net, lift net, cast net, clap net and traps. When using nets or traps to take fingerlings or adult fish out of ponds, care must be taken during seining since muddy water can stress the fish. To reduce mortality due to muddy seining, fish could be attracted with crumbles of feeds, light source or dewatering. Other practical handling and care of fish can be hinted as:

- ☆ Always use wet hand/contact materials *e.g.* nets
- ☆ Handle the fish gently
- ☆ Keep fish in water while counting
- ☆ Avoid overcrowding the fish in containers
- ☆ Avoid standing fish in the sunlight and for a long time

- ☆ Fill containers for transport with pond water
- ☆ Transport fish during cool weather
- ☆ Work fast and travel quickly
- ☆ Avoid squeezing, throwing or dropping fish
- ☆ Always check temperature and temper fish

Tempering is a pre-stocking treatment of especially fish seed. It involves homogenizing the water in which the seeds are transported with the water in the pond/system into which they are stocked. It is a crucial step in handling and care of fish to avoid high mortality

The Effects of Stress on Quality

Under stressful conditions, fish must choose between putting energy into increasing or maintaining growth, continuing to survive by fighting disease and pathogens or putting energy into reproductive effort. The most common sacrifices involve reproductive output including the fecundity of the animal and quality of the eggs and sperm.

Eggs

Under stressful conditions, female broodstock will choose between allocating energy for growth and survival or energy for reproduction. The most common negative trade-off involves reproductive performance including diminished egg quality and viability. The stickiness of common carp eggs can be removed by adding fertilization solution/tannin solution.

Fertilization Solution: (30 gms of carbamide/Urea +40 gms of Nacl) mixed in 10 litres of water. Add the solution @10 to 20 per cent of the volume of the eggs and stare it.

Tanin Solution: (5 to 8 gms of Tanin in 10 litres of water) after treatment it should be washed in freshwater for incubation and hatching.

Sperm

Stress in male brood stock causes major reductions in the sperm motility (duration of activity) and velocity (swimming speed) resulting in a reduced ability to fertilise the eggs.

Progeny

Stress in brood animals results in reduced survival and viability in developing and developed progeny. Parental stress results in skeletal deformities, abnormal skin pigmentation, organ failure and scale defects. Progeny with deformities generally show increased mortality rates, reduced growth and increased susceptibility to disease.

Stress Management

Stress in fish is a habitual, undesirable aspect of production. Stress results from biotic and/or abiotic challenges that act in changing or modifying the animal's natural or homeostatic state. Common fish stressors involve alterations in the fish's immediate environment such as:

- Chemical changes: contaminants, low oxygen and acidification
- Physical changes: handling, capture, confinement and transport
- Perceived changes: startling or predators

The above stressors are categorised as either acute or chronic depending on their duration and frequency. Acute fish stressors are events which the animal experiences for a short period of time such as handling. Chronic fish stressors are defined as a constant or recurring exposure that causes a prolonged physiological response. Fish stressors of both types have severe negative effects on the animal's growth and health.

The Physiological Response to Stress in Fish

Generally, Stress in fish is a non-specific response classified as three distinct responses:

1. Primary response: The primary response is characterized by the activation and the secretion of the hormones, corticosteroids (cortisol) and catecholamines into the blood.
2. Secondary response: The release of hormones triggers the secondary response which involves the release of glucose into the blood for energy production, followed by increases in heart rate, gill blood flow and metabolic rate which causes changes to blood lactate and hematocrit.
3. Tertiary response: Changes in blood physiology ultimately cascade into a whole body change or the tertiary stress responses. Changes associated with the tertiary stress response include reduced growth rate, decreased disease resistance, altered behavior and reduced survivability.

Fish Health Management

The Effects of Stress on Health

Fish health management is a term used to describe management practices that are designed to prevent fish disease. Successful health management begins with prevention of disease rather than its treatment. Controlling stress during health management operations such as health checks, vaccinations and bath treatments will maximize the ability of your fish to withstand infection or re-infection by ensuring the immune system is strong.

Health Checks

Routine monitoring of your animals for bacterial, viral and parasitic diseases is critical to maintain good fish health. Sampling fish for disease symptoms involves removing a sub-sample from the population and screening the individuals for disease. When removing fish from the water they must always be sedated. AQUI S®sedation will ensure fish survivability and growth post-handling. It is usually the process of capturing the animal that causes the greatest stress response. In order to maximise the benefit of using AQUI S® we recommend that fish are carefully crowded using low stress handling techniques then sedated with AQUI S®. Once fish are handleable, *i.e.* showing no response to being removed from water, health checks can be carried out.

Vaccinations

It is common practice on most aquaculture farms to vaccinate the fish at least once in their production cycle. Vaccination is achieved by either immersion for a period of time (generally 30 seconds to 30 minutes) or by injection, usually intraperitoneally (in the abdominal cavity). Sedating fish with AQUI S® before these handling procedures improves fish survivability and the effectiveness of the vaccine.

Bath Treatments

Bath treatments use compounds that are often toxic or irritating to the fish and result in significant stress in the animal. Bath treatments vary in duration from seconds to hours with short treatments often causing as much or more stress than longer treatments. Sedating the animals with AQUI S® before and during bath treatments to a level which ensures they maintain swimming action and equilibrium, is critical to bath success and fish survival. In these instances AQUI S® sedation will ensure a strong immune system and limited to no loss in growth post bath.

Transport Management

Effects of Stress during Transportation

Fish respond to transport with several adverse reactions. In general these reactions lead to a rapid deterioration in water quality which ultimately leads to deterioration in fish quality and increased mortality. When fish are stressed and confined to high densities they respond with an increase in metabolic rate, blood flow, gill permeability and ion exchange. This results in increased oxygen (O_2) consumption and increased carbon dioxide (CO_2) excretion. In addition, fish increase production and excretion of waste products such as feces and mucus. The waste products accumulate in the holding system leading to often dangerous levels of toxic ammonia (NH_3).

Types of Live Transport

There are two basic transport systems for live fish - the enclosed transport and the semi-enclosed transport.

How Can AQUI-S® Anaesthetic Help Your Transport?

Sedating fish with AQUI S® prior to and during transport will reduce stress and decrease the metabolic activity of the fish. This will lead to a lower O_2 demand and a reduction in CO_2 and waste production. As a result, a non-toxic environment will be maintained for longer in the transport vessel. AQUI S® has proven to be effective for short duration transport (<2 hrs) such as local smolts to sea or longer transports (>40 hrs) such as live lobster, eel or cod to distant international markets. The mode and duration of transport will dictate the method of sedation and the dose of AQUI S® required

Algal Bloom Control

Fertilization, at times, may result in the development of excessive microscopic plants which are less desirable or may even become undesirable when they form bloom. Depending on the intensity of such blooms, Diuron @ 0.1 to 0.3 ppm (1 to 3 kg per hectare-metre of water) may be used as control measure. The required quantity of herbicide may be mixed with water and spread uniformly over the pond surface. However, blooms in such small ponds are most conveniently controlled in about a week by thickly covering the water surface with duck weeds. The weeds may afterwards be removed from the pond. If grass carp exists in the system the same will serve as their food and thus need not be removed.

Nusery Pond Management

- ✰ The spawn are raised to fry stage in nursery ponds in about a fortnight during monsoon months.
- ✰ The preparation of ponds includes de weeding by manual labour, eradication of unwanted fishes by application of mahua oilcake @ 2000-2500 kg per hectare-metre. or tea seed cake @ 75-150 ppm or ammonia @ 20-25 ppm or bleaching powder @25-30 ppm in perennial ponds. This is done in summer months about five weeks before the anticipated date of stocking. Lime is applied @ 250-300 kg/ha about two weeks after application of piscicide, followed by manuring with cow-dung @ 10,000 kg/ha about two weeks before anticipated date of stocking. Harmful aquatic insects are controlled by soap oil emulsion treatment @ 56 kg vegetable oil: 18 kg cheap washing soap/ha about 12-24 hours before stocking. 3-day old spawn usually in the morning hours after about 5 weeks of piscicide application are stocked @ 25-35 lakhs/ha after ensuring complete detoxification of the piscicide.
- ✰ Supplementary feed mixture of groundnut and mustard oilcake and rice bran (1: 1 by weight) are fed daily @ 0.56 kg for first five days and 1.12 kg for next seven days for everyone lakh of spawn. No feeding is required thereafter.
- ✰ Harvesting of fry (25-30 mm size) is done on the 14th day by fine meshed drag net (1.5 mm).
- ✰ 3 to 4 such crops are possible to be obtained from the same pond during one season.

Fry Stocking

After testing for complete detoxification, 25-30 mm fry are stocked in various combinations at densities ranging from 2 to 3 lakhs/ha in any of the following ratios.

Species	*Ratio*
Catla + Rohu +Mrigal	2 : 4 : 4
Silver carp + Grass carp	1 : 1
Silver carp + Grass Carp + Common carp	4 : 3 : 3
Silver carp + Grass carp + Common carp	5 : 1.25 : 3.75
Catla + Rohu + Mrigal + Common carp	3 : 4 : 1 : 2
Catla + Rohu + Mrigal + Grass carp	4 : 3 : 1.5 : 1.5
Catla + Rohu + Mrigal + Grass carp	3 : 3 : 3 : 1
Silver carp + Grass carp + Common carp + Rohu	3 : 1.5 : 2.5 : 3
Silver carp + Grass carp + Common carp + Rohu	4 : 2 : 2 : 2

Rearing Pond Management

- ✰ The steps relating to de weeding, eradication of unwanted fishes, application of lime are same as in nursery pond management.
- ✰ Organic manuring is done with cowdung @ 2500 kg/ha in four equal monthly instalments. The initial dose is applied about 15 days earlier to stocking.
- ✰ Inorganic fertilization is done with urea @ 100 kg/ha or ammonium sulphate @ 200 kg/ha and single super phosphate @ 100 kg/ha or triple super phosphate @ 35 kg/ha in three instalments. The first instalment of inorganic fertilizers is given on second day of stocking and thereafter at monthly intervals, alternating with organic manuring.
- ✰ After ascertaining complete detoxification of piscicide, stocking of fry is done @ 2-3 lakhs/ha.
- ✰ Supplementary feeding with oi1cake and rice bran mixture @ 6 kg/lakh/day for the first month, 10 kg/lakh/day for the second month and 15 kg/lakh/day for the third month may be done commencing from the day of stocking.
- ✰ Duck weeds, preferably *Lemna* and *Wolffia* need be supplied as per consumption · of grass carp, if stocked.
- ✰ Fingerlings of 100-150 mm size are expected in about 3 months rearing period when they are ready for harvesting.

Fingerling Harvesting

The growth and well being of the growing fry during the rearing period need to be periodically checked. This is conveniently done by drag netting. Healthy fingerlings of 100-150mm size are obtained in three months rearing period.

Supplementary feeding is stopped a day before the date of catching. Harvesting is done during cool morning hours by repeated drag netting. Survival of about 70-90 per cent (average 80 per cent) is generally obtained in about 3 months rearing.

How to Improve the Brooders

1. Selective breeding (Final outcome: Jayanti Rohu)
2. Development of disease resistant stocks (SPF/SPR/SPT).
3. Transgenesis (GMO).
4. Hybridisation and Cross breeding (Hybrid Vigour)

Chapter 27

Fish Seed Certification Problems and its Future Perspective

Introduction

Fish Seed is the basic and most critical input for sustainable aquaculture. The response of all other inputs depends on quality of seeds to a large extent. It is estimated that the direct contribution of quality seed alone to the total production is about 15 – 20 per cent depending upon the crop and it can be further raised up to 45 per cent with efficient management of other inputs. Aquaculture is the fastest growing food production sector, globally. The adage of "what you sow is what you reap" certainly holds true. Quality seed is a fundamental pre-requisite for sustainable and successful aquaculture, be it small-scale or commercial farming. Inadequate supply of quality seed is often suggested as a major constraint for aquaculture in many parts of the world. That's why Fisheries and aquaculture are oftentimes both governed by fisheries policy. In recent past, there is progress in the recognizing aquaculture in government policies. Like mandatory registration of all hatcheries with the Department of Fisheries;

- ✰ Mixed spawning of carps is an illegal act;
- ✰ Growing of only pure strains of cultivable varieties of carp seed;
- ✰ Certification of seed by the competent authority;
- ✰ But this proper fish seed certification process does not exist in many countries in the region.

Fish Seed

Fish seed denotes spawn to fry and to advanced fingerlings. It is a collective name of hatchlings, post larvae, fry and fingerlings. Advanced fingerlings are used in beel and reservoir management.

Quality Seed

Good quality seed is defined as those having better food conversion efficiency (FCR), high growth rate potential, better ability to changing environmental conditions and resist diseases (Padhi and Mandal, 1999)

Good quality seed is the one that benefits the hatcheries, nurseries, intermediaries in the seed supply network and finally the producers. Seed which are active, healthy in appearance, uniform in size, with potential for high survival, better growth, less or no disease.

Present Problems of Aqua Sector: Responsible for Poor Quality Seed

1. Poor Pond Hygiene

Mortality of seed is attributed mainly to prevalence of **protozoan parasites**, particularly in ponds which are not properly managed and hygiene.Some common disease-causing organism which encountered in hatchery are **Protozoan parasites** like *Ichthyophthirius multifilis, Trichodina* complex and **ectodermal ciliates** like *Epistylis and Vorticella;* myxosporidians; Worm parasites like *Dactylogyrus and Gyrodactylus; Argulus; larger* ectoparasites, secondary invaders, systemic bacterial pathogens and viruses *etc.*

2. Inbreeding

Poor performance of the resultant seed had been linked to inbreeding of carps in India (Eknath and Doyle,1985). Since carps are highly fecund, hatchery operators tend to maintain a low effective population size (number of broodstock that contributes genetic material to the next progeny) and do not exchange broodfish between hatcheries.

3. Poor Management of Brood Fish and Seed

Competition among fish seed producers to meet demand sometimes leads to poor management of broodfish and fish seed which may negatively affect seed quality. According to Reddy (1995), most fish hatcheries in India are concerned more about the quantity rather than the quality of fish seed and produce them without following any selection norms.

4. Transportation Stress

During transportation, fish seed are subjected to confined environment, higher metabolic load, stress, strain and exhaustion. As a result, seed becomes susceptible to diseases and parasites.

Conditioning of Fish Seed

Conditioning is nothing but acclimatizing seed to a restricted environment prior to packing and transportation. During this period, seed are stocked at a very high stocking density in a *hapa or a pond* with running water, but without provision for food so that the weak seed dies and only the healthy fry survives. Only seed that survives the 'stress test' is selected for transport to a required destination. This type of conditioning of fish seed is not a common practiced in India.

Fish Seed Certification

Fisheries and aquaculture are oftentimes both governed by fisheries policy. In recent past, there is progress in the recognizing aquaculture in government policies. Like mandatory registration of all hatcheries with the Department of Fisheries

1. All hatchery owners must be registered with the Department of Fisheries.
2. Mixed spawning of carps by hatchery or hapa breeders is illegal and is an offence.
3. The Department of Fisheries is to exercise necessary initiative to collect data on fish seed production.
4. The officials of DOF are to take necessary action to form a network of seed producers in their respective areas.
5. The farmers may be educated to grow only pure strains of cultivable varieties of carps as recommended by DOF.
6. The research and extension wings of DOF may be strengthened with competent persons and necessary research conducted on different aspects of fish seed production.
7. Fish seed markets may have to be established in each district through a network of seed producers for the benefit of both producers and buyers. Seed may be certified by the competent authority before being sold.
8. The DOF may instruct all seed producers to use a standard method (perforated steel bati) to count fish spawn, fry and fingerlings.
9. Entrepreneurs in the private sector may be encouraged to establish feed plants to produce pelleted feed for broodfish and larvae to improve their quality.
10. Non-functional hatcheries may be leased out to technically qualified persons to operate them profitably.
11. Utmost care must be taken while releasing hatchery produced carp seed in natural water bodies like bheels, as most hatcheries are engaged in mixed spawning.

Introduction of Alien Species

Introductions and transfers of seed from outside region (known as Alien sp) may lead to competition with endemic fauna, genetic introgression with local fauna, and introduction of pathogens and parasites. Improved control and regulation of the worldwide transfer of shrimp post larvae and spawners are a vital step needed to reduce the spread of disease between regions, countries and continents. Only native species should be cultured to avoid the introduction of alien species and

minimise the spread of disease. Introductions of fish seed, including any "Genetically modified organisms" (GMO) and Exotic Species, have a role in the supply of fish seed for aquaculture. Any movement of fish seed between natural ecological boundaries/watersheds may involve risk associated with disease spread, biodiversity and management of genetic resources.

Status of Fish Seed Certification in India

India has an aquatic animal export and import business. Some legislation relating to the import and export of live aquatic animals does exist in India, but it is not complete. Some sort of health certification and quarantine procedure does exist in India and the aquatic animal health quarantine has been described by Punyabrata Das (1997). The NBFGR, located at Lucknow, Uttar Pradesh, has been recognized as the focal agency for formulating legislation on aquatic animal health certification and quarantine. Proper fish seed certification process does not exist in many countries in the region. However, there is no definite fish seed certification in India in National level.

Status of Fish Seed Certification in World

Some countries of the world followed some norms relating to fish seed certification. But several attempts towards seed certification has been taken by different countries. The following table shows some of the attempts.

Approach	*Measure*	*Country*
Institutional approach Certified	Supply of quality brood stock from government institutions	China
Fish seed certification and Quarantine	genetically improved varieties	China
Husbandry practices	Adoption of good husbandry practices	Pakistan, Sri Lanka, Bangladesh
Quality criteria	Standards to measure quality based on age, size and color, Body shape and behavior, Growth rate, survival, percentage deformity, behavior, Body shape, abnormalities, body and scale damage, mortality soon after stocking	Indonesia, Thailand, Sri Lanka, Viet Nam, Bangladesh
Health management	Periodic checks on diseases and infections, prophylactic treatment, Health certification	Sri Lanka, Thailand, Bangladesh, India, Thailand
Maintenance of genetic quality	Broodstock management, selective breeding. Awareness on genetic issues among hatchery operators and farmer	Thailand, Philippines, Cambodia, Viet Nam, Bangladesh, Philippines

Future Prospects and Recommendation

1. A live gene bank for producing and supplying pure seed of IMCs and exotic carps may be established since the quality of carp seed produced

in several hatcheries are reportedly not satisfactory due to continuous inbreeding.

2. Fish hatchery operators should be trained on better broodfish management, hatchery management and nursery management to produce quality fish seed.
3. Government or financial institutions should sponsor setting up of field laboratories for assessing and monitoring fish seed quality.
4. More emphasis should be laid on multiple spawning of carps so as to ensure the availability of seed over a longer duration in a year.
5. Greater support (technical as well as financial) from government agencies is needed for sustainable fish seed production.
6. Apart from the routine production of carp seed, emphasis needs to be given also to produce seed of valuable species like catfish and murrels, which command a good price in several parts of the country.
7. Mixed spawning should be prevented to protect the genetic purity of our precious carps gene pools.
8. Attempts should be made for efficient marketing and distribution of fish seed from surplus states to maximum deficit states.
9. The Government of India should explore the possibility of having a uniform fish seed grading system and pricing for the entire country.

Conclusion

Globally aquaculture is active, expanding and diversifying. Increasing aquaculture development will likely lead to more demand for quality seed. Aquaculture has suffered enormous losses due to trans-boundary diseases and increasing risks are expected in future as aquaculture expands leading to increased movement of seed and brood stocks across national and international boundaries. Some countries in the Asia-Pacific region still face significant challenges in the practical implementation of health management strategies, specifically in the areas of diagnosis, surveillance, risk analysis, and quarantine and certification programs. These challenges are mainly due to inadequate national capacity. There is, therefore, a need to continue to strengthen regional and international cooperation in aquatic animal health.

Chapter 28

Pond Bottom Soil: Its Problems and Remedies

Introduction

The pond bottom soil and the accumulated sediments are integral parts of ponds. Concentrations of nutrients, organic matter and microorganism density in the pond bottom are several orders of magnitude greater than in the water.However, suitable soil quality problems are common in aquaculture ponds, and therefore, many methods are used for improving pond soils.

General Features of Pond Bottom Soil

- ☆ Compared to original compacted pond soil, sediment contains more moisture.
- ☆ It possesses higher percentages of silt and clay.
- ☆ It has lower bulk density (<1.4 g/cm^3), greater porosity, specific surface area, and cation exchange capacity.
- ☆ It contains greater concentrations of organic matter and nutrients.
- ☆ Sediment organic matters are highly decomposed as evidenced by low proportions (5–15 per cent) of carbon and nitrogen.

Importance of Pond Bottom Soil

- ☆ The role of bottom soil in determining productivity of a pond is well documented.
- ☆ The production of various primary food organisms depends largely on the availability of different nutrients. Dynamics of availability of most of these nutrients, in turn, is determined by the condition prevailing in the bottom soil.

☆ Considering this significance, bottom soil is designated as the chemical laboratory of the pond.

Problems of different Types of Pond Bottom Soil

Acid Sulphate Soil

Acid Sulphate Soils occur naturally in both coastal (tidal) and inland or upland (freshwater)settings. Acid sulfate soils are extremely acid due to pyrite oxidation. Often, they are also unripe, sometimes also saline. Some occur naturally but most have developed because of drainage of previously coastalaluminum and peat.

Problems

Acid Sulphate Soils are harmless, but when excavated or drained, the sulfides within the soil react with the oxygen in the air, forming sulfuric acid. This acid, together with associated toxic elements (heavy metals and other contaminants), can kill aquatic plants and animals, contaminate drinking water and food such as oysters.

The main problems with acid sulfate soils are:

a. They tend to be waterlogged in their natural state and must be drained before cultivation, and
b. Draining beyond the pyrite layer will generate excessive acidity which is detrimental to fish growth.

Remedies

A procedure for rapid reclamation of ponds with acid Sulphate soils involves drying and filling of the soil to oxidize pyrite, filling the pond with water and holding till water pH drops to below 4, and then draining the pond, repeating the procedure, until the pH stabilizes at a pH above 5, and then liming the pond with 500 kg of calcium carbonate ($CaCO_3$)/ha.

Clay Soil

Clay soil have over 25 percent clay. Also known as heavy soils.These are potentially fertile as they hold nutrients bound to the clay minerals in the soil. They also hold a high proportion of water due to the capillary attraction of the tiny spaces between the numerous clay particles. They drain slowly and take longer to warm up in spring than sandy soils. Clay soils are easily compacted when trodden on while wet and they bake hard in summer, often cracking noticeably.

Problems

1. While sandy soils remain stable as water passes through them, soils rich in clay undergo significant changes based on their moisture content.
2. When clay type soils dry out, they shrink significantly so much that the ground can become covered with cracks. But after a heavy rain, the cracks are gone and the soil is fully saturated with water. Becauseclay soil absorbs so much water, it expands during wet weather.
3. Too clayey soil keeps all the nutrients adsorbed on to it.

Remedies

Adding of compost

Amend soil with 6 inches of good compost is very good for this type of soil. Work it right into the top of a clay soil and it will improve infiltration and will probably improve the amount of air.

Sandy Soil

A Sandy soil contain a high proportion of sand particles, with little silt or clay to modify the grainy nature of the soil. Sandy soils occur in patches in many districts, and can be very coarse, like builders' sand, or very fine and powdery. One of the commonest types of sandy soil occurs in heathland, such as the Surrey heaths and Norfolk breckland.This very poor soil is largely used for forestry rather than farming.

Problem

1. Sand particles are very much larger than silt and clay particles.This means the gaps between them are bigger too, so water flows away very quickly.
2. Summer drought is almost inevitable. Only an exceptionally wet summer will provide sufficient water to prevent it.
3. In sandy soils, most nutrients are washed out quickly.Even in sandy soils, phosphates are usually retained, but potash and nitrogen go very easily, and calcium is often in short supply. Therefore, sandy soils are usually infertile and acid.

Remedies

If the pond is constructed on sandy soils, heavy doses of organic manure are essential to control seepage loss of water. In general, the dose of raw or composted farmyard manure varies from 10,000 to 15,000 kg/ha/year.

Deteriorated Pond Bottom Soil (Black Soil)

How does Pond Bottom get Deteriorate?

Unconsumed feed, dead fish, molts, fecal matter and other organic matters(dead algae,aquatic weeds) settle at the pond bottom. Degradation or deterioration of this organic matter releases obnoxious gases. These are primarily due to generation of ammonia,hydrogen sulphide,methane generated due the degradation of the fecal matter.

Problem

1. Generation of ammonia,hydrogen sulphideand methane causes anaerobic conditions at the bottom of the pond.
2. Dissolved Oxygen at the pond bottom goes down drastically.
3. Bottom dwelling fishes face serious respiratory problems even death.
4. As prawn is a bottom dwelling animal, lack of oxygen at the bottom of the pond affects the growth rate of the prawn or even in some cases causes mortality.

5. Pond bottom management is one of the most important requirements for successful shrimp production.Poor pond management is one of the main causes for bacterial disease in shrimp and it is thought also to increase the risk of white Spot disease.

Identification of Deteriorated Pond Bottom Soil

- ✰ Check if the bottom soil is black or smells bad. If the soil is black or smells it means pond bottom has deteriorated. If the bottom soil is not black or smelling bad, then it can be considered as ok.
- ✰ Rapid consumption of feed from the feed trays can be an indication of poor pond bottom and not necessarily an indication that more feed should be used.

Remedies

1. One month after stocking, the pond bottom around the feeding area, trench, center, corners and outlet gate should be observed weekly.
2. If the soil color is black and smells bad, try to spread the feed further away from the dike(middle feeding). Avoid feeding in areas where the soil is black. Reduce the feeding rate for couple of days and change 5-10 cm water.
3. During water exchange, the feeding area and other areas where black soil occurs should be mildly and carefully agitated to dislodge the soil from the pond bottom. This will facilitate its drainage from the pond.
4. Good algal bloom should be produced before stocking and maintained. Preventing growthof benthic algae and Hydrilla in the pond helps to prevent pond bottom deterioration.
5. Chain dragging can help to dislodge the benthic algae but this practice should not be applied to more than ¼ of the pond bottom in any one day. Remove the benthic algae when it floats to the surface.
6. If the water is too green, exchange water to prevent collapse of the bloom and subsequent settling of dead algae on the pond bottom.
7. During routine application of Agri. lime to the pond, spread more lime in areas where the pond bottom is bad.
8. Zeolite can also be used as it is a hydrated alkali-aluminum silicate that absorbs ammonia, hydrogen sulphide, and other toxic gases.The recommended application of zeolite is 100-120 kg/1000 m^2.
9. After the harvest, remove black sludge either in wet condition manually or after drying using machines. If possible remove the black soil away from the pond site to a place where it does not cause any environmental problems.
10. If bottom black soil cannot be removed away from the pond site and instead should be deposited on the pond bunds, make sure to cover it with good soil and compact it properly to avoid leaching back to the ponds.

11. Diffused aeration is the best way to aerate deep ponds. Because the air diffuser lays on the bottom to achieve total pond aeration from top to bottom regardless of depth.

How to Treat a Leaky Pond Bottom

Pond bottom becomes leaky due to various reasons.Mainly it happens because of presence of sandy type of soil as well as soil with more gravel contents. The following processes is done to renovate the leaky pond bottom.

Sealing the Pond with Organic Matter

Light Organic Layer

1. A simple method that greatly accelerates the sealing of a pond bottom is to spread enough cattle manure over it (at least 2 $m^3/100\ m^2$) before filling the pond with water.
2. Repeat this procedure several times if necessary, at intervals of eight to 12 months, until seepage losses become acceptable.

Gleization

If the bottom soil is too permeable to be sealed by this method, an alternative is to create an impervious biological plastic layer in the bottom and on the sides of the pond. Such an impervious layer is called a gley, and the process of its formation is called gleization. Proceed as follow:

1. Prepare the pond bottom by clearing it of all vegetation, sticks, stones, rocks and the like. Fill all cracks, crevices and holes with well-compacted impervious soil.
2. Completely cover the cleaned surface with **moist animal manure**, preferably pig dung, spread in an even layer about 10 cm thick.
3. Cover the manure well with a layer of **vegetable material**, preferably broad leaves such as banana leaves. You can also use dried grass, rice straw, soaked cardboard or paper, *etc.*
4. Cover with a layer of soil about 10 cm thick.
5. Moisten and compact very well.
6. Wait two to three weeks before slowly filling up the pond with water.

Sealing the Pond Bottom with Puddling

1. A simple way to reduce water seepage, particularly if the pond bottom is very dry, hard and has open cracks in it, is to break the soil structure of the pond bottom before filling the pond with water. This is called puddling.
   - ✰ Saturate the soil of the pond bottom with water.
   - ✰ Let the water soak into the soil just enough to permit working.
   - ✰ Break the soil structure by puddling with a hoe or plough.

2. If a puddler is available, flood the pond bottom and puddle the soil under water for best results. A plough can also be used underwater.

Sealing the Pond Bottom with Compacting

1. If the bottom soil contains a wide range of particle sizes, from coarse sand to fine silt and clay, and at least 10 percent clay it can be made relatively impervious by good compaction alone. Proceed as follows.
2. Prepare the pond bottom by clearing it of all vegetation, sticks, stones, rocks and the like. Fill all cracks, crevices and holes with well-compacted impervious soil.
3. Loosen the soil to a depth of 20 to 25 cm. If possible, try to borrow a disc harrow or a rototiller from a neighboring farmer. A plough could also be used.
4. As necessary, dry or moisten the loosened soil so that it reaches its optimum moisture content for compaction, about 13 percent
5. Compact the soil surface into a dense, tight layer. You should preferably use four to six passes of a sheep foot roller. You may also use the tyres of a heavy tractor or a crawler-tractor.
6. Check compaction carefully. If necessary, readjust the soil moisture content and improve compaction.

Sealing the Pond Bottom with a Soil Blanket

Laying out a Soil Blanket

If the bottom soil contains a great proportion of coarse particles but lacks enough clay and silt to prevent excessive seepage, you can seal it with a soil blanket. Proceed as follows;

1. Prepare the pond bottom by clearing it of all vegetation, sticks, stones, rock and the like. Fill all cracks, crevices and holes with well-compacted impervious soil.
2. Build a soil layer about 15 cm thick over the entire bottom area and over the wet sides of the dikes. The soil you use should be well graded and contain at least 20 percent clay.
3. Moisten this soil layer so as to reach optimum moisture content for compaction.
4. Compact the soil layer well, preferably by four to six passes of a sheep foot roller.
5. Bring in additional clay soil to build a second layer about 15 cm thick over the first one.
6. Moisten and compact as before.
7. Protect the soil blanket against rupture.

Laying out a Subsurface Soil Blanket

For smaller ponds one should use a method which provides better protection for the soil blanket. Proceed as follows.

1. Prepare the pond bottom by clearing it of all vegetation,sticks,stones, rocks and the like.
2. Excavate the pond bottomdeeper to a depth of about 30cm.Store the soil close to the pond.
3. Build a well-compacted layer of clayey soil at least 5 cm thick all over the surface.
4. To protect this layer, cover it with the original soil and compact the soil cover.

Sealing the Pond Bottom with a Synthetic Membrane

Thin synthetic membranes can also be used to reduce excessive seepage, particularly in coarse-grained soils.Three kinds of synthetic membrane material are most commonly used:

1. Black polyethylene;
2. Polyvinyl chloride (PVC)
3. Butyl rubber.

To ensure zero seepage, the synthetic membrane must entirely cover the pond bottom and the wet sides of thedikes. To line a pond with a synthetic membrane, proceed as follows.

1. Prepare the entire pond area by clearing it of all vegetation, sticks, stones, rocks *etc.*
2. Fill holes and crevices with soil.
3. If the soil material to be lined is too stony or of very coarse texture, cover it with a cushion layer of finer material about 10 cm thick.
4. Lay a first strip of the membrane along the width of the pond and along the dike.
5. Lay a second strip of the membrane in the same way but allowing a 15-cm overlap for joining it to the first strip.
6. Repeat this process with as many membrane strips as necessary to completely cover the pond bottom and the wet sides of the dikes.

Sealing the Pond with Bentonite

If your pond bottom soil contains a high percentage of coarse grained particles and not enough clay, adding bentonite to it is another method of reducing excessive seepage. Bentonite is a fine-textured colloidal clay with as much as 90 percent of montmorillonite. When exposed to water, dried bentonite absorbs several times its own weight of water; at complete saturation, it swells as much as eight to twenty times its original dry volume. Sodium bentonite can be applied directly to the soil in the pond bottom or it can be sprinkled on the surface of the water and allowed to settle to the bottom. There are three basic methods: the blanket method, the mixed blanket method and the sprinkle method. Sodium bentonite is easily applied. In the case of small ponds, it can be applied with simple hand tools and for larger ponds, with ordinary farm equipment.

Some General Pond Soil Treatments

Liming

Agricultural limestone should be spread uniformly over bottoms of empty ponds, or alternatively, it may be spread uniformly over water surfaces. Agricultural limestone should be applied at the beginning of the crop, and it should be applied at least one week before fertilization is initiated.

Benefits

The reason for liming aquaculture ponds is to neutralize soil acidity and increase total alkalinity and total hardness concentrations in water. This can enhance conditions for productivity of food organisms and increase aquatic animal production.

Drying

The time required to dry a bottom soil depends upon soil texture, air temperature, wind conditions, rainfall, and infiltration of water from adjacent ponds or shallow water tables. Light-textured soils dry faster than heavy-textured soils. Warm, dry weather and windy conditions hasten drying,while rainy weather or infiltration of water into ponds retards drying. The decomposition rate in soil will increase up to the optimum moisture content and then decline if soils are dried more. Excessive drying makes soil too dry for microbial activity, so a drying period of 2-3 weeks is usually adequate.

Benefits

The purpose of drying pond bottoms is to reduce the moisture content of soil so that air can enter pore spaces among soil particles. Better aeration will improve the supply of oxygen and enhance aerobic decomposition of organic matter.By drying for two to three weeks, most of the labile organic matter remaining in the bottom soil from the previous crop will decompose and reduced inorganic compounds will be oxidized (Boyd and Pippopinyo, 1994).The main benefit of this practice is to reduce oxygen demand of bottom soil as much as possible before beginning the next crop.

Tilling

Pond bottoms should not be tilled when they are too wet to support tillage machinery. Depth of tillage usually should be 5 to 10 cm, so a disc harrow can be used. Rototillers require much more energy than disc harrows, and they are destructive of soil texture. Mold board plows, often called turning plows, can be used to turn soil over. A mold board plow should not be used for routine tilling for it requires more energy than a disc harrow.

Benefits

Tilling bottom soils can enhance drying to increase aeration and accelerate organic matter decomposition and oxidation of reduced compounds. Soil amendments such as agricultural limestone or burnt lime can be mixed into soil by tilling. Accumulations of organic matter of other substances in the surface layer of soil also can be mixed with deeper soils to reduce concentrations of the substances in the surface layer.

Sediment Removal

Soft sediment should be removed periodically before it reaches a troublesome thickness. Sediment can be excavated with a variety of equipment ranging from shovels to bulldozers.

Benefits

Accumulation of soft sediment in ponds is undesirable for several reasons. It fills deeper areas and can cause ponds to lose volume. Soft sediment can trap feed pellets and fertilizer granules. Anaerobic zones often occur in soft sediment, and soft sediment is not good habitat for benthic organisms. Fish harvest also is hampered by soft sediment because it impedes seining operations.So removal of sediments is required to overcome these problems.

Fertilization

Urea can be spread over pond bottoms at 200 to 400 kg ha^{-1} at the beginning of the fallow period to accelerate decomposition of organic soil.Sodium, potassium, and sodium nitrate can be applied to wet soil to encourage organic matter decomposition by denitrifying bacteria and to oxidize ferrous iron, manganous manganese, and hydrogen sulfide. The usual application rate is 20 to 40 g/m^2 over wet areas. Nitrate fertilizers are more expensive than urea and are not recommended where soils can be adequately dried. Chicken and other animal manures have been applied at 1,000 to 2,000 kg/ha to pond bottoms with concentrations of organic carbon below 0.5 to 1.0 per cent.However, application of a higher quality organic matter such as plant meals *e.g.*, rice bran, soybean meal, and crushed corn or low protein content animal feed at 500 to 1,000 kg/ha is more efficient.

Disinfection

Bottom soils can harbor aquatic animal pathogens or their vectors between crops and cause diseases in the succeeding crop. It is common practice to attempt to disinfect pond bottoms following disease outbreaks. Drying can eliminate most disease organisms, but the combination of drying and application of chemical disinfectant is more effective. The two most common treatments are chlorination with calcium hypochlorate to kill organisms by chlorine contact and application of lime (calcium oxide or calcium hydroxide) to cause a high soil pH and kill disease organisms and their vectors. Calcium hypochlorite is expensive, so lime treatment is more feasible. Application of 1,000 kg/ha of lime is the minimum amount necessary to raise pH high enough for disinfection, and 1,500 to 2,000 kg/ha is a more reliable rate.

Probiotics

Many products are promoted to enhance beneficial chemical and biological processes and to improve soil quality. These products include cultures of living bacteria, enzyme preparations, composted or fermented residues, plant extracts, and other concoctions. Nevertheless, they are not harmful to the culture species, surrounding environment, workers, or quality of aquaculture products.

Conclusion

Water quality management has been considered one of the most important aspects of pond aquaculture for many years, but less attention has been given to the management of pond bottom soil quality. There is increasing evidence that the condition of pond bottoms and the exchange of substances between soil and water strongly influence water quality. More attention is being devoted to the study of pond soils, and practical aqua culturists are beginning to seek information on pond bottom management.

Chapter 29
Problems of Snakes, Livestock and other Aquatic Animals

The Water Snake

The water snake is a common reptile in fish pond. It is very common in warmer climatic condition where as it is rarely seen in colder region. The water snake may be small or big in size. It has big mouth to swallow even the bigger prey. It has two sharp fangs to hold the prey. The water snake is not a poisonous one, however, people get afraid of it as its bite is very painful. The water snake likes to hunt the fingerlings as well as bigger fish too. The snake can be seen preying even in the day light, however, it likes to prey in the night. The water snake causes a considerable loss of fingerlings as well as yearlings of even 6″ - 8″. Therefore, once the snake is observed in and around the fish pond, immediately preventive measures, need to be undertaken,

Under the preventive measures, firstly the bushes around the fish pond need to be cleared. These bushes arc considered as shelter place of a snake. By clearing the bushes also, it minimizes the water snake around the fish pond and secondly an effective snake trap need to be prepared and operate it to trap the water snake.

The Snake Trap

The snake trap is a simple barrel type of structure. It can be made with locally available small mesh sized wire gauze. It is so simple that one can easily prepare it. To make the snake trap only two items are required. One is small mesh sized wire gauze found in a roll with a width of 4′ at the cost of Rs. 30 per running feet and secondly only 20 -25 feet of fine wire at the cost of Rs. 5.00 only.

How to make a Snake Trap

Take 4 feet of above mentioned sized wire gauze from the local market

(i) And cut out one feet width out of it.

(ii) The bigger sized wire gauze of (3′*3′) rolls back to give a shape of barrel

(iii) Both the end of barrel need to be overlapped. The overlapped portion of the barrel is stitched with fine wire from one end to other, so that the barrel shape stands firmly. This completes the first stage of Snake Trap.

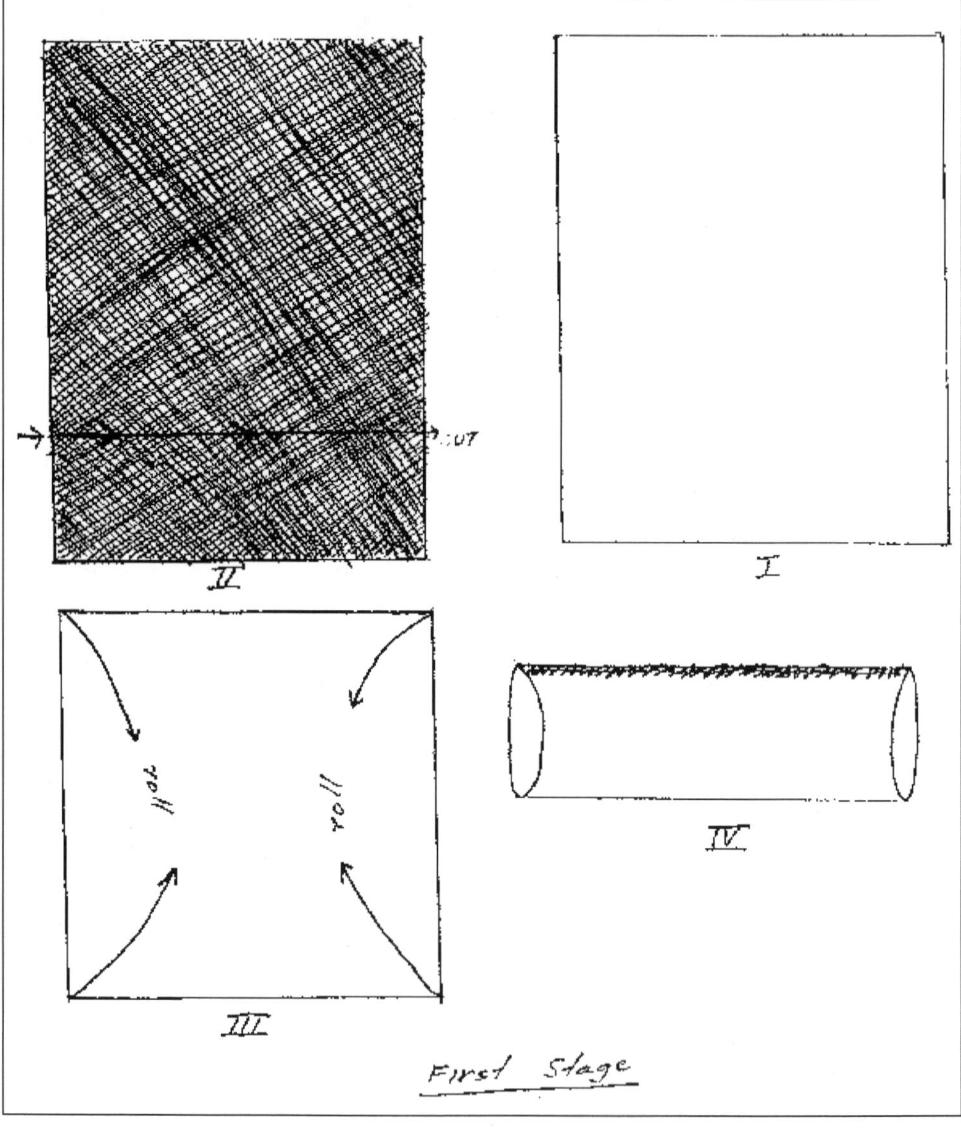

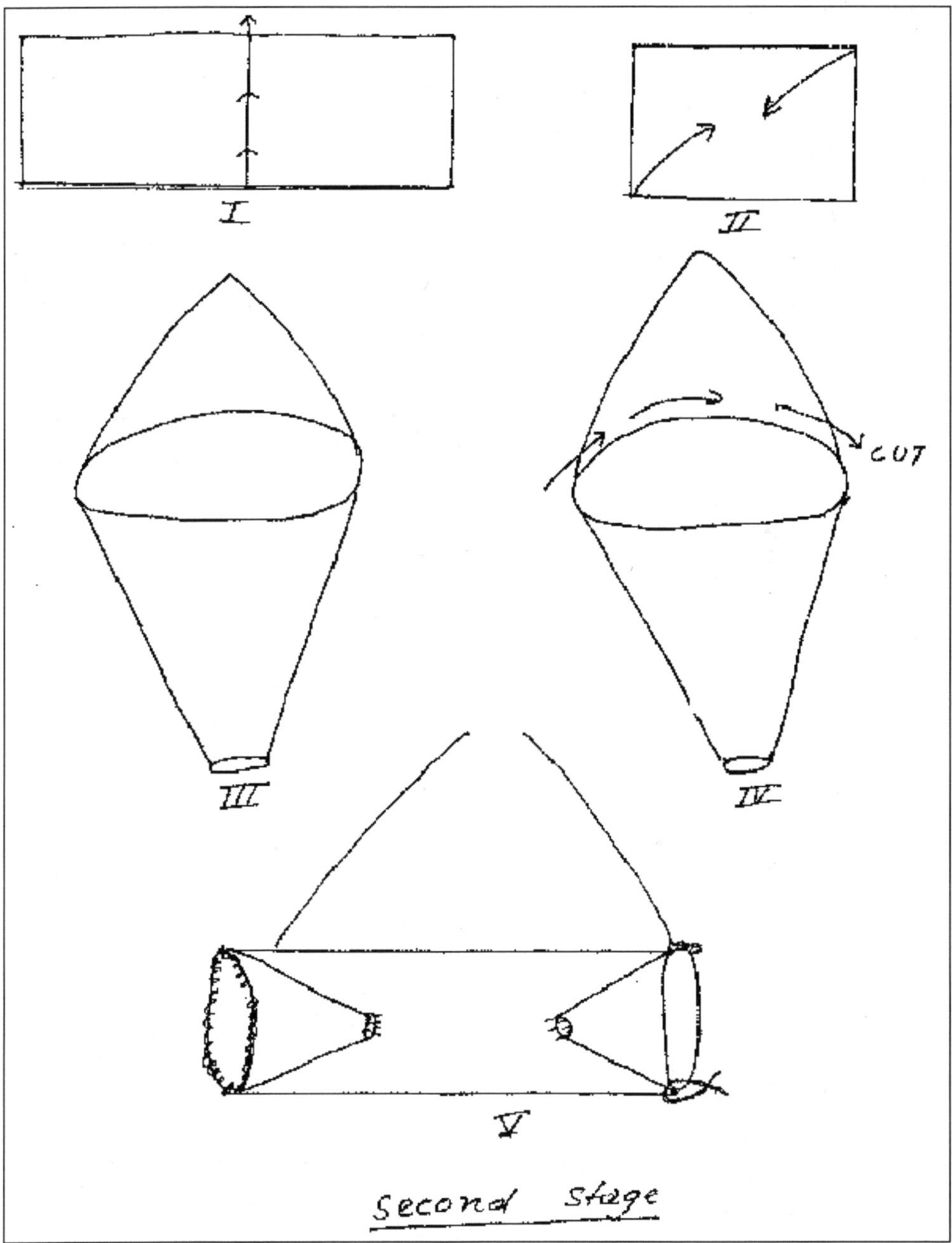

To complete the second stage, take the one feet wide cut piece of wire gauze and further cut it into two pieces

(i) Take one cut small piece of wire gauze and fold it diagonally

(ii) In such a way that the one end remains with big circle with a wing while the other ends will have 1-1.5 inches hole only

(iii) Thus, it turns out as a "Funnel". The over lapping side is again stitched with fine wire so that it remains firm. The wing like structure of bigger circle is cut in such a way that the given structure turns out as "funnel "

(iv) The second piece of cut wire gauze is also made the same sized funnel. Two funnel shaped structures are inserted into both the end of the barrel shaped structure, so that the narrow hole remains inside the barrel, while the wider portion Fits in with the mouth of the funnel.

In case of some deference in the size of the barrel and funnel, one need to adjust so that the bigger circle of funnel gets adjusted exactly with barrel opening. While fixing the funnel with the barrel, the end with 1" hole is directed towards the inner side of the barrel. It is better to check whether the wire at the 1" hole are sharp or pointed enough or not? If not, it is advised to cut out one mesh of wire and make them pointed. One of the wider end of the funnel and the opening of the barrel are tightly woven around the circle with fine wire, so that the funnel does not come out. This funnel shaped structure is fixed in second end of the barrel at two or three points by fine wire, so that in case of need, this funnel structure can be taken out with least effort, otherwise the funnel remains intact with the barrel. At the both end of the snake trap either with fine wire or nylon rope is tied in such a way that the snake trap can be uplifted holding the wire or rope. Now the snake trap is completely ready and is ready for its operation.

How to Operate

The operation of a snake trap has been found effective even in the day light, however, it is usually used at evening or night operation. The snake trap is fixed along the dike of the pond in such a way that half of the funnel hole of the snake trap is dipped under water and other half remains above the water. It is regulated with the help of wire or nylon rope tied on snake trap and a peg fixed on the dike of the pond. After fixing a snake trap, it is covered with some grasses or twinges of bushes to camouflage it. No bait is necessary in snake trap; however, some people prefers to put some live frog or wild fish as bait inside the trap to be sure to see a snake trapped.

In the night, when a snake comes out of its hole in search of its prey, it encounters with the trap and it attempts to peep inside the snake trap through the hole of runnel and gets trapped. As the snake try to peep inside the trap through the hole of funnel and does not see anything in it and tries to go back but the pointed wire of the funnel tip pricks the ventral surface of the snake body and forces to get into the trap. One should not surprise to see two or more snakes in the same trap next morning.

Next morning, examine the snake trap, if snake is not found in the trap, one can leave the trap as it is or place it again at evening at another place. But in case, a snake is trapped, the snake remains alive. The water snake is non-poisonous; however, no one like to handle a live snake. Therefore, the snake trap with snake need to be dipped into the water, just loosing the wire or rope tied on the snake trap. The whole snake trap gets easily dipped in the water due to the snake's body weight. After

15 to 20 minutes the snake gets drowned in water and die. Once it is confirming that the snake is dead take out the snake trap from water and take out one of the funnel which is not fixed properly but simple tighten two or three points so that the dead snake is taken out of the snake trap. The dead snake is buried in a ditch.

Again, fix the funnel with the snake trap and prepare it for next operation. This way the water snake problem is minimized to greater extent and stocked fish is saved to grow and gain higher economic returns from fish culture.

Snakes

The wet margins of ponds are ideal habitat for water snakes, both poisonous and nonpoisonous. Water dwelling snakes do eat fish, but too few to be detected by the manager. It is senseless to slaughter snakes as they provide a service by controlling pest populations. Unless the manager is skilled in distinguishing poisonous from nonpoisonous varieties, snakes should be observed and enjoyed from a distance.

Beavers and other Problematic Mammals

Beavers, otters, and muskrats may cause problems by interrupting water supplies, destroying vegetation, or by burrowing into to dams or pond banks. Trapping is the best way to reduce mammal populations, but all game laws must be observed. There are open seasons at certain times on otter, beaver, and muskrat, and depredation permits may be obtained during closed seasons with proper justification. The SCDNR has small game and furbearer specialists that are experts in dealing with problem mammals. You are encouraged to take advantage of their experience prior to attempting to take any nuisance mammal.

Livestock

Cattle and other livestock having free access to a fish pond may cause damage in several ways. Their wading stirs up sediments and causes muddy water, as well as bank damage. Wading will also hamper fish spawning in the spring and summer by damaging eggs and nests and by frightening fish from their nests.

Pond owners in South Carolina frequently place hog lots adjacent to their fish ponds to allow the hogs access to the water. This practice is fine for swine, but poses a direct threat to pond fishes. Hog lot waste is very high in organic matter and can cause a D.O. depletion if washed into a pond suddenly by a heavy rain. There is also a possible link between animal manure and fish disease.

If stock must be given access to your fish pond, it is a good idea to allow them to use only a small portion of the pond. This can be accomplished by the appropriate use of fencing. Hog lot runoff can be diverted from the pond with berms or diversion ditches.

Chapter 30

Shrimp Toilet: The Concept, Design and Utility– A Solution to *L. vennamei* Culture Problems

Introduction

Aquaculture in India, over the recent years has driven commercial and profitable venture. The practice of shrimp farming is an important arising industry in which extensive research is required to determine its biological and economic feasibility. India with its 1.24 million ha of brackish water area is the second shrimp producer in the world (Panda *et al.*, 2016). Selection for a suitable site by concerning layout, engineering, and management practices are considered while establishing a shrimp farm. The success of commercial aquaculture is attributed to the availability of vast untapped surface water resources, standardization of many new productions and associated techniques of input and output subsystems, extension of technology and practical knowledge to the farmers. Shrimp farming has always been predominantly susceptible to disease issues. Currently, shrimp aquaculture is facing different challenges and concerns to achieve sustainability allied to diseases outbreaks, lack of availability of quality seed, high feed costs, unauthorized farming, international price fluctuations, less demand in the domestic market, and others. The success or failure of shrimp farming depends on the environmental conditions of the location of the farms.

Recently, the main problem associated with shrimp farming is the removal of sludge from the pond bottom. When feeding and water quality are not managed properly the uneaten food, shrimp feces and dead microbiota starts accumulating in the stagnant water areas and becomes sludge placing tremendous strain on the pond ecosystems. Pathogenic micro-organisms grow profusely on this nutrient-rich sludge.

In case when shrimp feed sinks into those areas, contaminated by the pathogenic micro-organisms and when eaten by the shrimp end up with disease (Dash *et al.*, 2016). Hence, pond bottom should be monitored periodically and shrimp farmers must take a holistic approach in managing the ponds particularly on water quality and pond bottom hygiene. This problem can be elucidated by the erection of shrimp toilets/pits in the mid of the pond. The present article is to review in detail about the concepts of shrimp toilets practiced in shrimp farming.

Concept of Shrimp Toilet

The uneaten excess feed, fecal matter by shrimp and myriad opportunistic micro-organisms growing in it ultimately reaches the center of the pond where most anaerobic conditions exist leading to formation of ammonia and hydrogen sulfide simultaneously leading to the stress on animal. This stress gives chance for early and easy infection. After years of culture operation, disposal of accumulated sludge is posing a problem. In order to remove these waste accumulations, shrimp toilets/pits are made in the center of the pond bottom. The sludge formation at the center is the root cause of progression of many infections in shrimp farming. These shrimp toilets utilizing 1 or 2 per cent of space are deep sumps constructed in the center of the pond to collect and trap settled wastes (www.aquashrimpmagazine). By the continuous movement of water by aeration all the waste materials like fecal matter and excess feed will be dragged in to well. The effluent water carrying sediments from these sumps is carried to a series of recirculation reservoirs, some of which also have tilapia, before being circulated back to the shrimp pond. The net result is that they have vastly reduced the loads of EMS/APHNS carrying bacteria, and as a result have returned to very high survival rates on the order of 80 per cent to 90 per cent (www.aquashrimpmagazine).

Design and Construction

Shrimp toilets or central drain installed in shrimp farms plays an important role in the efficient removal of sediments/wastes. It requires approximately 5-7 per cent of the pond surface area with a submersible or floating pump. This pump continuously removes the sediment from the shrimp toilet. Establishment of shrimp toilet includes 7- 10ft cement concrete with a slope to the center where there will be a small well of 2-3ft depth (www.aquaculturetimes.com). The design consists of perforated pipes laid horizontally at the center of the pond and connected to a pipe leading to the outlet. A screen of small mesh size is used to cover the drain for the first 50 days of culture and is removed to allow for easy removal of water when the shrimps are larger than the diameter of the pipe. The construction of the shrimp toilet with smooth, steep slope surface moves solid wastes to the center at a faster rate.

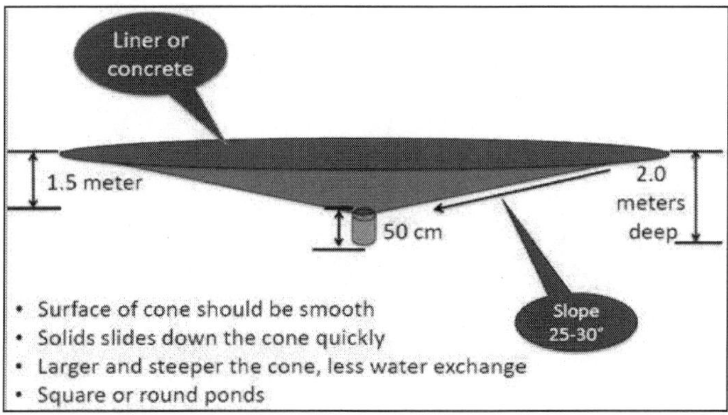

Fixation and Principles of Central Drain

Shrimp farms effectively use the central drain system with additional pipe structure affixed to the central drain pipe. The additional pipe sweeps the waste deposited in the middle of pond bottom in a circular motion and discharges into the drainage canal. Electric or mechanical suction pumps are employed. These techniques consume energy and require manpower but improve bottom quality and thus safeguard production. Some farm operators have invented rotational pond waste removal devices that do not require extra energy but simply utilize the water currents available within the pond. Different aeration equipment is commonly used to create circular currents that sweep shrimp pond waste into the middle of pond where it is deposited. The use of this device needs initial investment for the structure and slab at its base but it is worth the cost.

Figures Showing Design and Construction of Shrimp Toilets/Central Drain/Polyliner Coverings

Polyliners Covered on the Bottom of the Shrimp Pond.

Shrimp Culture Pond with Fixed Polyliners on the Bottom and Embankment.

Construction of Shrimp Pond with Polyliners.

Picture Showing Bundle of Imported Polyliners Brought for Fixing in the Farm.

Geomembrane Hot Wedge Welding Machine Used for Polyliner Fixing.

Discharge Pipes Used for Removal of the Wastes.

Polyliners Half Fixed in the Concavity Region of the Shrimp Pond.

Clear Cut Demarcation (Red arrow) of the Concavity Region Meant for Shrimp Toilet.

Concavity Region Filled with Water for Settlement.

Magnified View of Concavity Region Filled with Water (Red arrow) for Settlement.

Preparatory Phase during Shrimp Pond Construction.

Pathogenic Load Reduced by Shrimp Toilet Installation

The myriad opportunistic pathogenic micro-organisms that grow from the shrimp excretions and uneaten feed in the pond bottom can be removed by this method. The discovery that EMS grows on shrimp feed, on waste and on molts led to the realization that EMS could be treated not through exclusion but by limiting the growth of bacteria by installing "shrimp toilets" that reduces sludge accumulation. This method has the advantage that it can remove the waste and clean the pond bottom any time throughout the culture period.

Benefits for the Culture System

The major advantages of the shrimp toilet/central drain to the shrimp culture system are as follows:

- ✰ Diminishes the chance of infections
- ✰ Shrimps will be stress free
- ✰ Ammonia formation will be very less
- ✰ Controlled FCR
- ✰ Pollution free

Conclusion

Shrimp pond wastes affects greatly to growth and survival of shrimps and water quality of the pond. Frequent removal of shrimp wastes deposited in the pond bottom significantly reduces the organic nutrient concentration in water and can result in low levels of phytoplankton and low pond productivity. In this approach, the sludge is usually concentrated in the middle of culture pond in order to create larger clean space for the shrimp to inhabit around the edges. Effective management techniques are different from one farm to another depending upon personal preference, affordability, suitability and pond management techniques.

Chapter 31

Pond Automation: A Novel Ideal for Risk Management in Shrimp Culture

Introduction

This is a green energy and information based edge-computing technology for the ecosystem with an ultimate goal to transform aquaculture landscape across the globe. The technology is based on sensors which constantly monitors the water quality in the pond environment. The critical parameters like dissolved oxygen, ammonia, pH and temperature are being recorded from time to time by solar based auto cleaned sensors, which again has been connected with mobile application, so that the farmers/owners/technicians will get the accurate information at hand and necessary steps can be taken for monitoring the pond ecosystem. Also there is provision for forecasting the climatic condition for the next seven to ten days, which will help the farmers to be ready with time. The accuracy, longevity and tecnhoaccuracy of the instrument has made it popular among the shrimp farming community.

Top 7 Aqua Challenges

1. All the essential water quality parameters are not actually captured by Aqua Farmers
2. The water quality parameters are not captured immediately and transparently by the Farmers (delay in capturing, transporting samples and evaluating the water in the lab)
3. On time actions to treat the pond doesn't happen and a delay is associated with the same

4. The water quality parameters are captured only for the sample water collected from the top edge layers and not from the middle layer/bottom layer/center of the pond, which is the industry expectation.
5. Farmer has no control over the information about weather pattern changes, which impacts the water quality parameters
6. Farmer has no information/solution about the appropriate timings to feed the shrimps
7. Farmer at present uses a manual process to enhance the quality of water

Top 7 Farmer Approaches

1. The farmers collect the water samples and submit in the labs, once the results are available the lab executive provides the medicine to the farmer for their pond. The farmer has no knowledge of what chemicals are added and what values from testing represent.
2. The farmers are unaware of the water quality and usually visit the labs once or twice in a month. Sometimes they only visit to get the water tested at the time when the shrimp's mortality rate is increased or they see shrimp diseases present in their pond
3. Since the farmers become aware of the diseases when it's too late and then rush to the lab for water testing, the actions required to treat any change in value of the parameters does not exist at this point. Due to this delay the farmers bear loses due to low quality shrimp produce
4. The farmers usually take the water for sampling from the edge of the pond
5. Presently, there is no way for the farmer to capture the weather forecast metrics on/ahead of time. The DO levels changes due to change in temperature/weather conditions.
6. At present, the farmer feeds only during mornings or evenings but not specific to any timings
7. Currently the famers either uses bore water or a mix if available, these results in high salinity of water which the seeds cannot sustain. The farmer then puts a net with some X no of seeds to determine the mortality rate in a small area and then add chemicals as needed to bring the salinity of the water down. Once it reaches a 90 per cent mortality rate the releases the seeds in the whole pond.

Pain Points of the Farmers

1. Water Quality
2. Delayed results and Delayed actions
3. Feeding and Feeding material
4. Climate impact and required actions
5. High mortality rate

6. No transparency towards their pond
7. High risk of disease outbreak

System, Impact, Scalability and Achievement

- ☆ Green Technology yielding energy cost savings of 30 per cent -40 per cent
- ☆ Real-time water quality monitoring
- ☆ Reduces risk factors significantly and safeguards the farmer's investment by **80 per cent**
- ☆ Increases farmer's efficiency by 90 per cent
- ☆ Increases immunity towards diseases

Smart Aqua Farming

1. Solution based approach
2. Engaging Lab experts with their farmers
3. 24x7 monitoring
4. Alerting on radical changes in temperature and other parameters
5. Automation of Aerators saving high cost of diesel and electricity
6. Predictive Analysis on severe weather changes

How it Helps

1. Constantly monitors water quality
2. Proactive approach instead of reactive
3. Feeding time and material proposed by experts
4. Control on the climate changes to their pond
5. Results in low mortality rate
6. Complete transparency round the clock
7. Reduced risk of virus affecting the pond

Pain Points of the Farmers

Water Quality

- ☆ Delayed results and delayed actions
- ☆ Feeding and Feeding material
- ☆ Climate impact and Required actions
- ☆ High mortality rate
- ☆ No Transparency towards their pond
- ☆ High risk of virus

How it Helps

- ☆ Constantly monitoring water quality
- ☆ Proactive approach instead of reactive
- ☆ Feeding/Feeding material proposed by experts
- ☆ Control on the change impact of temperature due to climate
- ☆ Low mortality rate
- ☆ Complete transparency round the clock
- ☆ Reduced risk of virus affecting the pond
- ☆ Alerting on

For Farmers

- ☆ Complete transparency
- ☆ Optimum use of medicine and food
- ☆ 24x7 monitoring
- ☆ Access to their pond health anytime
- ☆ Able to achieve maximum FCR
- ☆ High profits

For Lab Technicians

- ☆ Able to manage multiple ponds using single dashboard
- ☆ 24x7 monitoring
- ☆ Able to take timely actions where needed
- ☆ Alerts and Notifications
- ☆ Save time and provide more efficient support
- ☆ Able to evaluate the changes in multiple pond without
- ☆ Visiting the pond and provides suitable remedies

Revenue Scale

- ☆ Expenditure 25000 per month on each Aerator
- ☆ With this technology a farmer able save 40 per cent revenue on fuel and electricity bills.
- ☆ 40 per cent of 25000 is 10000 on each aerator per month
- ☆ Per acre 4 aerators (10000 x 4) = 40000 INR
- ☆ Per crop its 120,000/month and in a year 360,000 INR for 3 crop cycles

Chapter 32
Frequently Asked Questions in Aqua Science

1. What are the major differences in Major and Minor Carps?

Ans.

Major Carp	*Minor Carp*
i) Number of Undivided dorsal Fin rays > 11	i) Number of Undivided dorsal fin rays ≤11
ii) Growth is faster and size is larger	ii) Growth is sluggish and size is small
iii) The eggs of major Carps are bigger compared to minor carps Ex. *Catla, Rohu, Mrigal L. calabasu, L. gonius,*	iii) The eggs of minor carps are smaller compared to major carps Ex. *Cirhinus sarana, C. reba, L. bata, L. angra, L. pangusia*

2. What are the requirement of the following animals/birds to fertilize 1 ha of water area?

Ans. Ducks → 200-300 Numbers/ha. (Khaki Kambel)

4-5 Nos/m^2 (Dry run)

3-4 Nos/m^2 (wet run)

Poultry Birds → 500-600 Nos/ha.

Pigs → 30-40 Piglets/ha.

Cows → 1 cow/7 m^2 area.

3. What are the advantages and disadvantages of presence of organic matter in the pond bottom?

Ans. Advantages

i) Increases the Productivity.

ii) Increases the arability of the pond bed soil.

iii) Reduces the sepage loss.

Disadvantages

i) Store house of all pathogenic forms of species.

ii) It forms noxious gases, polluting the entire system.

iii) It limits the interaction of soil and water, reducing the productivity level.

4. What should be the distribution pattern of area before construction?

Ans. The total area is to be divided in the following way:

80 per cent of the Total Area	→	Socking pond
8 per cent of the Total Area	→	Rearing Pond
4 per cent of the Total Area	→	Nursery Pond
6.8 per cent of the Total Area	→	Dykes
0.8 per cent of the Total Area	→	Segregation Pond
0.4 per cent of the Total Area	→	Brooder Tank

5. What is Redox potential? What is its importance? A pond with fish and another pond without fish, in which case redox potential is more and why?

Ans. The oxidising and reducing power of solution/system can be expressed in terms of Redox potential

The Redox poternial is expressed in terms of volts *i.e.* 'Eh'

The Redox potential depends upon the amount of oxygen dissolved in water and also the cycles of various elements taking place in water.

At pH = 7 ⇒ Eh = 0.4 to 0.5 Volts

If the water is well aerated Eh = 0.52 Volt.

If the pond mud has Eh = 0.2 ⇒ the O_2 is very less.

When redox potential reduces to 0.2 volt all the minerals were released into the water.

The pond having fish have more redox potential than the pond without any fish, because, The fishes will release the excreata, which helps in production of planktons. The death and decay of the plankton increase the Carbon content of water body, ultimately the acid combining capacity is more. So the redox potential is more.

6. Which form of Nitrogen is best for acid and alkaline soil and why?

Ans. In Neutral and acid soils Amonical form of N_2 has to be applied.

In Alkaline soil Nitrate form of Nitrogen has to be applied.

This combination is suitable, because in these above particular form leaching or loss of N_2 will be minimum.

7. What is the difference between male and female carp?

Ans.

Major Carp	*Minor Carp*
i) Scales are rough	i) Scales are smooth
ii) Base of the pectoral fin is rough sandy like structure/sand paper consistency	ii) Base of the pectoral fin is smooth
iii) Body is slim and slender	iii) The abdomen is buldging
iv) Vent is white/transparent and not protruding	iv) Vent is reddish and protruding
v) Gill operculum gives rough/ sandy touch	v) Gill operculum gives smooth Touch/feeling
vi) On slight pressure on abdomen milt will secretion Milt well ooze out	vi) On slight pressure, no such will come out

8. What are difference between Penaeid and Non Penaeid prawns.

Ans.

Non Penaeid	*Penaeid*
i) Pleuron of the 2nd Abdominal segment overlaps the pleuron of 1st and 3rd	i) The pleuron of 2nd abdominal segment overlaps the pleuron of 3rd abdominal segment only
ii) 1st two pereopods are chelate	ii) 1st three pereopods are chelate
iii) Petasma and thelycum are absent	iii) Petasma and thelycum are present
iii) Eggs are carried by the female in the abdomen	iii) Eggs are laid simply in water

9. What are difference between male and female prawns.

Ans. *Penaeid*

Male	*Female*
i) Petasma is present.	i) Thelycum is present.
ii) In endopodite 1st pleopod is modified absent	ii) In the endopodite is pleopod is to form petesma
iii) In 2nd Pleopod Appendix masculina is present	iii) Appendix masculina is absent

Non-penaeid

Male	*Female*
i) In the 2nd pleopod appendix masculine is present	i) In the 2nd pleopod appendix masculine is absent
ii) Male gonopore is present in the 5th walking leg base	ii) Female gonopore is present at of the 5th walking leg.

10. Which are exclusive species of prawns (male/female) get matured in Coastal water?

Ans. Male → *P. merguensis* (banana prawn) and *M. brevicornis* (Yellow prawn), the maturation takes place in coastal water.

Female → M. bennette (Australian Gresy back Prawn) get matured in Costal water.

11. What are 3 trophic phases of a newly formed Reservoir? (70)

Ans.
1. Initial fertility (1st stage)
2. Trophic depression (2nd stage)
3. Final Fertility (3rd stage)

12. How can you calculate the number of fingerlings required for a pond/ ha/annum?

Ans. C = aby where

C →Yield at harvesting

A → Survival rate

B → Size of harvesting

Y → Number of fingerlings.

13. What is Backward hatcheries, and how it is helpful to the farmers?

Ans. These are not the original hatcheries. These are mainly constructed at the back side of the house. The purpose of these hatcheries in to raise and rear, the spawns or hatchlings into fry or fingerling stage. In this way, there is the possibility of availability of seeds in off season. Also besides fulfilling the requirement of the person concerned, it may be possible to go for selling.

14. What do you mean by Eco-hatchery?

Ans. This circular type cemented hatchery *i.e.* Chinese hatchery having running water facilities with high production capacity of fish seed. The technique was devised by the scientists of mainland, China. This hatchery is of different dimensions starting from one meter diameter. It has two units *i.e.* breeding pool and hatching pool.

In West Bengal, this hatchery is popularly known as Eco-hatchery. The cost of setting of an Eco-hatchery with breeding pool, Over head tank, deep tube well, hatching pools ad pumping systerns is about 3-7 lakh depending on capacity of productions.

15. Why fisheries are not developed in India?

Ans. 1. Gap between lab to land (Generation Gap)

2. Lack of adequate amount of Capital.
3. Lack of adequate infrastructural facilities.
4. Lack of encouragement of Government.
5. Treating fishery under Agricultural ministry. (It should have unique feature)
6. Inadequate extension programmes or set up.

16. What will be the effect to fishes, if the pH is very low or very high?

Ans. If the pH is very low or water is acidic

1. Plankton Production becomes low.
2. The fishes become susceptible to diseases.
3. Resistance of fishes to toxic materials become very less.
4. The fishes loose appetite.
5. The growth remain stunted if the pH is very high.
6. Skin of the fishes will Corrode or Corrosion of skins.
7. Respiration is hampered.

17. When the pH of pond water is very low and very high, give the reason?

Ans. The pH of the pond water is very low is the early morning because in night time, there is no photosynthesis, so there is no oxygen production. What ever reserve O_2 was there, is consumed by the existing organisms, so CO_2 concentration is more, which implies less pH value.

In the similar manner, pH value is very high during, mid afternoon. Because, at this time, there is peak in photosynthesis, So O_2 production is maximum which implied high pH condition or Alkaline Condition.

Relationship:

High O_2 Conc. $\rightarrow$ low Co_2 Conc. $\rightarrow$ high pH value $\rightarrow$ Alkaline Condition.

High CO_2 Conc. $\rightarrow$ low O_2 Conc. $\rightarrow$ low pH value $\rightarrow$ Acidic Condition.

18. What are the difficulties in Inland fishery and why it is not developing?

Ans. 1. Existence of different sets of regulation for observation and management in Inland fishery.

2. Control of Inland fishery rights, Inland water by different agency, tourism and fishery revenue.
3. Inadequate Communication and transport.

4. Presence of obstruction and bridges in rivers.
5. Lack of suitable techniques for estimation of fish production (Tagging help to estimate fish stock).
6. Lack or adequate seed.

19. Fish culture is more efficient compared to culture of any other land animals. Justify it?

Ans. 1. Fish is a poikilothermic animal, so need not spend any energy for temperature regulation.

2. It does not spend any energy to maintain it's body position in water column.

The specific gravity of fish (1.02 to 1.06) is nearly same with that of water.

20. What do you mean by satellite farming?

Ans. It is a very complex and interrelated farming system, where the original farm, being at the centre,provides seed,feed, fertilizer and others ch crical inputs including labour charge to the farming communities, with the purpose of taking their final product at the Govt. approved rate and and process it to agricultural products/byproducts. For ex. Shakti Sugar, Dharani Sugar *etc.*

21. In which way Agriculture and Fisheries are related?

Ans. Agriculture and Fisheries are two sister concerns or fisheries is an allied subject of Agriculture.

1. The pond mud which is very much harmful to the fishes can be very well utilized manure for rice cultivations.
2. The pond embankment can be used for cultivation of flower plants, coconut, papaya *etc.*
3. The agriculture wastes can be used for preparation of compost, which can be used as manure for fish pond.
4. Also in Rice cum fish culture, the relation is quite obvious.

22. What are the drugs/medicines commonly prescribed for prevention of EUS?

Ans. 1. Micro-encapsulated feed containing 30 per cent of protein forfeited with Nalidixic acid and erythromycin observed the recovery of EUS affected fishes.

2. Homeopathic drugs like 'Heaper Sulphur' and 'Arnica Spray' claimed to achieve encouraging results within 7 days in EUS affected fish.
3. Oral administration injections or bath of tetracycline 75 mg/kg feed also give encouraging results.

4) Paddy straw ash at used at the rate of 500kg/ha/m, which is claimed to have given effective results within 10-15 days.

5. The application of turmeric powder 4 kg/acre along with unslaked lime at split doses indicated the positive results.

6. The supplementation of hostacycline and vitamin mineral mixture with the feed at the rate of 2 per cent body weight gives good result.

7. Use of quick lime at the rate of 450 kg/ha and Kmo_4 at 3 Kg/ha as a control measure against EUS infection.

8. Oxytetracycline at the dose of 75mg/Kg/day for control of EUS within 10-15 days.

9. Lime at the rate of 250 Kg/ha to be repeated after 3 weeks, Kmo_4 at the rate of 0.5 Kg/ha, bleaching powder at the rate of 5-10 Kg/ha, Common salt and teramycin at the rate of 100 mg/kg of feed, Sodium bicarbonate at the rate of 100 Kg/ha and salt at the rate of 300 Kg/ha *etc.* for control of ulcer disease.

10. Also penicillin injection can be used as a curable measure of ulcer disease of fishes.

11. CIFAX at the rate of 0.01 PPM used as chemotherapeutic measure for ulcer disease in fishes (1 lit per ha meter water)

23. What are the beneficial effects of alternate farming of shrimp and paddy?

Ans. The experts of MPEDA have suggested cultivation of shrimp in alteration with paddy in low saline coastal soils. This could enable preservation of ecosystem and soil quality to increase the paddy harvest and reduce the possibility of deadly viral attack.

The experts said the paddy stem acted as biofilters for the water bed and by providing substrata for the healthy growth of shrimps. Paddy could be grown in the season and shrimps during others in the same ponds along the coastal belt, particularly in the low saline soil with freshwater flow. The virus affected shrimp farmers suggested the adaptation of crop alteration technique immediately. The use of harmful pesticides, insecticides and other chemicals could be eliminated if farmers adopted this practice and used saline and pest resistant verities of paddy as practiced in Kerala. They also suggested the effective of management of water quality and bunding of shrimp ponds. Alternative crop practices would reduce the incidence of deadly virus attack, such practices will improve soil fertility and yields in coastal areas, with good freshwater in take during the rainy season and low salinity, the districts were well suited for the new technique. They said Kerala did not experience any viral attack due to crop alternation technique.

24. What is allergy and how it occurs?

Ans. Allergy is an antigen antibody reaction in the body mechanism. Usually antigen (foreign body) when mixes with blood antibody is found, which ultimately forms a complex, but incase of allergic condition after insertation of the antigen, the antibody will not form. So it will not bind or form colloidal complex and this causes allergy.

25. Justify the compatibility of Mrigal and Comon Carp in Composite fish farming?

Ans. In Composite fish farming both Common Carp and Mrigal remain at the bottom layer of the pond. Both prefer to detritic matter as well as animal matter as their food. But Common Carp has more affinity towards animal matter (Carnivonus in nature) where as Mrigal has more affinity towards plant matter. So there is no keen competion between the two different species in a single niche for food.

26. Which design of pond is mostly desirable for pisciculture and why? Finger shaped, Circular, Rectangular, Conical, Trapezoidal *etc.*

Ans. Out of all these types finger shaped ponds are the most desirable ones, because, this will produce the natural environment more closely than any other. In finger shaped pond, though the water area seems to be less, it's natural simulation enhances the growth in shortest possible time, (Follow Practical Manual on Agriculture by H.B. Willson, USA).

27. What are the methods to measure the age of fishes. ?

i) Known age method ii) Scale method iii) Otolith Method iv) Bone Method iv) Length Frequency Method vi) Fin ray of Pectoral Spine Method.

28. What is the dose of MOC, Deris Root Powder and Urea and bleaching powder in eradicating the predatory fishes. ?

Ans.

	MOC	*Derris Root Powder*	*Bleaching Powder and Urea*
1.	Dose → 75 PPM (Sufficient) 200-250 PPM (Some workers advised)	4-20 PPM	Urea →100 kg Bleaching →175 kg Powder (Both are added per hectre meter of water)
2.	Active Ingredients → Mowrin or Resparin or Saponin	Rotenone	Urea+Ca(ocl)Cl (Calcium cyramide)

3. Cause of death → Haemolysis of RBC	Damages the Respiratory System	Failure in Nervous System
4. Advantages → Poisoning effects remain upto 21 days then MOC acts as a manure	Poisoning effects remain upto 4-12 days.	Poisoning effects remaining upto 7 days. After poisoning effect is over it acts as fertilizer.

29. What are the hormones secreted from the fish Pituitary Gland ?

Ans. PG is otherwise called as Hypophysis.

Two Parts I) Neurohypophysis – Origin is from brain

2) Adenohypophysis – Ectodermal growth of brain

(1) Neurohypophysis secrets the following hormones:-

i) Oxytocin

ii) Vasopressin

iii) Dieuretic

iv) Antidiruretic

2) Adcrohypophycis Consists of 2 parts and produces the following hormones.

A) Pars distalis (PD) B) Pars Intermedia (PI)

A) PD is again divided into 2 parts

i) Rostral Pars distalis (RPD)

ii) Proximal Pars distalis (PPD)

RPD → 1) Prolactin. (Which regulates osmoregulation)

2) ACTH Cells (Function to produce steroid hormones)

PPD → 1) Somatotrophs or Growth hormone

2) Gradotrophs (FSH, LH)

3) Thyrotrophs (TSH- Thyroid Stimulating Hormone)

B) PI →

1) MSH Cell (Melanocyte Stimulating Hormone producing cells)

2) PAS^+ Cells (Periodic acid skift)

30. What is bundh breeding and describe in brief with the probable factors responsible for spawning in confined bunds?

Ans. The Indian Major Carps (IMC) bread in special type of perennial or seasonal ponds or impoundments locally known as bundhs *i.e.* breeding in ponds where riverine condition is simulated.

Two types i) Wet bundh ii) Dry bundh

Wet bundh – Bundh is perenial *i.e.* water is present throughout the year.

Dry bundh – Seasonal type only during rainy season the water is filled up. In all otherseasons pond is dried

Area of operation – It is practiced in Midnapur and Bankura district of West Bengal and

Nawabgang in Chatrapur district of Madhyapradesh.

Probable Factors

i) Presence of Petrichor – Petrichor is an oil,(biogeocoplexes) obtained by steam distillation of silicate minerals and rocks, which stimulate the fishes to spawn.

ii) Food

iii) Turbidity

iv) Sudden rise in water level.

v) Water temperature (23-24°C)

vi) Current velocity of water (0.5 to 1.7 mt/sec)

31. How do you control continuous turbidity in ponds whether it is new or old?

Ans. The continuous turbidity is due to high content of sodium (Na) salts. It can be controlled by introduction of weeds in the pond water.

32. What are the doses of Cowdung and Biogas slurry?

Ans. Cow dung:

The double filtered cow dung (liquidated condition) should be used in order to avoid unnecessary deposition of undigested particle present in the dung.

dose- 1000-15000kg/ha or 20-30 baskets/Acre/month

This dose should be applied in installments

Biogas slurry is advantageous over cow dung,as it is completely decomposed and the harmful gases such as CH4, has been released. The addition of eichornia weeds in biogas plant will increase the production of gas by 30 per cent and the quality of slurry will be better.

Dose: 12.5 tonnes/Acres

(In different installments according to the need)

33. What is the special characteristic of viruses and which type of cells are most preferred by viruses for its proliferation and why?

Ans. Viruses require the living cell for it's growth living cells are must for growth and proliferation of viruses.The neural cells are the most preferred sites for viruses to grow and multiply. Because these cells do not undergo meosis or mitosis. Only these cells increase in length and breadth wise and the viruses spread and settle nicely through the nerve endings *i.e.* Axons and dendrons.

But incase of other cells, they undergo mitosis and meosis division. So during that time there is every chance of removal of the unwanted organisms *i.e.* the viruses will be thrown out. So they prefer the neural cells. Soon after they get the entry to the body of the organisms they find the chance to enter into the neual system the most safest place for the viruses.

34. Why do not the mosquitoes die even after sucking blood of different groups?

Ans. Blood groups in human beings are classified on the basis of antibodies (protein particles that fight against foreign bodies). The four different blood groups (A,B,AB and O) are specified by the presence and absence of particular antibodies and antigens.

When the blood of two non-compatible groups combine inside our body on antigen-antibody reaction (called transfusion reaction) takes place. This results in haemolysis (breakdown of red blood cells) and clumping of the blood corpuscles. This leads to total stoppage of blood circulation, making the person die.

After mosquito sucks in blood from human body it remains in the stomach of the insect and does not take part in any physiological or metabolic processes.

It acts as a thermal source within its body to facilitate the egg development in abdominal cavity of female mosquito (male mosquitoes do not bite). So a mosquito does not have any problem even though it sucks in blood of different groups.

35. Whether the baby in the mother's womb is considered under parasitism, explain it?

Ans. The very definition of parasitism explains the undesired/unspecific association of the host and the pathogen But in this case the baby well take its place the womb only by the desireness of the mother. Again it is a natural biological phenomena, which again explains the desired association of male and female.

But in a broader sense it can be grouped under true parasitism.

36. The taste of a fish depends on which factors?

Ans. It is the amino acids which are responsible for the taste or flavour of the fish. The amino acids break down into ATP molecules and finally the ATP molecules break down to ionosine Monophosphate which is a flavouring compound.

The taste of fresh from one pond or location may be different compared to another pond or location,it is due to the availability of food i.e protein nature. Which finally break down giving rise to the flavour or taste of the fish. In case of marine fish, the fishy smell is due to the presence of TMAO

(Trimethyle Amino Oxide), which finally break down to TMA (Trimethyle Amine) which gives this fishy smell in marine fishes.

37. What is the difference between microspores and myxospores?

Ans. **Myxospores**: Myxospores are known for it's multicellular spores possess nematocyst like polar capsules. The sporoplasm is in the posterior (or middle) part of the spore and may be a binucleate or two uninucleate or two uninucleate sporoplasms one inside the other. The sporoplasm contains dense bodies (sporoplamosomes), evident in electron microscope. Thick walled pollar capsules which are retractile, contain coiled polar filaments. The polar filament is a hollow tube which is spirally twisted along it's length and is capable of rapid extrusion when it evarts inside out. Spore size is usually larger(8 to 10 μm)

Microspores: Microspores are strictly intracellular protozoan parasites which are characterized by the production of spores. The spore has an elaborate extrusion apparatus with a conspicuous extrusible pollar tube. They infect animals from protozoa to man. They are the usually parasites of the invertebrates. Most microsporidians are small parasites usually (3-10 μm) long. They are lacking mitochondria. Their nuclei are either single, isolated or paired (diplokarya consisting of two closely apposed nuclei). The spores are small and mostly ellipsoidal structure or egg shaped. They have a very complicated structure. They have an elaborate hatching apparatus. The most conspicuous part, is the polar tube. The tube extends obliquely from the cap to the posterior half of the spore, where it forms a coil beneath the spore wall. The sporoplasm contains either single nucleus or a diplokaryon.

38. What is the remedy for algal fouling over the body of prawns in farming practices?

Ans. Usually algal fouling over the body of the prawns is due to Eutrofication condition of the water body, so it is better to exchange the water if possible 100 per cent. After exchange of water, if the algae still persists then go far (BKC) Benzal Konium Chloride treatment at it's prescribed doses.

39 What is the microbiological difference of a old and new pond?

Ans. In new pond the microbiological load is comparatively lower than the old pond. In new pond the content of organic matter is less, the low content of organic matter (humus), provides poor shelter for microorganisms. In old ponds the high organic matter provides better shelter to the microorganisms. So in old ponds, a large variety of microorganism population is usually found.

40. Why vaccination programme is not successful is shrimp aquaculture?

Ans. In shrimps immune system is very poorly developed, more particularly, being an invertebrate, immunoglobulins or antibodies are absent or merely present. So the aim of vaccination to increase the immune power in shrimps is not possible.

41. What is the difference between immunostimulant and vaccine?

Ans. Vaccines are nothing but dead or attenuated microorganisms designed to produce the specific antibodies in the host body

An immunostimulant is a chemical/drug/stressor/or action that elevates the non-specific defence mechanisms or the specific immune response.

Immunostimulant can be given by themselves to activate non-specific defence mechanisms, or they way be administered with a vaccine to activate non-specific defence mechanisms, or they may be administered with a vaccine to activate non-specific defence mechanisms as well as heightening a specific immune response.

42. What is difference between antibiotic, vaccine, and antimicrobials?

Ans. **Antibiotic-** are the chemicals or drugs of broad spectrum to control the bacterias

Vaccines- are nothing but dead or attenuated microorganisms designed to produce the specific antibodies in the host body

Antimicrobials - are the broad spectrum drugs or chemicals mainly meant for controlling the microbes

43. Which antibiotic is of choice to the farmers?

It is the sulfa drugs, mainly dominated over the other antibiotics because

1. It is cheaper
2. Less risk
3. Broad spectrum *etc.*

44. What are the control measures for viral diseases?

Ans. There are no therapeutic measures for viral diseases but prophylactic measures can be taken up to prevent the occurrence of the disease.

As prophylactic means – In aquatic ecosystem disease spreads faster compare to terrestrial unit. So all the sources like water, equipment, feed, fartiliser, nets *etc.* should be disinfected and checked, to meet the prophylactic means.

More over the diseased animals should be taken up immediately, burnt and buried. If the condition is more acute, the total stock may be destroyed.

45. How you will detect the virus?

Ans. Externally the viral attack can be known by:

Different virus shows different types of symptoms as well as behavioral changes in the fishes/shrimps. According to the external signs, we can proceed one by one step, up to the diagnostic procedure for viruses.

The virus growth in the cell can be determined by any or the following ways.

1. Detection of inclusion bodies.
2. Electron microscopy
3. Fluroscent antibody and immuno peroxidase staning
4. Haemadsorption test
5. Viral interference test.

46. What is the difference between bacteria and virus?

Ans. Viruses are those which require live cells to grow and multiply, having cell wall, cell membrane and the nucleus. But the bacterias are those which can grow in both live and dead cells, without having any definite cell wall cell membrane and nucleus.

47. What is difference between bacteria and parasite?

Ans. **Bacteria**: it is a parasite which produces directly and usually at high rates in or on the host usually small with short generation time, inducing immunity and of transient nature.

Parasite: it is an organism without direct reproduction in or on the host (usually large with long generation time) inducing immune response and that usually depends on infection intensity.

48. What is El Nino and what La Nina?

Ans. The hot water current in the pacific ocean near the equator is seen every year, known as El Nino. It is a Spanish word, means "the boy"

La Nina- the cold water current seen in the oceans occasionally is known as La Nina, means "the girl"

El Nino creates hurricane and violent storm where as La Nina creates poor rain and drought

49. What is the difference between RNA virus and DNA virus?

Ans. Viruses are primitive organisms *i.e.* lower group in the animal kingdom. So it contains either RNA or DNA in their cell. But not both type of nucleic acid in the same organism.

The pathogenicity of the RNA virus or DNA virus is considered depending on the species of particular virus, not that RNA virus is more pathogenic than DNA virus.

50. Why the Aquatic animals grow throughout the life where as terrestrial animals do not?

Ans. Number of reasons can be attributed towards this answer. Aquatic animals more particularly fishes are poikilothermic and need not have to spend more energy to respire. swim or towards other body physiological functions.

But another thing is that a mackerel can not be of 60kg size or as like that of whale, simply as because the growth also controlled by species genetic code.

51. When we should go for liming and why?

Ans. Usually liming is done in order to bring an buffering condition in the water body.

The water bodies where organic matter load is less, we should go for liming in the night, because very less amount of co_2 is available in the water surface, it we add lime in the day time (CaOH),it will make use of the CO_2 forming $Ca(CO_3)$ + free H^+, so the buffering capacity of water surface will be lost.

But in sewage fed fisheries or in Bhery fisheries where organic matter load is more, CO_2 availability in the surface water is plenty i.e a toxic conditions. So we should apply lime in this case in the day to absorb some amount of CO_2 or to reduce the CO_2 level in the surface water, so as to restore a buffering of buffered zone in the surface

So lime in both way to reduce CO_2 level and to restore CO_2 level is useful.

52. What are the other way help we are getting from putting the bamboo poles in the water body?

Ans. These bamboo poles not only helps to avoid poaching, reducing the parasitic load by abrasion but also it helps the sedentary organisms to settle over them, which helps the column feeders like Rohu, to take those perphytons. So the bamboo sticks in ponds are other way helpful in this manner.

53. How can you control heavy planktonic growth in the pond?

Ans. If the pond is heavily loaded with phyto ad zoo plankton it can be controlled by putting the weeds like. Azolla, spirulina, pistia etc. If excess growth of planktons are there then these weeds may be put in packets into the water body, within 5 to 7 days. This is very good method to control plakton population in the pond. These weeds take care of the nutrients which help for the growth of planktons. Once the nutrients will not be available to them the growth will be restricted.

54. What should be the ideal size of the pond, give reasons for it?

Ans. The ideal size of the pond should be 5hactre.

Because

1. Management will be easier
2. The water parameters will be more stable

Now a days in order to get more production by better management, the fish growers are putting bonds or embankments and making smaller units, but these are of less advantageous in long run. In profit wise, they may get a good quantum in a short periods, but in long run, the bigger water bodies will give more production and more profit. Too small ponds are also not desirable it well lessen the production though management will be easier

55. How tilapia are helpful in culture systems like Bheries?

Ans. Tilapia is commonly known as poorman's chicken. In the bheries the proliferative breeding is not harming the target species, because the target species are stocked in a vary high density so these tilapias do not get sufficient space for their growth and survival. More over the spawn and fry of tilapia are consumed by carps as planktons. So in this case tilapia growth controlled naturally. It is no way harming the carp farming (experience of farmed as well as Govt. officers)

More over tilapia control the aquatic vegetation to a great extent.

56. How the pond dyke vegetables helps in eradicating the pests of Rice fields, more particularly rice-cum fresh culture plots?

Ans. Vegetables growing in the pond dykes attract the beneficial insects which invade/counter the harmful insects of rice fields. In this process, the pests of rice field can be biologically controlled without using any pesticides into it.

57. Why the small size fish production is advisable in swage fed systems or in heavy metal accumulating zones?

Ans. The small size fish production is advisable, because it will reduce the biomagnifications of heavy metals and pesticides. The more the age and weight of fish, the more the accumulation as well as biomagnifications of the elements. So in order to avoid such problems small size fish production is advisable

1. Secondly it can encourage continuous stocking and harvesting, which gives some daily return.
2. Thirdly, pouching problems can be avoided to some extent.
3. Fourthly it is economically profitable in comparison to the table fish production.

58. Why the smell from sewage fed farms usually come? Where is it's origin?

Ans. The characteristic smell of sewage fed farms is due to the presence of planktons like Gomphonema, microcystis (Recently renamed as Anacystis), and anabaena etc. These planktons on living condition as well as on death gives the characteristic bad odour.

59. Why Argulus are found with the fishes, where organic matter in the pond is more?

Ans. Angulus usually prefers acidic environment, and as because organic matter leads to low PH, so these parasites proliferates in the ponds having more organic matter.

60. Why in the Bheries, Argulus infection is more with *Catla catla* than *L. rohita* or *C. mrigalla*?

Ans. Bheries are having putrefied environments, loaded with high amount of sewage materials. So dissolved oxygen content is more only in surface region than buttom and column. More over, the adult parasites are infective one than the nauplius. These adult ones require oxygen for their survival. So they prefer surface environment than column and bottom also for the same reason Catla which are found in the surface regions are more affected in Bheries than Rohu and Mrigal, Where as argulus infection will be more in Rohu, in pond envinonment, because, the of dissolved oxygen content is more in column region than suface, where Rohu species are found. It's clean environment.

61. How argulus population can be checked?

Ans. Argulus parasites needs a definite substratum for laying eggs. These substrata may be palm leaves, asbestos sheets, coconut leaves, Eichornia weed etc. These substrata should be removed n from pond environment in order to have control over it. Drugs or chemicals should be applied in multiple times so as completely eradicate these parasites.

62. What precautions one has to take before application of chemicals/drugs in argulised ponds?

Ans. The commonly used chemicals to control Argulus population are Nuvan/ BHC/Gamaxene/Cleaner. These chemicals has cidal effect on the crustaceans. So the ponds or bheries where, prawns/shrimps has been stocked along with carps, there these above applications are not advisable. Because it will not only kill the Argulus but also the prawns. So it is better to segregate the prawns and then apply the chemicals to control argulus.

63. What do you mean by PCR tested seed ? How it is beneficial?

Ans. The seeds has to be tested through PCR cycle. Among the seeds muscle from some seeds are being taken and tested with PCR, and if it will give positive reaction to WSV, their the it may be discarded and if at all it will show negative reaction, their the lot may accepted for culture.

This as beneficial with regard to get the viral free stocks for culture purpose.

Index

N

O

P

Q

R

S

Color Plates

Page 2

Photo courtesy TFH Publications

Page 4

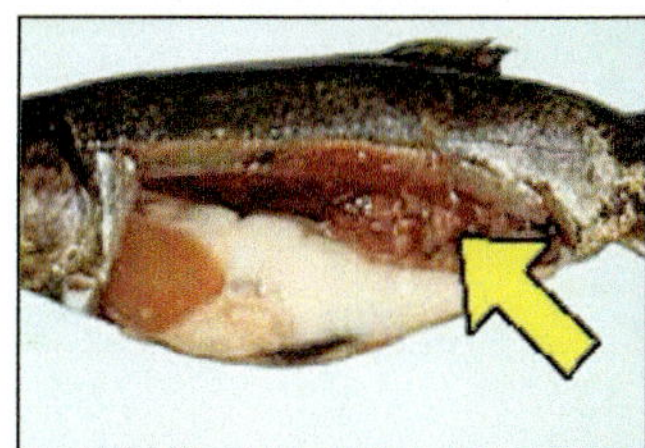

Page 5

Page 7

Page 29

Page 48

Page 50

Page 58

Page 60

Page 65

Pg. 68

Page 73

Page 73

Page 75

Page 113

Page 117

Page 118

Page 120

Page 121

Page 122

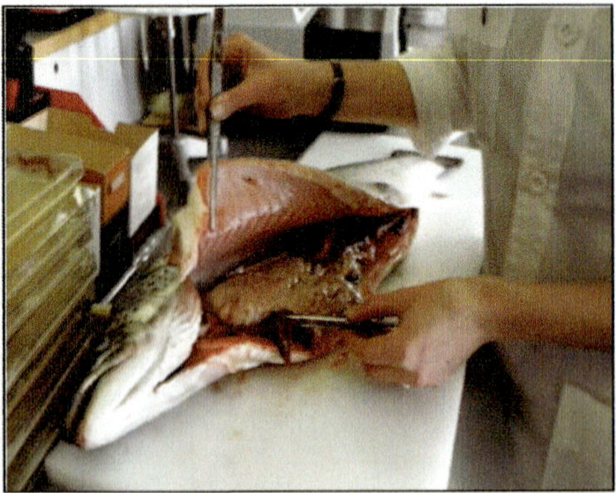

Page 151

Page 159

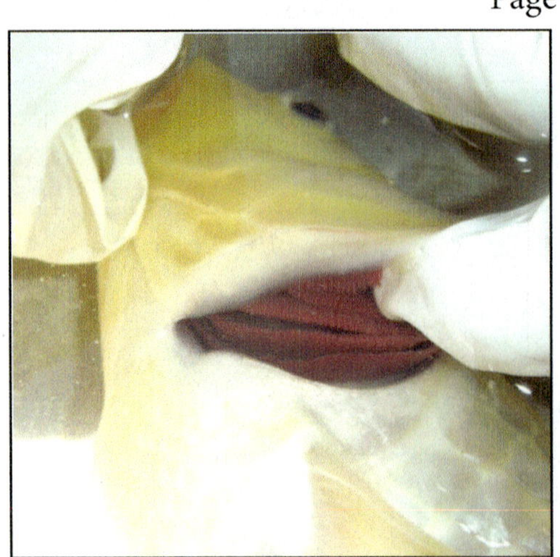

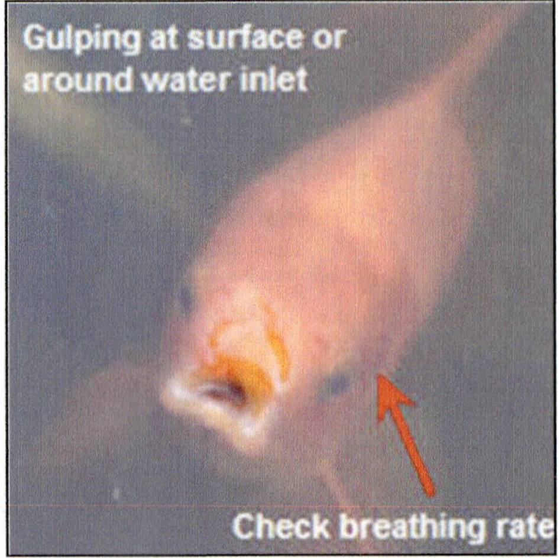

Page 164

Page 200

Page 218

Page 219

Page 220

Page 221

Page 222

Page 223

Page 225

Page 227

Page 227

Page 227

Page 229

Page 230

Page 250

Page 251

Page 252

Page 253

Page 254

Page 255

Page 256

Page 257

Page 257

Page 258

Page 271

Page 272

Page 273

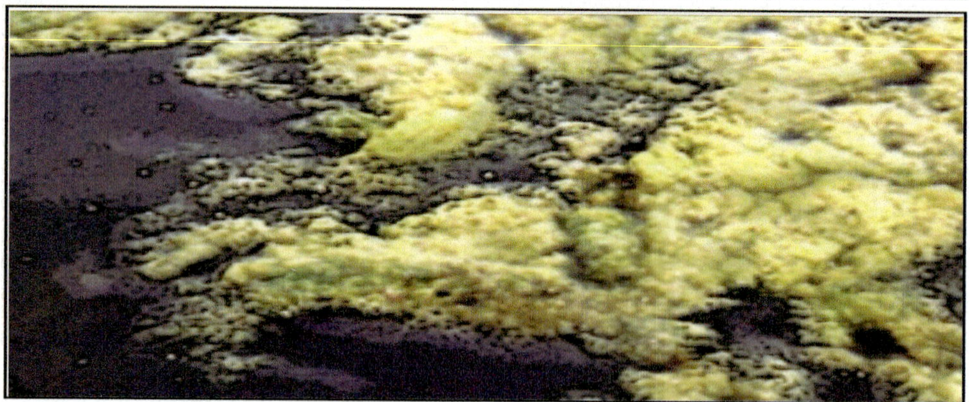

Page 273

Page 274

As 33
74.922
Arsenic

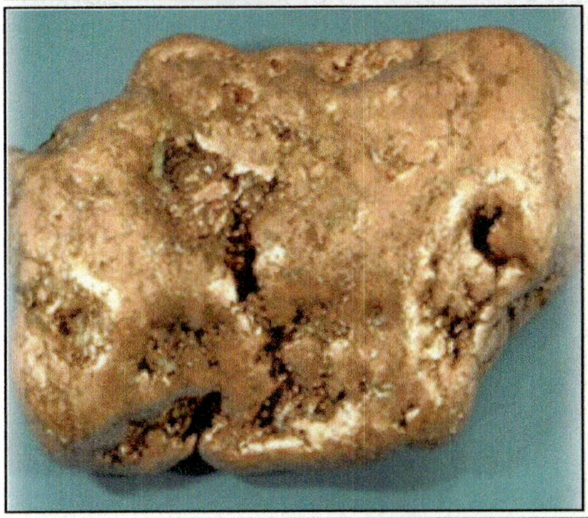

Page 284

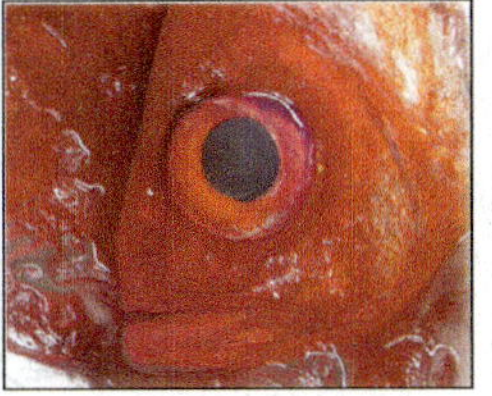

Page 287

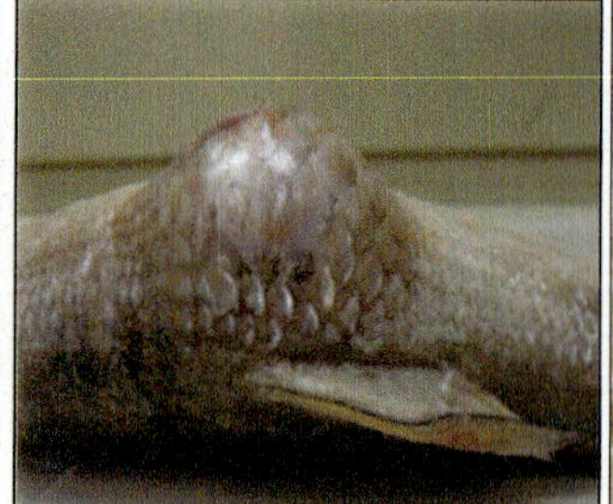

Page 288

Page 289

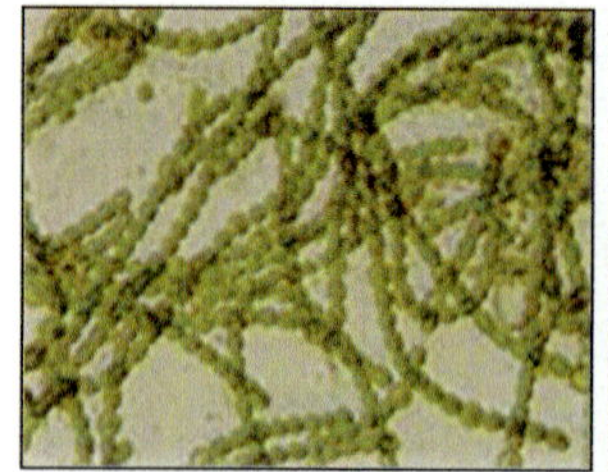
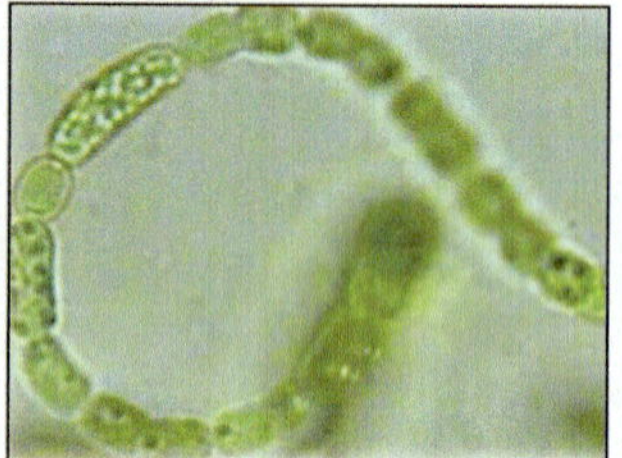
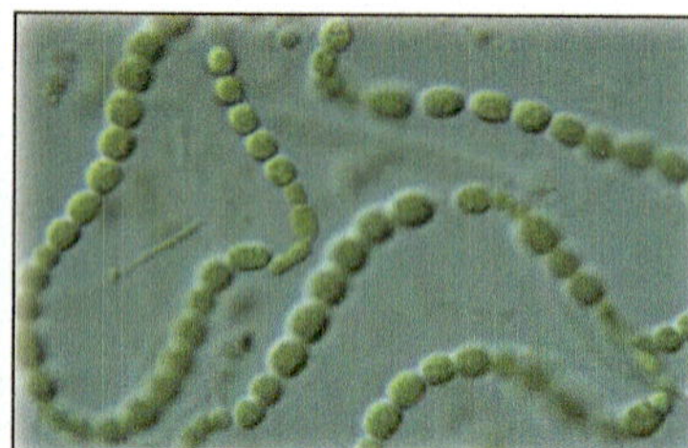

Page 298

Page 301

Page 302

Page 303

Page 306

Page 310

Page 310

Page 311

Page 352

Page 352

SMC-900 Geomembrane Hot Wedge Welding Machine
Simon Machinery (China) Co., Ltd.
Email: sales@simon-machinery.com

Discharge pipes

Pg. 352

07 03 2017
07 03 2017

Pg. 353